D1483622

*Michael Eckert*

**The Dawn of Fluid Dynamics**

A Discipline between Science
and Technology

## Related Titles

R. Ansorge
**Mathematical Models of Fluiddynamics**
**Modelling, Theory, Basic Numerical Facts - An Introduction**
187 pages with 30 figures
2003
Hardcover
ISBN 3-527-40397-3

J. Renn (ed.)
**Albert Einstein - Chief Engineer of the Universe**
**100 Authors for Einstein. Essays**
approx. 480 pages
2005
Hardcover
ISBN 3-527-40574-7

D. Brian
**Einstein - A Life**
526 pages
1996
Softcover
ISBN 0-471-19362-3

*Michael Eckert*

# The Dawn of Fluid Dynamics

A Discipline between Science and Technology

**WILEY-VCH**

WILEY-VCH Verlag GmbH & Co. KGaA

**The author of this book**

*Dr. Michael Eckert*
Deutsches Museum München
email: M.Eckert@deutsches-museum.de

**Cover illustration**
"Wake downstream of a thin plate soaked in a
water flow" by Henri Werlé, with kind permission
from ONERA, http://www.onera.fr

**Library of Congress Card No.: applied for
British Library Cataloging-in-Publication
Data:** A catalogue record for this book is
available from the British Library.
**Bibliographic information published by Die
Deutsche Bibliothek**
Die Deutsche Bibliothek lists this publication in
the Deutsche Nationalbibliografie; detailed
bibliographic data is available in the Internet at
http://dnb.ddb.de.

© 2006 WILEY-VCH Verlag GmbH & Co. KGaA,
Weinheim

**Typesetting**   Uwe Krieg, Berlin
**Printing**   betz-druck GmbH, Darmstadt
**Binding**   Litges & Dopf Buchbinderei GmbH,
Heppenheim

Printed in the Federal Republic of Germany
Printed on acid-free paper

**ISBN-13:**  978-3-527-40513-8
**ISBN-10:**  3-527-40513-5

# Contents

*The Dawn of Fluid Dynamics: A Discipline between Science and Technology.* Michael Eckert
Copyright © 2006 WILEY-VCH Verlag GmbH & Co. KGaA, Weinheim
ISBN: 3-527-40513-5

# Preface

A leading representative of fluid dynamics defined this discipline as "part of applied mathematics, of physics, of many branches of engineering, certainly civil, mechanical, chemical, and aeronautical engineering, and of naval architecture and geophysics, with astrophysics and biological and physiological fluid dynamics to be added." [1, p. 4]

Fluid mechanics has not always been as versatile as this definition suggests. Fifty years ago, astrophysical, biological, and physiological fluid dynamics was still in the future. A hundred years ago, aeronautical engineering did not yet exist; when the first airplanes appeared in the sky before the First World War, the science that became known as aerodynamics was still in its infancy. By the end of the 19th century, fluid mechanics meant hydrodynamics or hydraulics: the former usually dealt with the aspects of "ideal," i.e., frictionless, fluids, based on Euler's equations of motion; the latter was concerned with the real flow of water in pipes and canals. Hydrodynamics belonged to the domain of mathematics and theoretical physics; hydraulics, by contrast, was a technology based on empirical rules rather than scientific principles. Theoretical hydrodynamics and practical hydraulics pursued their own diverging courses; there was only a minimal overlap, and when applied to specific problems, the results could contradict one another [2].

This book is concerned with the history of fluid dynamics in the twentieth century before the Second World War. This was the era when fluid dynamics evolved into a powerful engineering science. A future study will account for the subsequent period, when this discipline acquired the multifaceted character to which the above quote alluded. The crucial era for bridging the proverbial gap between theory and practice, however, was the earlier period, i.e., the first four decades of the twentieth century. We may call these decades the age of Prandtl, because no other individual contributed more to the formation of modern fluid dynamics. We may even pinpoint the year and the event with which this process began: it started in 1904, when Ludwig Prandtl presented at a conference the boundary layer theory for fluids with little friction. Prandtl's publication was regarded as "one of the most extraordinary papers

*The Dawn of Fluid Dynamics: A Discipline between Science and Technology.* Michael Eckert
Copyright © 2006 WILEY-VCH Verlag GmbH & Co. KGaA, Weinheim
ISBN: 3-527-40513-5

of this century, and probably of many centuries" [1]—it "marked an epoch in the history of fluid mechanics, opening the way for understanding the motion of real fluids" [3].

In order to avoid any misunderstanding: this is not a biography of Prandtl, however desirable an account of Prandtl's life might be. Nor is it a hero story; I do not claim that the emergence of modern fluid dynamics is due solely to Prandtl. If Prandtl and his Göttingen circle's work is pursued here in more detail than that of other key figures of this discipline, it is because the narrative needs a thread to link its parts, and Prandtl's contributions provide enough coherence for this purpose. The history of fluid dynamics in the age of Prandtl, as presented in the following account, is particularly a narrative about how science and technology interacted with another in the twentieth century. How does one account for such a complex process? In contrast to sociological approaches I pursue the history of fluid dynamics *not* within a theoretical model of science–technology interactions. Nevertheless, the relationship of theory and practice, science and engineering, or whatever rhetoric is used to refer to these antagonistic and yet so similar twins, implicitly runs as a recurrent theme through all chapters of this book. I share with philosophers, sociologists, and other analysts of science studies the concern to better grasp science–technology interactions, but I cannot see how to present the history of fluid dynamics from the perspective of an abstract model. My own approach is descriptive rather than analytical; I approach the history of fluid dynamics from the perspective of a narrator who is more interested in a rich portrayal of historical contexts than in gathering elements for an epistemological analysis. This approach requires deviations here and there from the main alley, so to speak, in order to clarify pertinent contexts, but I am conscious not to lose the narrative thread and regard as pertinent only what contributes to a better understanding of the theory–practice issue. I postpone further reflections to the epilogue, when this issue may be better discussed in view of the empirical material presented throughout the remainder of the book.

Many people and institutions have contributed to this work. Instead of acknowledging their help here individually in the form of a long list of names, I refer readers to the notes in the appendix, where readers may better appreciate how archives and authors of other studies helped to add flesh to the skeleton of my narrative. The only exceptions concern my colleagues from the Deutsches Museum and the Munich Center for the History of Science and Technology, whom I owe thank for years of fruitful collaboration and stimulating discussions, and the Deutsche Forschungsgemeinschaft for funding the Research Group 393, which formed the framework of this study.

*Michael Eckert*, Munich, May 2005

# 1
# Diverging Trends before the Twentieth Century

The flow of water or air around an obstacle is such a familiar phenomenon that we tend to underrate its importance in the history of science and technology. Throughout the centuries, the behavior of a body in a fluid was a fundamental theoretical problem and an obvious practical concern. The motion of celestial bodies, ships, projectiles, and other phenomena involved conceptions of fluid dynamics. Although the development of science from Aristotle to Einstein is usually presented without excursions into the history of fluid dynamics, concepts about motion inevitably involve assumptions about fluid resistance.

## 1.1
## Galileo's Abstraction

In Aristotle's natural philosophy, the medium through which a motion proceeded played a paradoxical role. In order to sustain the motion, a motive agency was required. Aristotle (384–322 BC) imagined that this motive agency resided in the medium: "We must, therefore," Aristotle wrote in Book VIII of his *Physics*, "hold that the original movent gives the power of causing motion to air, or water, or anything else which is naturally adapted for being a movent as well as for being moved" [4, p. 506]. At the start of the motion of a projectile, the medium would be displaced by the projectile, and together with this displacement, a motive force would be passed along the trajectory. Thus, the medium acquired the power to propel the projectile. At the same time the medium would resist the motion: "If air is twice as tenuous as water," Aristotle argued, "the same moving body will spend twice as much time in travelling a certain path in water as in travelling the same path in air" [5, p. 21].

Aristotle dominated pre-modern natural philosophy – but some of his views also served as bones of contention. How could the same medium at the same time propel and resist the motion of a projectile? Most famous among those who criticized this concept was Jean Buridan (1300–1358), who argued that the propulsive property resided in the projectile itself rather than

*The Dawn of Fluid Dynamics: A Discipline between Science and Technology.* Michael Eckert
Copyright © 2006 WILEY-VCH Verlag GmbH & Co. KGaA, Weinheim
ISBN: 3-527-40513-5

in the medium. He called this property impetus: "Whenever some agency sets a body in motion," Buridan wrote, "it imparts to it a certain impetus, a certain power which is able to move the body along in the direction imposed upon it at the outset (...) It is this impetus which moves a stone after it has been thrown until the motion is at an end. But because of the resistance of the air and also because of the heaviness, which inclines the motion of the stone in a direction different from that in which the impetus is effective, this impetus continually decreases" [5, pp. 49–50]. Now the medium through which the motion proceeded was left with just one property: resistance.

The impetus concept marked the emergence of the modern notions of inertia and momentum. But that did not happen at once. Even Galileo Galilei (1564–1642), with whom we associate the revolutionary turn from the medieval philosophy to the "new science" of motion, still mixed Aristotelian concepts with modern concepts of motion. Like his predecessors, Galileo struggled with the role of the medium through which a body moves. His famous *Dialogues Concerning Two New Sciences* reveals what problems were behind the effort to imagine how a body would move without the resistive property of the medium. Galileo lets Salviati ask, for example, "What would happen if bodies of different weight were placed in media with different resistances?" The answer was presented by comparing the motion in air and water: "I found," Salviati continues, "that the differences in speed were greater in those media which were more resistant, that is, less yielding. This difference was such that two bodies which differed scarcely at all in their speed through air would, in water, fall the one with a speed ten times as great as that of the other" [6, p. 68].

Obviously, motion in air would be closer to a motion without any influence of the medium. But there were problems in quantitatively measuring differences for bodies with different weights in air. "It occurred to me therefore," Galileo argues with the voice of Salviati, "to repeat many times the fall through a small height in such a way that I might accumulate all those small intervals of time that elapse between the arrival of the heavy and light bodies respectively at their common terminus." With the repetition of the free fall, he meant the repeated swings of a pendulum:

> "Accordingly I took two balls, one of lead and one of cork, the former more than a hundred times heavier than the latter, and suspended them by means of two equal fine threads, each four or five cubits long. Pulling each ball aside from the perpendicular, I let them go at the same instant, and they, falling along the circumferences of circles having these equal strings for semi-diameters, passed beyond the perpendicular and returned along the same path. This free vibration repeated a hundred times showed clearly that the heavy body maintains so nearly the period of the light body that neither in a hundred swings nor even in a thousand will

the former anticipate the latter by as much as a single moment, so perfectly do they keep step. We can also observe the effect of the medium which, by the resistance which it offers to motion, diminishes the vibration of the cork more than that of the lead, but without altering the frequency of either; even when the arc traversed by the cork did not exceed five or six degrees while that of the lead was fifty or sixty, the swings were performed in equal times" [6, pp. 84–85].

In order to find out how the resistance of air depends on the velocity, Galileo compared the swings of pendulums with equal weights but different amplitudes. He found that the air resistance is proportional to the velocity of the moving body [6, p. 254].

Already Galileo's contemporaries noticed that these conclusions could not have resulted from actual experiments. Marin Mersenne (1588–1648) compared the swings of equal pendulums with different amplitudes: he found that one which started swinging with an amplitude of two feet differed from one with an amplitude of one foot already after thirty periods of oscillation by as much as one full period. In 1639, a year after the publication of the *Dialogues Concerning Two New Sciences*, he remarked that if Galileo had performed real pendulum experiments and only waited for thirty or forty swings, he would have noticed the difference [7]. Recent pendulum experiments confirmed Mersenne's critique [8, 11].

This and other observations of Galileo stirred considerable debate among historians of science – to what extent did Galileo actually perform experiments? Only his pendulum experiments with small amplitude are presumed "real"; those with larger amplitudes are regarded as "imaginary" or "hypothetical," i.e., they were not performed in reality, but (contrary to mere thought experiments) are based on extrapolation from empirical observations [7]. Earlier interpretations tended to categorize Galileo's style of research into one of two extremes: either as deductive in the tradition of Platonic and idealistic natural philosophy, in which the experiment only plays a role as a confirmation of insights gained by mere thinking; or as inductive, with the experiment as the origin of new knowledge. According to more recent historical studies, however, Galileo's science was more complex and does not fit neatly into one category or the other alone [9].

The question whether Galileo actually performed free fall experiments from the leaning tower of Pisa attracted particular scrutiny [10]. As with the pendulum experiments, the problem of resistance plays an important role here, too. "Aristotle says that 'an iron ball of one hundred pounds falling from a height of one hundred cubits reaches the ground before a one-pound ball has fallen a single cubit.' I say," Salviati responds to such an obvious discrepancy with reality, "that they arrive at the same time. You find, on making the experiment, that the larger outstrips the smaller by two finger-breadths" [6, p. 64]. However, a modern calculation, which takes into account the air resistance, yields a difference of 1.05 m for the free fall of a 100-lb. iron sphere (with a radius of 11.13 cm) and a 1-lb. iron sphere (with a radius of 2.4 cm) over a distance of 100 cubits (58.4 m). The lighter sphere would be more than one meter behind the heavier one – certainly much more than the "two finger-breadths" in Galileo's argument [11]. If Galileo really performed the tower experiment, why didn't he notice this discrepancy? The puzzle can be resolved by a psycho-physical argument: when an experimenter intends to release simultaneously two different weights from his outstretched hands, the palm with the lighter weight tends to open a bit earlier than the palm with the heavier weight; this difference could have compensated for the difference due to the air resistance [12], [13, Supplement 3].

But Galileo, presumably, was rather motivated by a theoretical argument. The medium had to be "thrust aside by the falling body," Salviati argued. "This quiet, yielding, fluid medium opposes motion through it with a resistance which is proportional to the rapidity with which the medium must give way to the passage of the body." By such reasoning, Galileo related the displaced mass of the medium to the resistance: "And since it is known that the effect of the medium is to diminish the weight of the body by the weight of the medium displaced, we may accomplish our purpose by diminishing in just this proportion the speeds of the falling bodies, which in a non-resisting medium we have assumed to be equal" [6, pp. 74–75].

In other words, despite a flawed concept of fluid resistance in terms of buoyancy, Galileo arrived at his goal: the abstraction of a motion in a non-resisting medium. With a vanishing buoyancy, the resistance would vanish too. In this case, with no mass to be displaced, all bodies would fall in the same manner. Galileo's law of free fall certainly has to be rated among the most important accomplishments in the history of science, but it is erroneous to infer from Galileo's abstraction that he "had a correct notion of air resistance," as a widely read book on the history of aerodynamics has claimed [14, p. 8]. Galileo did not aim at a theory of aerodynamics; his predominant concern was Aristotle's natural philosophy. The abstraction of a motion in a non-resisting medium, perceived as a motion in which no medium had to be displaced, touched upon another ancient philosophical belief: Aristotle believed in the impossibility of a vacuum; for Galileo, it was the domain in which the laws of free fall hold. Maybe it is not an exaggeration to state that Galileo's elaborations on the medium through which a body moves only served to justify his abstraction of a motion in empty space.

Against this background it does not come as a surprise that it was a pupil of Galileo, Evangelista Torricelli (1608–1647), who is credited with presenting the first experimental evidence of a vacuum. Torricelli emptied glass tubes filled with mercury into a container, such that the openings of the tubes were not exposed to the air. Inside the inverted tube, above a remaining column of mercury, there was left an empty space, a "Torricellian vacuum." The height of the mercury column in the tube was found to depend on the ambient air pressure. Torricelli undertook these experiments with another pupil of Galileo (Vincenzio Viviani). Like Galileo himself, his pupils also were primarily interested in refuting Aristotelian dogmas. "Many have said [that vacuum] cannot happen," Torricelli wrote to another follower of Galileo after his experiment; yet, it "may occur with no difficulty, and with no resistance from nature." Thus, he refuted the dogma of a "horror vacui." He concluded, with a now famous quote: "We live submerged at the bottom of an ocean of elementary air which is known by incontestable experiments to have weight." [15, p. 84].

After Torricelli's experiment the old debate among "vacuists" and "plenists" seemed to be decided in favor of the "vacuists," but René Descartes (1586–1650) renewed the belief of a universal filling of space. He denied that Galileo's extrapolation of free fall in empty space was based on sound arguments. According to Descartes' doctrine, all natural phenomena resulted from the motions of infinitely fine weightless particles of an ether that pervaded the entire universe. The particles of ordinary matter, such as air or water, were supposed to have weight, so that their displacement by a moving body would retard its motion. In order to prevent a temporary depletion behind a moving object, the displacement of matter involved a flow around the object, which Descartes imagined as vortical. He extended his doctrine to the entire universe. The solar system was supposed to be an enormous vortex of matter, in which the planets orbited as smaller vortices around the center [16].

Descartes did not produce quantitative results – neither for his cosmogony nor for the domain of earthly physics. Once, he communicated in a letter a formula about the retardation of a free falling body in a medium, whereby the speed approached a limit in the form of an infinite geometrical series, but he did not provide a physical argument for this result [23, p. 110]. Nevertheless, he exerted a remarkable influence on seventeenth century natural philosophy. Christiaan Huygens (1629–1695) pursued several of Descartes' ideas, such as the concept of an attracting force due to vortical motion around a center. Such a force would keep a planet embedded in a vortex in his orbit around the sun. In order to illustrate this force, Huygens arranged a little sphere in a cylindrical vessel filled with water such that it was free to move in a radial direction only. When the vessel was rotated around its axis, the sphere moved inwards against the centrifugal force [16, pp. 76–77].

**1.2**

**Hogs' Bladders in St. Paul's Cathedral**

Descartes' concepts of motion also influenced Isaac Newton (1643–1727), but as an opponent rather than as a follower. Alluding to Descartes' *Principia Philosophiae*, Newton titled his own three-volume treatise on mechanics *Philosophiae Naturalis Principia Mathematica*. The first volume with "Newton's laws of motion" for a body in a vacuum is celebrated as the foundation of classical mechanics. It is less known that Newton also spent a lot of time developing the laws of motion for a body in a fluid. The entire second volume is dedicated to this problem. It was regarded as "the most original part of the whole work, though also largely incorrect" [17, p. 167]. As for Galileo and Descartes, the debate among "vacuists" and "plenists" was also a major issue for Newton. One of his pupils, Henry Pemberton, wrote in 1728 a book titled *A View of Sir Isaac Newton's Philosophy* about the second volume of Newton's *Principia*: "By this theory of the resistance of fluids, and these experiments our author decides the question so long agitated among natural philosophers whether the space is absolutely full of matter. The Aristotelians and Cartesians both assert this plenitude; the Atomists have maintained the contrary. Our author has chosen to determine this question by his theory of resistance" [18, p. 314].

If the universe were filled with a material substance, as taught by Descartes and his school, then the planets would encounter a resistance along their orbits around the sun. Descartes' vortex conception could not escape that fundamental problem and therefore would have given rise to contradictions if it had been formulated in a quantitative manner. Newton presented an alternative concept with his theory of universal gravitation, which assumed an empty space between the celestial bodies–or a "bodiless" medium that would not exert a noticeable resistance: "And therefore the celestial spaces, thro' which the globes of the Planets and Comets are perpetually passing towards all parts, with the utmost freedom, and without the least sensible diminution of their motion, must be utterly void of any corporeal fluid, excepting perhaps some extremely rare vapours, and the rays of light." This was Newton's conclusion at the end of the section "Of the motion of fluids and the resistance made to projected bodies" [19, vol. 2, proposition 40, pp. 161–162].

From the outset, Newton assumed: "In mediums void of all tenacity, the resistances made to bodies are in the duplicate ratio of the velocities." Galileo's relation "that the resistance is in the ratio of the velocity," according to Newton, was "more a mathematical hypothesis than a physical one" [19, vol. 2, proposition 4, p. 11]. Others had already made the same assumption of a quadratic velocity dependence, which seemed to be more in agreement with empirical observations (see below); but Newton was the first natural philosopher who attempted to justify this relation on the basis of a physical model.

His concept is too complex for a short summary. It may suffice to hint at Newton's argument for an "elastic fluid" like air, which he conceived as a gas of particles. Based on certain assumptions about the mutual collisions of these particles, Newton obtained quantitative results about the resistance of such a fluid. "But whether elastic fluids do really consist of particles so repelling each other," he concluded, "is a physical question. We have here demonstrated mathematically the property of fluids consisting of particles of this kind, that hence philosophers may take occasion to discuss that question" [19, vol. 2, proposition 23, theorem 18, p. 79]. Newton explicitly envisioned different sources of resistance, "as from the expansion of the particles after the manner of wool, or the boughs of trees, or any other cause, by which the particles are hindered from moving freely among themselves; the resistance, by reason of the lesser fluidity of the medium, will be greater than in the corollaries above" [19, vol. 2, proposition 34, theorem 17, p. 117]. He also developed a notion of viscosity: "The resistance, arising from the want of lubricity in the parts of fluid, is, ceteris paribus, proportional to the velocity with which the parts of the fluid are separated from each other" [19, vol. 2, proposition 51, p. 184].[1]

Newton did not content himself with establishing theorems. "In order to investigate the resistances of fluids from experiments, I procured a square wooden vessel (...) this I filled with rain-water: and having provided globes made up of wax, and lead included therein, I noted the times of descents." Thus, Newton described the beginning of a series of experiments on fluid resistance. He used a pendulum with an oscillation period of a half-second for the measurement of time, and meticulously compared the various outcomes with his theoretical formulae: "Three equal globes, weighing 141 grains in air and 4 3/8 in water, being let fall several times, fell in the times of 61, 62, 63, 64 and 65 oscillations, describing a space of 182 inches," he described one of these experiments. "And by the theory they ought to have fallen in 64 1/2 oscillations, nearly." He noticed that sometimes "the globes in falling oscillate a little" and believed that for this reason the resistance was "somewhat greater than in the duplicate ratio of the velocity." But in general he regarded the outcome as an experimental verification of his square law formula for the resistance of a "globe moving though a perfectly fluid compressed medium." After a series of 12 experiments he concluded "that the theory agrees with the phaenomena of bodies falling in water; it remains that we examine the phaenomena of bodies falling in air" [19, vol. 2, proposition 40, pp. 145–155].

**1)** In modern terms, this is equivalent to a linear relation between shear stress and strain rate: we call fluids with such viscous behavior "Newtonian." However, Newton did not investigate the relation between stress and strain. See [20, pp. 258–259].

In order to verify his theory of the resistance of a spherical body moving in air, Newton, like Galileo, first performed pendulum experiments. He suspended a sphere by a fine thread on a hook, then varied the diameters and materials of the sphere, as well as the lengths of the thread. According to his theory, the resistance was proportional to the square of the velocity of the sphere, and this is what he "nearly" observed. But he could not account for the additional resistance of the thread "which was certainly considerable." He also compared the oscillations of the pendulum in air with those in water, but found the outcome not reliable because the vessel in which the water was contained was not large enough so that "by its narrowness [the vessel] obstructed the motion of the water as it yielded to the oscillating globe." Even less conclusive were pendulum experiments in mercury. "I intended to have repeated these experiments with larger vessels, and in melted metals, and other liquors both cold and hot: but I had not leisure to try all," Newton admitted [19, vol. 2, proposition 31, pp. 95–110].

Newton hoped to obtain more reliable measurements of air resistance with free fall experiments: "From the top of St. Paul's church in London in June 1710 there were let fall together two glass globes, one full of quicksilver, the other of air." The two spheres traversed a height of 220 English feet (67 m) before they shattered into pieces on the cathedral's floor. They were released by a sophisticated trapdoor-mechanism which ensured their simultaneous begin of fall. The time was measured by a pendulum with a period of oscillation of one second. The experiment was repeated several times with varying weights. The spheres filled with mercury had a diameter of 0.8 inches; those filled with air were between 5.0 and 5.2 inches in diameter. The time of free fall was 4 s for the heavier spheres and between 8 and 8.5 s for the lighter ones. (In a vacuum the time would have been 3.7 s.) In order to compare these results with his theory, Newton compared the experimental height of free fall with the distance they would have traversed within the measured time according to his formula. Both distances differed by less than 11 feet [19, vol. 2, proposition 40, pp. 155–157].

Newton mentioned in his *Principia* yet another series of free fall experiments from a somewhat greater height in St. Paul's cathedral: "Anno 1719 in the month of July, Dr. Desaguliers made some experiments of this kind again, by forming hogs' bladders into spherical orbs; which was done by means of a concave wooden sphere, which the bladders, being wetted well fist, were put into. After that, being blown full of air, they were obliged to fill up the spherical cavity that contained them; and then, when dry, were taken out. These were let fall from the lantern on the top of the cupola of the same church; namely from a height of 272 feet." For comparison, a leaden sphere was let fall down at the same time. The air filled hogs' bladders had diameters of about 5 inches and required about 20 s to fall down; the leaden spheres, by contrast, reached the ground in 4 1/4 s. Newton also reported about phenomena which delayed the free fall by as much as a whole second sometimes, because "the bladders did not always fall directly down, but sometimes fluttered a little in the air, and waved to and fro as they were descending." One bladder "was wrinkled, and by its wrinkles was a little retarded." Nevertheless, he found that the results agreed much better with his theory than nine years ago: "Our theory therefore exhibits rightly, within a very little, all the resistance that globes moving either in air or in water meet with; which appears to be proportional to the densities of the fluids in globes of equal velocities and magnitudes" [19, vol. 2, proposition 40, pp. 157–159].

In modern terminology, Newton's formula for the resistance of a fluid is expressed as $\sim \rho D^2 v^2$, with $\rho$ representing the density of the fluid, $D$ the diameter of the sphere, and $v$ its velocity. This has become known as "Newton's square law" and has been established as a valid description of fluid resistance for a wide range of flow regimes. However, although Newton's experiments seemed to corroborate this law, they bear little evidence for Newton's theory because only one quantity, the time of free fall, was observed. The particle model gave rise to contradictory results when applied to bodies of different shape in fluids such as air and water. In air, a "rare medium, consisting of equal particles freely disposed at equal distances from each other," the resistance of a sphere would be half of that of a cylinder with the same radius moving in the direction of it axis. In water, "a compressed, infinite, and non-elastic fluid," would both experience the same resistance [19, vol. 2, propositions 34 and 37, pp. 117, 135, and 141]. If Newton had compared experimental results of spheres and cylinders, he would have noticed a contradiction with his theoretical results. Similarly, if he had calculated by the same reasoning the air resistance of a flat plane oriented at an oblique angle to the flow of air, he would have found a result proportional to the square of the sine of the angle of incidence. Newton did not perform such a calculation, but among aerodynamicists "Newton's sine square law" became famous as an erroneous formula for the lift of a wing. At the beginning of the nineteenth century, this formula was even used to demonstrate the impossibility of flying, and later aerodynamicists blamed Newton for having delayed aviation at least for half a century [18, p. 311].

## 1.3
## Ballistics

Beyond its pertinence to natural philosophy, the resistance of a body in a medium had always been a practical problem. Since antiquity, understanding the trajectory of projectiles was an obvious challenge for natural philosophers as much as for practically minded men. It was part of a science named "bal-

listics" (derived from the Greek word βαλλειν, to throw). Niccolò Tartaglia
(1499–1557), a mathematician with some experience in military affairs, described the knowledge of his epoch on ballistics in a treatise *Nova scientia*.
This work exerted some influence on Galileo, who spent considerable time in
his youth coming to grips with ballistics problems. In Tartaglia's treatise, one
could read, for example, that a projectile traveled the farthest when fired at
an angle of 45 degrees; but the trajectory was no parabola: on a horizontal
plane, the distance between the vertex of the projectile and the site of its impact on the ground was always shorter than the distance between the origin
of its trajectory and the vertex, as depicted in Fig. 1.1. Initially, he adopted
the Aristotelian belief that a trajectory starts out straight, but in a subsequent
work, he argued that the trajectories are curved everywhere [23,24].

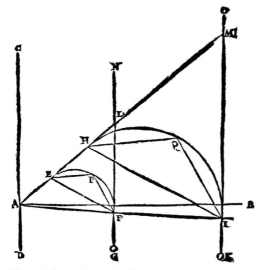

**Fig. 1.1** Tartaglia imagined that a projectile's trajectory starts out
straight due to the "violent" motion impressed by the shot; it is followed
by a curved mixed motion, and finally becomes "natural" [21, p. 38].

As recent studies of Galileo's manuscripts have shown [25], Galileo described the trajectory of a projectile in a vacuum as a parabola *before* he arrived at his law of free fall – not the other way around, as a deductive approach would suggest. The parabola emerged in 1592, when Galileo lectured
on military technology at the University of Padua. Based on his conviction
that the air resistance did not exert an appreciable effect, Galileo assumed a
symmetric trajectory – in contrast to Tartaglia's more realistic descriptions of
asymmetric trajectories. But when Torricelli derived ballistic tables based on
parabolic trajectories in 1644, an artillery officer uttered doubts: he wrote to
Torricelli that if it were not for the authority of the great Galileo, whom he
revered, he would not believe that the motion of projectiles is parabolic. Torri-

celli admitted that if there are discrepancies, one should find out what caused them, but with the experimental and theoretical tools available at the time – a hundred years before the advent of calculus – such efforts were futile. The correspondence between the practical artillery officer and the theorist (Torricelli was court mathematician at the Medicis) ended without a tangible result shortly before Torricelli died in 1647 [26].

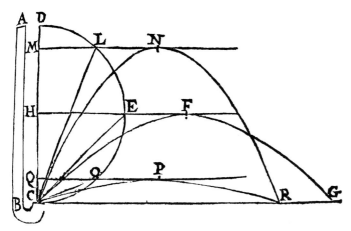

**Fig. 1.2** Like the jets of a fountain [22, p. 325], ballistic trajectories were assumed to be parabolic.

Throughout the seventeenth century it was fashionable for practical gunners to assume, like Galileo and Torricelli, that the parabola is the true trajectory of a projectile – despite air resistance. Francois Blondel (1617–1686), a field marshal of the Royal French Army, published a treatise on *L'Art de Jetter les Bombes* in which he addressed the problem of air resistance but assumed that it could be neglected. He pointed to fountains with their nearly parabolic jets as evidence for this assumption – see Fig. 1.2. Another treatise on *The Genuine Use and Effects of the Gunne*, published in 1674, also claimed that air resistance is negligible and, therefore, a parabola describes the real trajectory of a projectile. It is ironic that those who believed that air resistance does exert a considerable influence and that the resulting trajectory is different from a parabola were not the gunners with experience with "real" trajectories, but men like Huygens and Newton, who based their arguments on mathematics rather than practical observations. Newton derived from the square law for air resistance that a parabolic trajectory would require that the air density not be constant but become negative along part of its trajectory. A hyperbolic trajectory would not result in such a blatant contradiction. Therefore, "It is evident that the line which a projectile describes in a uniformly resisting medium, approaches nearer to these hyperbola's than to a parabola" [23, pp. 120–127, 140–141].

In retrospect, it is not astonishing why ballistic theory and practice diverged to such an extent before the eighteenth century. Only in 1742, with the treatise by Benjamin Robins (1707–1751), *New Principles of Gunnery*, did a method become known by which it was possible to measure the velocity of a projectile in the beginning of its trajectory: the ballistic pendulum (see Fig. 1.3).

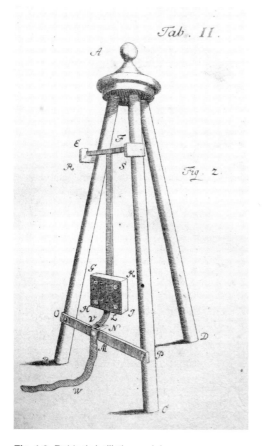

**Fig. 1.3** Robins's ballistic pendulum.

Robins's innovative contributions to experimental ballistics made him famous as "Father Gunnery." He experimented with projectiles that left his gun with velocities as high as 1,700 feet per second (559 m/s). If such a projectile, when fired at an angle of 45 degrees, would follow a parabolic trajectory, it would hit the ground 17 miles away – in contrast to an actual range of only about half a mile. Projectiles with such a high starting velocity obviously experienced an enormous resistance if their range was so much shorter. "The track described by the flight of shot or shells is neither a parabola," Robins concluded, "nor nearly a parabola, unless they are projected with small velocities." Robins also invented a whirling arm technique to measure the air resistance of objects with small velocities. Based on his experiments, he found "that all the theories of resistance hitherto established, are extremely defective" [27, pp. 153–154]. But Robins's mathematical abilities were limited. It was left to Leonhard Euler (1707–1783), in his German translation of Robins's *New Principles of Gunnery*, to elaborate a theory of ballistic trajectories [20, pp. 211–220].

Besides Euler, mathematicians and natural philosophers like Johann Bernoulli (1667–1748) and Jean le Rond d'Alembert (1717–1783) became deeply engaged in ballistic calculations. The problem to find a projectile's trajectory became a proving ground for the newly developed calculus.[2]

## 1.4
## D'Alembert's Paradox

The same eighteenth century thinkers who had recognized that air resistance posed a serious problem for calculating the trajectory of a projectile also formulated the laws of motion for ideal, i.e., inviscid, fluids, and found a strange result. D'Alembert published a treatise in 1768 titled "Paradoxe proposé aux Géometres sur la Résistance des Fluides" in which he asserted that a body moving through an ideal fluid does not experience a resistive force. D'Alembert, from his efforts in ballistics, knew about the practical importance of air resistance, but neither he nor other theorists were able to derive Newton's square law or any other law of fluid resistance from the laws of mechanics. In retrospect, d'Alembert's paradox does not appear so paradoxical because it was derived under the assumption of an inviscid fluid. Yet, it is difficult to understand why the displacement of the fluid does not involve a force. Euler had expressed this strange result many years before d'Alembert, after whom the paradox finally became named. If each fluid particle flowed around the body in such a way that it maintained the direction it had when it was in front of the body, argued Euler in one of his comments to Robins's *New Principles of Gunnery*, then there is no net force "and the body would not experience any resistance" [20, p. 245].

D'Alembert's paradox, therefore, should have entered the history of fluid mechanics more appropriately as the "Euler–d'Alembert paradox." Both had approached these problems as theorists. Their mutual relation was often one of fierce rivalry, which gave rise to some legendary stories. D'Alembert's rep-

---

**2)** The major problem, however, was due to yet unknown physical processes rather than an unavailability of mathematical tools. At the high (usually supersonic) velocities of projectiles fired from cannons and guns, the density of the air around the moving body is no longer constant. The study of air resistance in varying air density had to await twentieth century gas dynamics.

utation has been overrated, claimed one historian of mechanics, while Euler's role in this history was not appreciated enough.

Despite his theoretical leanings, Euler was very open-minded about practical problems. Nevertheless, his practical work was regarded largely as a failure. When Frederick the Great, king of Prussia, gave orders to decorate his Royal Garden of Sanssouci with water art, Euler became involved with hydraulic calculations about pumps and pipes required to raise water into an elevated water reservoir from where it was supposed to feed the fountains in the park. The king, however, never came to enjoy a fountain. He blamed Euler for having failed miserably: "My mill was constructed mathematically, and it could not raise one drop of water to a distance of fifty feet from the basin. Vanity of Vanities! Vanity of mathematics." Based on this passage, historians concluded "The mathematical genius Euler was a second-rate physicist," or "Euler's theory was not applicable for practical ends." This is how Euler is seen in the history of science – as a prime example of the proverbial schism between theory and practice. However, although it is true that the water art constructions in the Royal Garden of Sanssouci were abandoned unfinished in the lifetime of Frederick the Great, this fact was not Euler's fault, but was the result of the king's stinginess. He employed cheap laborers who had no experience with such work and who completely ignored Euler's hydraulic advice, which could have prevented the sad outcome. Euler conceived a theory of pipe flow that explained why the pipes always burst before water was raised to the elevated reservoir: as a consequence of the pumping action, which accelerated the water through the pipes, the walls of the pipes had to sustain a much higher water pressure than expected from the height difference between the pumps and the reservoir [28].

This was not the first incidence that a study in fluid flow was motivated by problems with water art. The science of moving water was among those specialties that were met with the greatest interest from Royal Academies. One outstanding work on hydraulics, which resulted from the patronage of the Paris Académie Royale des Sciences, is Edme Mariotte's *Traité de Mouvement des Eaux*, published in 1686. Mariotte and other academy members performed experiments investigating the speed with which water is ejected from a pipe, the principles of raising water, the height of water jets, and the resistance of a body as a function of the flow velocity. The motivation to undertake such experiments came from ambitious projects of water constructions, such as the canals all across France and the plans for the Royal Park at Versailles, where the world's most sophisticated water art was established for the pleasure of the Sun King and his court. The flow of water in open canals and in closed pipes became the subject matter of intensive study. The law of energy conservation in fluids, Bernoulli's equation, was formulated in the context of pipe flow by Johann Bernoulli (1667–1748) and his son Daniel Bernoulli (1700–

1782). Like Euler, the Bernoullis are mainly renowned for their mathematical work, but as is evident from the father's *Hydraulica* (1732) and the son's *Hydrodynamica* (1738), their work was motivated to a large extent by practical concerns of contemporary water art. In the age of Euler and the Bernoullis, the notions of hydrodynamics and hydraulics were used almost synonymously, often with an emphasis on the "art of raising water" and "the several machines employed for that purpose, as siphons, pumps, syringes, fountains, jets d'eau, fire-engines, etc.," according to a contemporary dictionary [29].

After he had established and solved the equations of fluid motion for the special case of pipe flow, Euler formulated the general equations of motion for inviscid fluids. They were published in 1755 under the title "Principes généraux du mouvement des fluides"; with "Euler's equations," as they were called, fluid mechanics was based on a firm theoretical foundation. Although these equations are valid for ideal fluids only, which inevitably involves d'Alembert's paradox, a number of practical problems can still be solved on that assumption.

## 1.5
### New Attempts to Account for Fluid Friction

In 1822, Claude Louis Marie Henri Navier (1785–1836) added a term to Euler's equations, which turned them into equations of motion for viscous fluids. A few years later, Siméon-Denis Poisson (1781–1840) arrived at the same result. Other contributors to this new formulation of the theory of fluid flow are Augustin Louis Cauchy (1789–1857) and Barré de Saint-Venant (1797–1886). But only in 1845 did George Gabriel Stokes (1819–1903) present a valid derivation for the Navier–Stokes equations, as they became known. The earlier theories of Navier and Poisson were based on hypotheses of atoms which, from a modern perspective, have to be dismissed as wrong, illustrating "a common phenomenon in the history of science: Falsehood $\Rightarrow$ Truth," commented a twentieth century expert on fluid mechanics and historian of mechanics on the gradual emergence of the Navier–Stokes equations, but then the treatise of Stokes appeared as "a burst of sunlight" [30, p. 316].

It is not accidental that it was mostly scientists in post-revolutionary France who paid so much attention to the mechanics of continuous media – not only fluid mechanics but also elasticity theory – in the early nineteenth century. This interest was rooted in the Laplacian program, in which all phenomena in nature were believed to be explainable in terms of an attraction or repulsion of particles. This program emerged in the tradition of Newton's natural philosophy: inspired by the model of celestial mechanics, central forces were believed to govern phenomena on a large scale as much as they do on the scale of atoms [31]. However, there was little unanimity on how to pursue this program: Navier was adhering the school of "analytical mechanics," in contrast to another faction which headed for a more "physical mechanics" approach. Institutionally these traditions were rooted in l'École Polytechnique and the special engineering schools, l'École des Mines and l'École des Ponts et Chaussées. Navier, for example, had studied at l'École Polytechnique and at l'École des Ponts et Chaussées before he became a professor himself at these institutions. From a sociological perspective, his career was described as an early example of a "hybrid career," where the realms of science and technology became entangled [32].

Stokes's effort to account for friction was also initially based on assumptions about "ultimate particles", but he became aware that his conclusions did not depend on such assumptions [33]. Like Navier, Stokes was primarily a theorist, but in contrast to Navier, he was not affiliated institutionally with engineering. As a professor at the University of Cambridge, Stokes had no official research interests devoted to experimental or technological studies. Nevertheless for Stokes "mathematics was the servant and assistant, not the master." His approach was described in an obituary: "His guiding star in science was natural philosophy. Sound, light, radiant heat, chemistry, were his fields of labour, which he cultivated by studying properties of matter with the aid of experimental and mathematical investigation" [34].

For Stokes, like for other nineteenth century natural philosophers, hydrodynamics was a specialty where fundamental questions about the constitution of matter sometimes went hand in hand with practical problems. This dual orientation, which led to the Navier–Stokes equations, is also apparent in the derivation of what is known as Stokes's law: a sphere of radius $a$ moving with a constant velocity $V$ in a fluid of viscosity $\mu$ experiences a resistance $6\pi\mu aV$. Stokes arrived at this result by simplifying the Navier–Stokes equation so that terms involving the square of the velocity were neglected. It was published in 1850 in a paper titled "On the Effect of the Internal Friction of Fluids on the Motion of Pendulums" [35, vol. 3, 1–141].

The relation to pendulums hints at the practical context that motivated this study: from Galileo via Huygens and Newton until the nineteenth century, the pendulum was the preferred instrument to measure time, but the precision that could be obtained theoretically was, in the true sense of the word, dampened by air resistance. When Stokes started to analyze the potential reasons why the swings of a pendulum would slow down, he investigated the buoyancy that the sphere at the end of a pendulum experiences in a medium as a primary cause. A second cause was the dynamic effect of the displacement of the medium, which resulted in an apparent increase of the inertia of the sphere. Stokes concluded "that the mass which we must suppose added to that of the pendulum is equal to half the mass of the fluid displaced." With

regard to friction, it was unclear to what extent the density played a role.[3]
Experiments commissioned by the Board of Longitude had shown that the resistance depended both on the density and composition of the gas in which the pendulum swung. In practice, medium-related influences for a pendulum designed for a certain period of oscillation were accounted for in terms of correction factors for the ideal length of a pendulum in vacuum. There were numerous theoretical and experimental studies in order to determine such correcting factors. Stokes cited studies performed by the German astronomer Friedrich Wilhelm Bessel (1784–1846) or the Frenchman Louis Gabriel Dubuat (1732–1787), whose research had been largely ignored by those interested in pendulum clocks, as Stokes argued, "probably because such persons were not likely to seek in a treatise on hydraulics for information connected with the subject of their researches. Dubuat had, in fact, rather applied the pendulum to hydrodynamics than hydrodynamics to the pendulum." The same may be said about Stokes. His goal was to derive an "index of friction," by which the experimentally determined correction factors for pendulums used for precise measurement of time could be understood in terms of hydrodynamics.

Stokes's law was of interest far beyond its original pendulum context. Stokes argued, for example, that the resistance of the water droplets in a cloud may be estimated from his law. "The terminal velocity thus obtained is so small in the case of small globules such as those of which we may conceive a cloud to be composed, that the apparent suspension of the clouds does not seem to present any difficulty," he argued. "The pendulum thus, in addition to its other uses, affords us some interesting information relating to the department of meteorology" [35, p. 10].

Stokes had also sketched another application which could be analyzed by the Navier–Stokes equations: he derived a formula for the velocity profile of a fluid in a tube. If one assumes that the velocity is zero at the inner wall of the tube (which Stokes mentioned as a possible assumption but did not pursue), one finds a parabolic increase of the velocity towards the tube's center. Integration over the tube's cross section yields the total flow as proportional to $r^4$ (with $r$ being the radius of the tube), or the resistance per unit length as proportional to $1/r^4$. This law was found earlier from experiments by the German hydraulic engineer Gotthilf Hagen (1797–1884) and the French physiologist Jean Louis Poiseuille (1797–1869); it became known as the Hagen–Poiseuille law. Hagen's experiments were performed with metal tubes with a diameter of a few centimeters and were motivated by practical considerations concerning the design of water pipelines. Poiseuille experimented with

**3)** Stokes assumed that the viscosity $\mu$ is proportional to $\rho\mu'$, where $\mu'$ is the "index of friction," and $\rho$ is the density of the medium; it was later shown by Maxwell that contrary to Stokes's assumption, the viscosity is independent of the density and therefore, the density does not enter into the formula of Stokes's law.

glass tubes with a diameter of only a tenth of a millimeter; he aimed at a better understanding of blood circulation [36, 37].

The theoretical explanation of the Hagen–Poiseuille law was published in 1860 in a physiological as well as a physical context, the former in the *Archiv für Anatomie, Physiologie und Medizin* and the latter in the *Annalen der Physik*. What is remarkable about these publications is that the result stemmed from such diverse disciplines – physiology and physics—which seems to indicate that after the Navier–Stokes equations were formulated and the first applications appeared, theory and practice would grow closer together. However, this was not the case. Hydrodynamics became an ever more theoretical science and hydraulics a specialty for practical men. The interest in the theory of ideal fluids did not fade away but further increased when mathematicians and physicists explored new avenues of fluid behavior in the second half of the nineteenth century.

## 1.6
## Revival of Ideal Fluid Theory

Despite d'Alembert's paradox, there is an influence upon the motion of a body in an ideal fluid that is due to the displacement of the fluid. In 1852, the mathematician Gustav Lejeune Dirichlet (1805–1859) investigated this influence through a novel analysis of Euler's equations. He wondered whether there were specific motions in which a resistance in an ideal fluid was theoretically possible. Dirichlet analyzed the case of a sphere in a uniformly accelerated fluid. He found that the sphere experiences a constant force proportional to the ratio of the densities of the fluid and the sphere, and to the accelerating force. This "resistance" was independent of the momentary flow velocity and disappeared with a vanishing acceleration, so that for the case of uniform motion, d'Alembert's paradox was established. Dirichlet's "resistance" had nothing to do with friction but was a mere inertial effect due to the displacement of fluid by the solid body, as analyzed by Stokes in his pendulum motion experiments. It was most conspicuous when expressed in terms of the kinetic energy: compared with motion in a vacuum, the kinetic energy of the sphere in the fluid was as if the motion involved an increased mass of the sphere. That mass corresponded to the mass of the fluid which had to be displaced by the sphere [38].

Although Dirichlet's result was derived from ideal fluid theory, it was important for the understanding of fluid resistance in real fluids because it showed how to discern forces due to inertial effects from friction. One is tempted to conclude that his result stemmed from efforts to learn more about the differences between ideal and real fluids, but that was not Dirichlet's mo-

tivation. His primary incentive was to refute Navier, who had expressed the
opinion that the known methods of integration are insufficient to solve the
partial differential equations of hydrodynamics, even in such cases where the
fluid extended to infinity and the body moving in it had the simplest shapes.
Dirichlet's analysis was meant to restore the trust in mathematics rather than
to show where ideal fluid theory could be a valid approach to real-world problems.

From the perspective of physics, too, ideal fluid theory held surprises in
store. Vortical motion, for example, was regarded as outside the scope of ideal
fluids because the mechanism to create vortices was believed to be the friction
between the fluid particles – a mechanism absent in ideal fluids. Hermann
von Helmholtz (1821–1894) regarded friction in fluids as one of the great riddles of mid-nineteenth century physics [39]. "The problem to define its influence and to find methods to measure it, is due to a large extent to the lack
of notion about the form of motion which is caused by fluid friction." This
is how Helmholtz introduced in a 1858 paper "On integrals of the hydrodynamic equations which correspond to vortex motion" [40]. After defining vortex motion (by defining the notions of vorticity, vortex lines, vortex filaments,
and vortex tubes), he derived three theorems from Euler's equations: 1) fluid
particles originally free of vorticity remain free of vorticity; 2) vorticity sticks
to the fluid particles on a vortex line; 3) the strength of a vortex tube remains
constant in time. In more colloquial terms, these theorems state that vortices
may not be created or destroyed in an ideal fluid, confirming the older view in
so far as vortices require an external cause and are alien to ideal fluids. However, vortices are not entirely alien to ideal fluids because if there are some
present at one time they persist forever. Helmholtz also mentioned that there
is an analogy of such vortical motion with magnetic fields: the electric current in a metallic wire may be compared to a fluid vortex filament, and the
magnetic field caused by the electric current in the wire may be considered
analogous to the rotating motion of the fluid particles around the filament.
This analogy rendered the theory important for the future development of
electromagnetism.

In 1868 Helmholtz studied a special case: what happens along the contact surface of two infinitely extended fluids if one fluid moves relative to
the other [41]? Based on his concept of vortical motion, Helmholtz regarded
the infinitesimal boundary between both fluids as a plane of parallel vortex
lines—a vortex sheet. Because the vortex lines, according to his theorem, stick
to the fluid particles, the vortex sheet would move with half of the relative
speed between both fluids like a ball-bearing. In a real fluid, the rotating motion of fluid particles would be communicated to neighboring particles; the
slightest motion perpendicular to the boundary would cause pressure differences (according to Bernoulli's law), which would further increase the defor-

mation such that more and more fluid particles would be caught in a vortical motion, and the surface of discontinuity would become a vortex layer of a finite thickness.

Based on Helmholtz's vortex sheet concept, Gustav Kirchhoff (1824–1887) developed a "Theorie freier Flüssigkeitsstrahlen" (theory of free fluid rays) in which he analyzed the case of a flat plate exposed to a flow under an oblique angle. William Strutt, better known as Lord Rayleigh (1842–1919), independently analyzed the same problem. According to Rayleigh–Kirchhoff theory, as it became known among aerodynamicists, surfaces of discontinuity extend from the edges of the plate, bounding a wake with dead-water at the back of the plate. Although the theory treated the fluid as inviscid, the plate experienced a force resulting from high pressure in front of and low pressure behind the plate. The normal component of this force could be interpreted as the lift of the plate, and the component parallel to the flow could be interpreted as the resistance. Both followed the square law as far as the dependence of the velocity was concerned, but in contrast to the discredited sine square law attributed to Newton's flawed particle concept, Kirchhoff–Rayleigh theory predicted a different dependence of the plate's lift on the angle of attack. The theory seemed to agree with experimental results "remarkably well," as Rayleigh believed. But this agreement was the consequence of an erroneous comparison between theory and experiment, as was later found, and Kirchhoff–Rayleigh theory entered the history of aerodynamics as another futile attempt to understand the lift of a wing [42, pp. 100–106].

With the focus on vortical motion, the gap between theory and practice became more pronounced. Stokes regarded it as an inappropriate attempt to transfer Helmholtz's vortex sheet concept from mathematics to physics—from ideal to real flow. William Thomson, better known as Lord Kelvin (1824–1907), discussed the concept critically in a paper, "On the doctrine of discontinuity of fluid motion, in connection with the resistance against a solid moving through a fluid." The discussion focused on the problem of the stability of vortex sheets. The analysis of Helmholtz–Kelvin instability, as it became known, became an active topic for research among theoretically minded fluid dynamicists for another century [43].

Although it was clear that Helmholtz's vortex theorems were valid only in ideal fluids, they were regarded as fundamentally important in wider areas. A consequence of the third theorem was that vortex lines in an infinitely extended fluid could not just start at one point and end at another; they had to be either infinitely extended or closed to a ring. "If there is a perfect fluid all through space, constituting the substance of all matter, a vortex-ring would be as permanent as the solid hard atoms assumed by Lucretius," Kelvin wrote to Helmholtz in a letter in 1867. Peter Guthrie Tait (1831–1901), a friend of Kelvin and a gifted popular lecturer, had visualized vortex rings with a simple experiment. He replaced one wall of a box by fabric and made a circular opening in the opposite wall; when he filled the box with smoke and pushed against the fabric, smoke rings where blown through the hole, as illustrated in Fig. 1.4. Kelvin was enthusiastic about Tait's demonstration: one could "easily make rings of a foot in diameter and an inch or so in section, and be able to follow them and see the constituent rotary motions," he reported to Helmholtz [44, p. 418].

**Fig. 1.4** Tait's smoke ring performance.

The idea of vortex atoms appealed to many physicists. The ultimate particles of matter could be conceived as vortex rings or an entangled combination of vortex rings made up of an ether with the properties of an ideal fluid. Natural philosophers in Victorian England cherished this idea for decades and ideal fluid theory became a universal basic science [45]. Stokes, Kelvin, and James Clerk Maxwell (1831–1879) were obsessed by the idea of a universal fluid whose properties could be explained by hydrodynamics. Joseph John Thomson (1856–1940), renowned as the discoverer of the electron, was an ardent advocate of the vortex atom in his youth. Still, in 1907, he regarded the atomistic conception based on the electron as "not nearly so fundamental as the vortex-atom theory of matter." For Albert Abraham Michelson (1852–1931), whose experiments finally became instrumental for the demise of the ether, the concept of the vortex atom was "one of the most promising hypotheses," and he hoped "that all phenomena of the physical universe are only different manifestations of the various modes of motion of one all-pervading ether" [46, pp. 472–473]. Even for mathematics, the vortex atom became a challenge. Entangled vortex rings were conceived as representations of molecules, and the study of such entanglement gave rise to new mathematical specialties – the theory of knots and topology [47].

**1.7**

**Reynolds's Investigations of "Direct or Sinuous" Flow**

Vortex rings also played a role in the discovery of turbulence as a heretofore unexplored source of fluid resistance: "Had not Professor Helmholtz some twenty years ago called attention to the smoke ring by the beautiful mathematical explanation which he gave of its motion," Osborne Reynolds (1842–1912) wrote in 1877, "it would in all probability still be regarded as a casual phenomenon, chiefly interesting from its beauty and rarity." But these vortex rings also provided "evidence of a general form of fluid motion," Reynolds recalled his investigations into this phenomenon. Two years earlier, he had speculated in the *Proceedings of the Literary and Philosophical Society of Manchester* how to prove an observation often made by sailors, "that rain soon knocks down the sea." He demonstrated with an experiment that the impact of drops on a water surface creates vortex rings. Not the splashes at the surface but the vertical mass exchange caused by the vortex rings calms the sea in a rain shower. By adding a colored surface layer, Reynolds was able to make this vertical mass transport visible. In a similar manner, he made visible the vortical motion that occurs in the wake of a solid body moving through the water: "Colouring the water behind the solid shows, that instead of passing through the water without disturbing it, there is very great disturbance in its wake. An interesting question is as to whether this disturbance originates with the motion of the solid, or only after the solid is in motion. This is settled by colouring the water immediately in front of the solid before it is started. Then on starting it the colour is seen to spread out in a film entirely over the surface of the solid, at first without the least disturbance, but this follows almost immediately" [48].

With these experiments, Reynolds became aware of the fundamentally important role played by vortical motion in various flow configurations. In 1880 he began to measure its influence on fluid resistance in tubes. The apparatus Reynolds used is depicted in 1.5. He injected a thin jet of colored liquid into the water, which flowed at an adjustable speed through a glass tube [49]. At small flow velocities the injected thread of colored liquid was smooth and straight, and the flow in which it was embedded could be assumed as direct or laminar. At a critical velocity it became sinuous, indicating the transition from the laminar state to a turbulent flow regime. Reynolds argued that this transition happens when a certain dimensionless quantity exceeds a critical value. He derived this quantity as $\rho U_m D / \mu$, where $U_m$ is the mean flow velocity, $D$ the diameter of the tube, $\rho$ the density of the fluid, and $\mu$ its viscosity. This quantity was later named the "Reynolds number" (usually the viscosity is replaced by the so-called kinematic viscosity, $\nu = \mu/\rho$) [50].

Turbulence had been studied both theoretically and experimentally before Reynolds [51]. The transition from laminar to turbulent flow was noted in Ha-

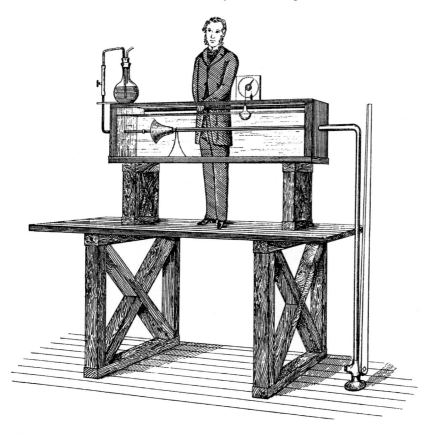

**Fig. 1.5** Reynolds's apparatus to observe vortical pipe flow.

gen's experiments as early as 1839, but Reynolds's approach to flow problems was nevertheless highly original. Analyzing the dimensions of all quantities involved in the Navier–Stokes equations, Reynolds singled out the dimensionless ratio that would bear his name as the crucial parameter on whose value it depends whether a flow develops vortices. The Reynolds number may be regarded as the ratio of inertial to viscous forces; it expresses the similarity law of fluid dynamics: flows around objects of different size that are geometrically similar are equivalent (i.e., the differential equations describing the flow behavior are identical) as long as the ratio of inertial to viscous forces, i.e., the Reynolds number, is the same. "Professor Reynolds has traced with much success the passage from the one state of things to the other, and has proved the applicability under these complicated conditions of the general laws of dynamical similarity as adapted to viscous fluids by Professor Stokes," acknowledged Lord Rayleigh (1842–1919), a pioneer of dimensional analysis (quoted in [52, p. 62]; on earlier uses of dimensional analysis see [53]).

Reynolds was too prolific to reduce his scientific legacy to the Reynolds number or the ingenious way he derived it. "I had no intention whatever of laying down the conditions of dynamical similarity," he once commented in a letter to Stokes when he became aware that Stokes interpreted his derivation in that sense. It was merely the comparative view of viscous and inertial forces he had in mind [51, p. 256]. He also did not regard himself as either a scientist or an engineer. "The results of this investigation have both a practical and philosophical aspect." The practical aspect was obvious, because the friction of water flowing through pipes was of enormous practical importance for hydraulic engineering. But Reynolds revealed that this was not his main concern: "The results as viewed in their philosophical aspect were the primary object of the investigation", because his results were addressing "the fundamental principles of fluid motion" [49, p. 51]. By "philosophical" Reynolds meant the theoretical problems of how two such fundamentally different modes of motion – laminar and turbulent—could be described by the same equations of motion, and what made a steady laminar flow unstable so that it became turbulent. In a popular lecture at the Royal Institution he compared the flow of water with military tactics: "For although only the disciplined motion is recognized in military tactics, troops have another manner of motion when anything disturbs their order. And this is precisely how it is with water: it will move in a perfectly direct disciplined manner under some circumstances, while under others it becomes a mass of eddies and cross streams, which may be well likened to the motion of a whirling, struggling mob where each individual particle is obstructing the others." The analogy went further: "The larger the army, and the more rapid the evolutions, the greater the chance of disorder; so with fluid, the larger the channel, and the greater the velocity, the more chance of eddies." Reynolds also performed experiments with a model of a ship moving in an illuminated water tank; he made the turbulent flow behind the ship visible with the motion of thin threads in the vortical wake: "It is these eddies which account for the discrepancy between the actual and theoretical resistance of ships" [54].

## 1.8
### Hydraulics and Aerodynamics: A Turn Towards Empiricism

Hydraulics began to thrive as an empirical science particularly in France. Late eighteenth-century inquiries concerning the motion of water in open canals and in closed conduits nourished the conviction that theory was little help for practical constructions. Antoine Chézy (1718–1798), at the request of the Paris Academy of Sciences, had established an empirical formula for the discharge of a canal in relation to its slope and its cross-section, for which no theory was available as a justification. Jean Charles Borda (1733–1799), an engineer concerned with harbor construction and hydraulic machinery, performed rotating-arm experiments to measure the drag of bodies immersed in water. Although he was able to verify the prevalent theoretical assumption of a proportional relation between the resistance and the square of the velocity, there was no theory to account for the variation of resistance with the body shape. Charles Bossut (1730–1814) performed towing experiments with boats and found that the resistance of the boats increased with the narrowness of the canal through which the boat was towed. Pierre Louis Georges Du Buat (1734–1809), who is regarded as the founder of the French hydraulic school, deplored in 1786 in the preface of his *Principes d'hydraulique, verifiés par un grand nombre d'experiences faites par ordre du gouvernement* "the uncertainty of the principles, the falsity of theory which is contradicted by experience, the paucity of observations made up till now, and the difficulty of making them well" [58, p. 130]. Nineteenth-century French hydraulics was already largely divorced from theoretical hydrodynamics, despite some "hybrid careers" like those of Navier or his pupil Saint-Venant or Joseph Boussinesq, whose research into turbulent flow were considered as pioneering from a theoretical and a practical point of view [51].

By the mid-nineteenth century, the growing gap between hydraulics and hydrodynamics could also be felt in other countries. In a textbook on *Hydromechanik* "analytical forms as well as natural-philosophical debates" were deliberately largely omitted, except where they proved indispensable. The book was authored in 1857 by a professor of the Technical University in Hanover "with the practical engineer as a reader in mind." When a second edition was published in 1880, the author wanted to change its title so that it appealed better to the intended audience, but the publisher insisted on publishing it with the old title under which it was well-known. The size of the second edition nearly doubled to 760 pages [57]. More than a hundred years after Bernoulli's and Euler's work, hydrodynamics and hydraulics were certainly no longer regarded as synonymous designations for a common science. Hydrodynamics had turned into a subject matter for mathematicians and theoretical physicists—hydraulics became technology.

Aerodynamics, too, became divorced from its theoretical foundations in hydrodynamics. Although aerodynamics was not yet regarded from the perspective of aviation before the twentieth century, this does not mean that it was without interest for practical purposes. Already in the 1857 edition of the textbook on *Hydromechanik*, there were chapters titled "Aerostatik" and "Aerodynamik"; the former provided the knowledge engineers required to build air pumps or steam vessels; the latter addressed the practical problems of windmills, the arrangement of sails, the construction of anemometers for meteorological observations, or ballistics. In all these areas of application, air resistance was the central problem. Aerodynamic theory could not provide a single formula that accounted for the various practical goals. Therefore, empirical formulae derived from experimental investigations were introduced for each special case. In 1896 a textbook on ballistics lists in chronological order 20 different "laws of air resistance," each one further divided into various formulae for different ranges of velocity. Many of these formulae could be expressed as proportional to the square of the velocity, but only if a so-called Siacci's factor was included, and this factor itself was a velocity-dependent function. The Siacci–Berardinelli law from 1892, for example, involved a Siacci factor that extended over the width of the page when it was explicitly formulated with all its coefficients and parameters [59, p. 53].

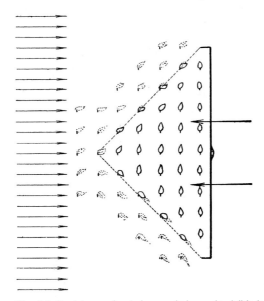

**Fig. 1.6** Resistance due to jammed air, made visible by candlelight [60, fig. 14].

Even the most cumbersome resistance formulae in hydraulics, such as in open-channel flow, did not represent the experimental data in such a weird mathematical form. No physical theory could provide a logical framework for justifying these empirical "laws." Ballistics illustrates to what extent practice had become divorced from theory, because it encompassed a broad range of experimental data on air resistance. In other areas of applied aerodynamics, the gap between theory and practice was expressed in a different manner. In 1896 an Austrian engineer published a book titled *Die Luftwiderstandsgesetze, der Fall durch die Luft und der Vogelflug* (The laws of air resistance, the fall through the air, and the flight of birds) in which the author developed the idea that the air in front of an obstacle becomes jammed and that the shape of this air barrier ("Stauhügel") is responsible for the magnitude of the resistance. If the obstacle is a square plate oriented perpendicular to the stream of air, the barrier of jammed air is assumed to be a pyramid; if it is a circular plate, the barrier is a cone, as shown in Fig. 1.6. The empirical proof was presented in the form of drawings of the inclination of a flame of a candle held in various positions in front of the obstacle [60]. This idea did not originate in the mind of a screwball; it found serious discussion, for example, in a review article published in 1902 on aerodynamics in the famous *Enzyclopädie der mathematischen Wissenschaften* [61, p. 153].

Concepts based on such speculations should not be derided as mere fantasies but rather as another expression of diverging trends between hydrodynamic theory and practical engineering, although it seems doubtful in this case whether the speculation of jammed air barriers was confirmed to such an extent by experiments as the author claimed. Another empirical aerodynamicist was Otto Lilienthal (1848–1896), a German engineer who became famous for his glider flights, as seen in Fig. 1.7. His accidental death gave him an aura of a martyr for the age-old dream of mankind to fly like birds. Less known are Lilienthal's experiments to measure air resistance. He was able to demonstrate that a curved plate experiences a lift even when it is oriented parallel to the stream of air. Lilienthal also introduced a novel method to display drag and lift data in so-called drag polars [42, pp. 141–153].

**Fig. 1.7** Lilienthal's glider flights were based upon experiments concerning the drag and lift of curved surfaces (Source: Deutsches Museum, Munich).

For Lilienthal, like for other pioneers of flight in the nineteenth century, aerodynamics was largely unrelated to the theoretical foundations of fluid mechanics. Nevertheless, driven by practical skill and enthusiasm for flight, they achieved considerable intuitive understanding for developing experimental techniques and deriving new aerodynamic insights from their observations. In 1870 the first primitive wind tunnel was made on behalf of a group of flight enthusiasts in Great Britain. In the 1880s Horatio Phillips (1845–1924) built an improved wind tunnel and performed airfoil tests, which he checked with an enormous whirling arm of 50 feet radius guided on rails and driven by a steam engine. In a letter to the British Journal *Engineering*, he reported that his experiments "have conclusively shown that by the use of convex surfaces, a 4 lb. weight may be sustained in the air by each sq. ft. of undersurface with a speed of current of about 40 m.p.h., and this without presenting any appreciable angle of undersurface to the current" [27, p. 165]. In America, Samuel Pierpont Langley (1834–1906) performed more whirling-arm investigations, which further nourished the belief in the feasibility of heavier-than-air flight. In his *Experiments in Aerodynamics*, published in 1891, Langley concluded "that these researches have led to the result that mechanical sustentation of heavy bodies in the air, combined with very great speeds, is not only possible, but (...) that we now have the power to sustain and propel them" [42, p. 165].

## 1.9
### Fluid Mechanics ca. 1900

The author of a textbook on hydrodynamics published in 1900 argued that "the actual processes are often in such unsatisfactory agreement with the theoretical conclusions that technology has adopted its own procedure to deal with hydrodynamical problems, which is usually called hydraulics. This latter specialty, however, lacks so much of a strict method, in its foundations as well as in its conclusions, that most of its results do not deserve a higher value than that of empirical formulae with a very limited range of validity" [55, p. III].

When Arnold Sommerfeld (1868–1951) presented a talk in the same year on recent investigations in hydraulics, he pointed to the case of fluid resistance in pipes as an example for the gap between hydrodynamics and hydraulics: "Physical theory predicts a frictional resistance proportional to the velocity and inversely proportional to the square of the diameter, according to the technical theory it is proportional to the square of the velocity and inversely proportional to the diameter. The physical theory agrees splendidly in capillary tubes; but if one calculates the frictional losses for a water pipeline one finds in certain circumstances values which are wrong by a factor of 100" [2]. Sommerfeld argued that the contradiction was due to laminar flow in the former case and turbulent flow in the latter, but could not offer a way out of the dilemma because there was no theory available for both cases. Sommerfeld's view was that of an aspiring young mathematician with a keen interest in theoretical physics, but the gap between hydraulics and hydrodynamics was noticed also by those who represented the practice. The author of a review article on hydraulics described in 1906 his specialty as "a domain of coefficients with a working method based often only upon the interpolation of empirical data." Hydraulic laws would only address special cases and not lend themselves to deepen our knowledge of how the phenomena are connected with one another. Nevertheless, "a theoretically unsatisfactory solution, even if it only turns out to be useful within the limits in which it is used in technology, is still better than no solution at all" [56, p. 327].

The rise of hydraulics and aerodynamics as practical specialties indicates that basic theory had little to offer for practical purposes – but it did not lead to a decline of their mother-discipline. Hydrodynamics flourished independently of its practical offspring as a basic science throughout the second half of the nineteenth century. Light, electricity, magnetism, heat, gravity, in one way or another, seemed to be ultimately explained by a universal ether whose properties could be understood in terms of hydrodynamics. Even when the belief in a mechanical ether faded away, hydrodynamics survived and further flourished into the twentieth century as part of the fledgling discipline of theoretical physics. Some theorists, like Vilhelm Bjerknes, hoped that new theories such as the fashionable electron theories would soon lose attention and hydrodynamics would rise up again as the predominant field of fundamental research: "I myself hold no doubt that this will become the future for representation for mathematical physics," he wrote to his father in 1901, whom he helped to publish a fundamental treatise on hydrodynamics. "In this manner hydrodynamic phenomena will receive a central position in mathematical physics." When this did not happen to the extent he had hoped, he specialized in geophysics and meteorology as new fields in which the introduction of hydrodynamics offered useful prospects. He regarded the atmosphere as "a big laboratory for hydrodynamics" [62, pp. 24–25, 45].

# 2
# The Beginnings of Fluid Dynamics in Göttingen, 1904–1914

Among those who experienced the rise of fluid dynamics in the twentieth century it was common to notice the wide gap between theory and practice around 1900: "At the end of the nineteenth century, fluid mechanics was split along two different directions, which were barely in touch with one another," one author introduced his textbook. Theoretical hydrodynamics on the one side was opposed to the more technical specialties, hydraulics and aerodynamics, on the other. "It is the great merit of L. Prandtl, to have shown a way how both diverging trends of fluid mechanics could be combined again" [63, p. 1]. This remark hinted at the boundary layer theory, conceived by Ludwig Prandtl (1875–1953) in 1904. This theory became the epitome of modern fluid dynamics. "The paper will certainly prove to be one of the most extraordinary papers of this century, and probably of many centuries," another pioneer of fluid dynamics remarked on Prandtl's publication [1, p. 11]. A Japanese aerodynamicist wrote: "This paper marked an epoch in the history of fluid mechanics, opening the way for understanding the motion of real fluids" [3, p. 87].

As is obvious from these quotes, boundary layer theory is regarded by the international community of fluid dynamicists as crucial to bridging the gap between theory and practice in this discipline. However, this is a retrospective evaluation. Outside Göttingen boundary layer theory was largely ignored for almost two decades after Prandtl's first publication in 1904. This delayed reception is another manifestation of the gap that separated theory and practice at the beginning of the twentieth century. The same is true for the so-called circulation theory of lift, which Wilhelm Martin Kutta (1867–1944) and Nikolai Joukowsky (1847–1921) introduced independently of one another in the decade before the First World War. This theory emerged as a mathematical concept and seemed remote from practical application before Prandtl transformed it into a full-fledged airfoil theory for wings of a finite span. Both boundary layer theory and airfoil theory became the subject matter of heated debates in the 1920s.

*The Dawn of Fluid Dynamics: A Discipline between Science and Technology.* Michael Eckert
Copyright © 2006 WILEY-VCH Verlag GmbH & Co. KGaA, Weinheim
ISBN: 3-527-40513-5

This chapter describes the emergence of these theories before the First World War, when Prandtl and his pupils developed them without much resonance from outside Göttingen.

## 2.1
### Prandtl's Route to Boundary Layer Theory

"I have posed myself the task to do a systematic research about the laws of motion for a fluid in such cases when the friction is assumed to be very small," Prandtl explained in August 1904 at the Third International Congress of Mathematics about the motivation for his boundary layer theory. To present the talk, "On fluid motion at very small friction," at such a congress may seem unusual, but Prandtl did not immediately address practical problems. He started with a remark on Dirichlet's motion and the paradoxical result that the ideal flow around a sphere may not be obtained as a limiting process of viscous flow with vanishing viscosity. If one goes to the limit of inviscid flow, Prandtl argued, "one obtains something quite different from Dirichlet's motion" [64, p. 576].

Dirichlet's motion dealt with the resistance due to the inertia of the displaced medium rather than to viscosity (see Chapter 1). Although a sphere moving in an inviscid fluid, because of the displacement of fluid, has a larger kinetic energy than in a vacuum, its speed does not decrease (d'Alembert's paradox) as in the case of viscous resistance. Dirichlet's motion and d'Alembert's paradox were familiar topics among mathematicians. Prandtl added to these another paradox: reducing the viscosity in the Navier–Stokes equations did not yield Dirichlet's solution. Or, put differently, in the limiting case of zero viscosity, the Navier–Stokes equations reduced to Euler's equations – but their solutions did not reduce to those of Euler's equations!

Alluding to the tradition of mathematical hydrodynamics from d'Alembert to Dirichlet appealed to the expectations of Prandtl's audience at this mathematics congress. But Prandtl's own interest in this problem was not mathematical. He was an engineer by training, educated at the Munich Technical University. Although he had finished his studies with a doctoral dissertation in mathematics at the University of Munich, mathematics was not his primary area of interest. Prandtl had been first a student and later an assistant of August Föppl (1854–1924), a famous professor of technical mechanics, and presented his dissertation to the mathematicians of Munich University only because technical universities at the end of the nineteenth century were still not granted the right to award doctoral degrees. Mechanics and mathematics, however, were regarded as neighboring disciplines, and so it was not unusual to submit a treatise in theoretical mechanics to mathematicians for acceptance as a doctoral thesis. Theoretical mechanics was regarded by mathematicians to some extent as their own territory. Mathematicians from universities also found employment as professors of mechanics at technical universities. At a time when technical universities were involved in a struggle of emancipation with the universities, this cross-fertilization caused tensions. Mathematicians from universities who were called as professors of mechanics to technical universities were regarded with suspicion by their colleagues from engineering departments who saw their own concerns not appropriately addressed by the academically trained mathematicians. On the other side, applied mathematicians invited derision from the pure mathematicians at the universities if they approached the interests of technology. "Grease!" was the despicable definition of the hydrodynamical theory of lubrication, for example, from which a pure mathematician kept some distance [65, 66].

Against this background, Prandtl's boundary layer presentation at the International Mathematical Congress, like Prandtl's career itself, appears to bridge theory and practice. His ascent began 1900 when he was an engineer at the Vereinigte Maschinenfabrik Augsburg und Maschinenbaugesellschaft Nürnberg (which became the Maschinenfabrik Augsburg-Nürnberg AG, MAN after 1908). In the following year Prandtl was called as professor of mechanics to the Technical University of Hanover. Three years later, in 1904, he became director of the new Institute for Technical Physics at the University of Göttingen, funded by a society of industrialists and academics, the Göttinger Vereinigung zur Förderung der angewandten Physik und Mathematik. The initiative for the establishment of applied university institutes in Göttingen came from Felix Klein (1849–1925), who became famous for his achievements in mathematics as much as for his entrepreneurial activities in science. Klein regarded Prandtl as the embodiment of the ideal of a theorists oriented toward practical problems, who combines "a strong power of intuition and great originality of thought with the expertise of the engineer and the mastery of the mathematical apparatus," as he once described him in a report, and above all, Prandtl displayed "pedagogical interest" [65, p. 232].

When Prandtl presented his boundary layer paper at the Heidelberg Congress, he had just accepted the call to Göttingen. Although the prestige of a professor at a university was regarded higher than that of a professor at a technical university, Prandtl's new position was a step down the career ladder: in Hanover, he had been full professor, in Göttingen he was associate professor. However, Prandtl was not concerned, because Klein must have confided in him that his position would soon be raised to the status of an ordinary professorship. Behind Prandtl's deeper concerns lurked the tension between purely academic and technical universities. After three years at a technical university and a year in industry he felt loyal with the technical universities' struggle for emancipation. "The gravest doubt emerged from my sense of belonging to technology," he wrote to Klein in May 1904. He went through "a hard inner struggle," as he confessed in another letter, but in the end he accepted the call to Göttingen. The Göttinger Vereinigung added a subsidy so that he had not to face a reduction of his salary. But what attracted him most was the prospect of running his own laboratory and having more time for research; above all he was looking forward to "the beautiful scientific Göttingen intercourse" [67, pp. 14 and 297]).

Besides Klein the Göttingen circle of theorists comprised the famous mathematician David Hilbert (1862–1943), the theoretical physicist Woldemar Voigt (1850–1919), and the astronomer Karl Schwarzschild (1873–1916). Furthermore, and simultaneously with Prandtl, the mathematician Carl Runge (1856–1925) was called to Göttingen as professor of applied mathematics—one of the first professorships in the world explicitly dedicated to this specialty. Runge and Prandtl had already been colleagues in Hanover; in Göttingen their disciplines were combined in 1905 under the same roof in a new institute for applied mathematics and mechanics. Nowhere could Prandtl have found more resonance with his own research interests.

Against this background it is understandable that Prandtl put the theoretical problem of Dirichlet's motion at the beginning of his presentation at the International Mathematical Congress in Heidelberg. He must have regarded the problem as a business card for introducing himself to the ambitious circle of theorists in Göttingen and elsewhere. His own motivation, however, was rooted in practice rather than theory. As he recalled later, the impetus came from a problem he had encountered as an engineer a few years earlier. He had to design an exhaust system for wood shavings and swarf: "In a larger installation of air conducting pipes at the Nürnberg machine factory I had arranged a tapered tube in order to restore pressure," Prandtl recalled many years later, "instead of a pressure retrieval, however, the flow of air became detached from the walls. Today I know that I only would have had to shape the cone more slender in order to succeed. I received a call to the Technical University of Hanover then, and the firm did not much care about the loss of pressure. But I could not forget the problem why a flow rather than streaming along the wall detaches itself from it, until three years later boundary layer theory provided the answer" [68, p. 1605]. In this recollection, presented many years after the events at the Nürnberg machine factory, Prandtl played down the practical importance of this problem. To "restore pressure" was not only an academic problem. In the talk "Shavings and Swarf Exhausters" presented in 1903 at a meeting of engineers in Hanover, Prandtl had explained that the operation of the exhaust system consumed an enormous amount of power, of which a considerable part was used only to compensate for the loss of pressure. Retrieving pressure meant avoiding energy loss. Prandtl had invented a power-saving pipe connect that reduced the losses considerably. He even obtained a patent for this invention [67, pp. 10–11].

At the Technical University of Hanover, Prandtl became charged with new tasks. In his main lectures he had to cover the entire field of mechanics. His special lectures also did not yet reveal a preference for fluid mechanics; he lectured on "selected chapters of technical mechanics," for example, or on "statics of constructions." In his publications and in his scientific correspondence from that period he focused on elasticity theory and vector calculus, then a novel formalism in mathematics [69]. From his unpublished notes, however, it is apparent that he pursued the problem of flow separation. He even constructed a water canal for visual observation of flow phenomena, shown in Fig. 2.1.

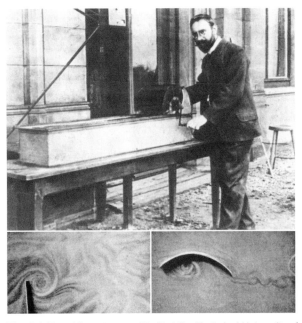

**Fig. 2.1** Prandtl's water canal built at the Technical University of Hanover for the illustration of flow separation.

In his presentation at the Heidelberg Congress, he used photographs obtained from this canal in order to illustrate the flow separation from the walls of an obstacle in the stream of water. It must have been unusual for the attending mathematicians to be presented with such experimental detail (see Fig. 2.2):

"The apparatus (displayed in the figure both in a plan view and in vertical section) consists of a 1.5-meter-long container with an intermediate bottom. The water is circulated by a paddle wheel and, guided by vanes 'a' and tranquilized by four sieves 'b,' enters the upper part relatively free from swirls; at 'c' the object is inserted. A mineral (iron mica) is suspended in the water which consists of tiny shining lamellae; thereby all deformed sites of the water, particularly all vortices become visible by a peculiar effulgence, caused by the orientation of the lamellae at the respective sites" [64, p. 580].

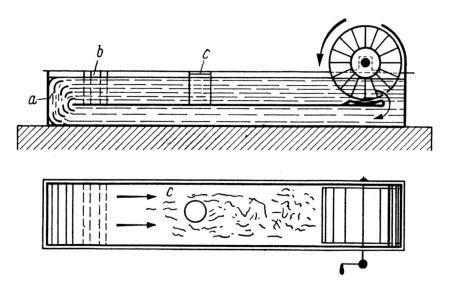

**Fig. 2.2** Vertical section and plan view of Prandtl's water canal.

About half of Prandtl's presentation was concerned with visualization and phenomenological descriptions of vortex formation in the wake of obstacles placed in the stream. The published paper presents two plates with 12 photographs. The boundary layer theory itself was only sketched qualitatively. Prandtl described the central idea behind his theory in these words:

> "By far the most important part of the problem concerns the behav
> ior of the fluid at the walls of solid bodies. The physical processes
> in the boundary layer between the fluid and the solid body is ad
> dressed in a sufficient manner if one assumes that the fluid does
> not slip at the walls, so that the velocity there is zero or equal to
> the velocity of the solid body. If the friction is very small and the
> path of the fluid along the wall not very long the velocity will at
> tain its free stream value already at a very close distance from the
> wall. Although friction is small, within the small transitional layer,
> the abrupt changes of velocity result in considerable effects" [64,
> p. 577].

In other words, Prandtl localized the influence of friction within a thin transitional layer – the boundary layer. Outside this layer friction could be neglected, i.e., ideal fluid theory (Euler's equations) was valid. He derived simplified partial differential equations from the Navier–Stokes equations and determined that within the layer, the no-slip condition applied to one side of the body's surface and free-flow conditions applied to the other. (Solving these boundary layer equations remained a challenge, but they were easier to solve than the Navier–Stokes equations.) The particular virtue of this approach, as Prandtl saw it in 1904, was that it offered the prospect to account for flow separation and vortex formation:

> "For the application the most important result of these inquiries is that in certain cases the flow separates from the wall at a position which is completely determined by the exterior conditions (see Fig. 2.3). A fluid layer set in rotational motion by the friction at the wall moves into the free fluid and, exerting a complete change of motion, plays there a similar role as Helmholtz's discontinuity sheets" [64, p. 578].

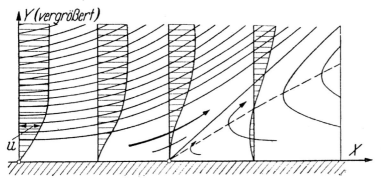

**Fig. 2.3** Velocity profile of the boundary layer, indicating the site where the flow separates from the wall. (The $X$-axis indicates the distance along the wall, the $Y$-axis the distance perpendicular from the wall).

The site of flow separation could be localized to the point where the gradient of the velocity changed in sign. Solving a specific flow problem would have required the numerical evaluation of the boundary layer equations for specific boundary conditions. Prandtl did not provide such a calculation, but he presented a physical argument about the mechanism involved in this process. The flow would only separate from the wall if the pressure in the direction of the flow increased. Together with such an increase of pressure, kinetic energy would be transformed in potential energy so that there was not enough kinetic energy left in the boundary layer to move forward into the region of higher pressure, and therefore, the flow would swerve away from the wall.

Prandtl's scant argument based on imagination rather than mathematics left much to be desired. When a colleague many years later asked him why he had presented such a fundamental concept in such terseness, Prandtl replied "that he had been given ten minutes for his lecture at the Congress and that, being still quite young, he had thought he could publish only what he had time to say" [1, p. 11]. "Your talk was the most beautiful one of the whole congress," Klein reportedly remarked to Prandtl at the end of his presentation. If Klein did praise Prandtl as such, and if he, as was furthermore suggested, "immediately recognized the momentousness of Prandtl's method" [70], then Klein was out on a limb. It took years before the scope of the new theory was recognized. Prandtl had performed preliminary calculations, as is evident from dozens of unpublished pages of manuscript, but he did not immediately pursue the matter further. Obviously boundary layer theory was not immediately regarded as such a breakthrough to bridge the gap between theory and practice in fluid mechanics as it appears to have been in retrospect.

## 2.2
### "Per Experimentum et Inductionem Omnia"

At about the same time when Prandtl took pictures of swirls in his water canal at the Technical University of Hanover, Friedrich Ahlborn (1858–1937), a biology teacher at a Hamburg high school, performed similar experiments in a somewhat larger water hod. Ahlborn experimented between display cases of impaled butterflies, stuffed animals and aquarium containers in the "zoological cabinet" of his school, but his water canal, shown in Fig. 2.4 was much more sophisticated than Prandtl's, and his photographs of flow phenomena showed finer details.

Ahlborn is largely unknown today. He was portrayed already half a century ago as a "forgotten pioneer of fluid dynamics" [71]. In his time, however, he was renowned for his research about the flight of seeds, bird flight, and flying fish. He was an accomplished experimental scientist and a respected member of the Hamburg Natural Science Association.

In November 1903, Ahlborn presented his observations of flow phenomena to the general business meeting of the Shipbuilding Society in Berlin. He introduced his talk with the remark that since Newton's days the problem of fluid resistance has not ceased to challenge the most distinguished physicists as well as the leading experts of ship building and hydraulic engineering, but there remained a "yawning gap between theory and experience, as well as between the results of the various theories among another." Then he expressed his belief that a real understanding into this matter was possible only if we did not make a priori judgments based on opinions and imaginations, but rather instruct ourselves by exact experimental methods. His motto was: "Per experimentum et inductionem omnia," everything is to be obtained inductively from experiments [72, p. 4].

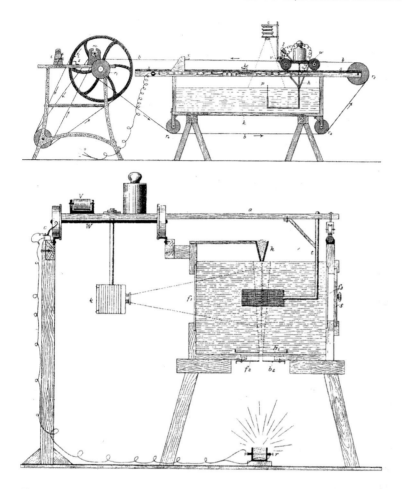

**Fig. 2.4** Ahlborn's apparatus used to photograph flow phenomena.

Ahlborn's motto is illustrated by his apparatus: in contrast to Prandtl's water canal where objects were placed in a circulating flow of water, Ahlborn moved his test bodies together with a camera on rails through still water. In order to visualize the flow, he sprayed spores of lycopodium into the water – see Fig. 2.5. Exposures could be obtained both from above and through a glass window from the side, with the illuminating flashlight at a vertical position. The bright traces of lycopodium "designate as natural stream lines the flow directions with utmost precision," Ahlborn explained, "while their length reveals the velocity of motion at each position. Because the duration of exposure is constant, about 1/25 seconds, the slowly flowing spores create shorter lines than the fast-flowing ones" [72, p. 7].

Prandtl was probably not aware of Ahlborn's experiments at the time of his own first efforts to visualize flow phenomena. But the Hamburg biol-

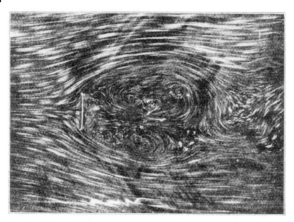

**Fig. 2.5** Streamlines in the wake of a plate according to Ahlborn. For comparison, see Fig. 2.1, Prandtl's flow images presented at the Heidelberg Congress in 1904.

ogist would not remain unknown to the Göttingen mechanics professor for very long. In 1905 the *Jahrbuch der Schiffbautechnischen Gesellschaft* published Ahlborn's treatises, "The vortex formation in the mechanism of resistance of water" and "The action of the ship propeller on the water," papers that must have found Prandtl's interest as much for their themes as for the journal in which they were published, because shipbuilding and hydraulic engineering was the closest area of application for new research on fluid resistance. In 1909 appeared another paper by Ahlborn titled "The mechanisms of resistance in water on plates and hulls of ships. The generation of waves." In the same year Ahlborn established contact with Prandtl by correspondence. He expressed his curiosity about the new aerodynamic research facilities in Göttingen, about which Prandtl had lectured in summer 1909 at a meeting of the *Verein Deutscher Ingenieure (VDI)*. Ahlborn proposed that "we act in concert according to a common plan" for future research on fluid and air resistance and that the details were to be negotiated at a personal meeting [73].

But the proposed concerted plan was not executed. In their subsequent correspondence there was no talk of common research on fluid resistance, neither in aerodynamics nor for hydraulic engineering. Instead Ahlborn conceived new plans for his research in Hamburg. On December 11, 1909, he presented slides and a film on flow phenomena in a lecture on "Hydrodynamical experimental investigations" before the Göttingen Physical Society, surveying his "very troublesome and lifelong experiments in the area of fluid motions." However, because of a lack of means, the continuation of his research was "seriously called into question," he argued. His research canal was too short for quantitative measurements because the sensitive measuring instruments would not reach a stationary state within the short range of motion. He hoped that the Göttingen physicists could provide "moral support" for his plan to found a large laboratory equipped with all required means for more quantitative hydrodynamical experiments in Hamburg [74].

Prandtl and the Göttingen physicists provided more than merely moral support. "Those of us concerned with hydrodynamics and aerodynamics have long been aware of professor Fr. Ahlborn's work and appreciate it very much," Prandtl wrote as spokesman of the Göttingen academics in an advisory opinion; Ahlborn's flow images produced by superb techniques were unsurpassed and very beneficial both for the theory and practice of fluid motion. For example, practical uses could be drawn "for the investigation of ship models, and furthermore for models of ship propellers, airships and wings of airplanes by revealing at what sites of a model resistances originate and how these sites should be modified." The Göttingen professors therefore "most warmly" supported the plan for a new laboratory [75]. Ahlborn used this expert opinion as an attachment in his proposal to the Hamburg senate. He was convinced that it "will be of utmost benefit for me here," as he thanked Prandtl [76]. It took two years before Ahlborn informed Prandtl that his project "now is in the hands of the Hamburg chamber of commerce, whose judgment will be decisive." But Ahlborn was disappointed. The Hamburg hydrodynamical laboratory did not materialize. Only during the First World War was Ahlborn provided with larger research facilities, when the military aviation establishment in Berlin-Adlershof founded a laboratory equipped with a water canal 20 m long for hydrodynamical flow research, designated as "Testing Department Captain Ahlborn." Apart from occasional reprints of Ahlborn's flow photographs in review articles and textbooks, however, Ahlborn's experimental hydrodynamics left few visible traces in the history of fluid dynamics.

Ahlborn's experimental orientation could have served as a challenge for Prandtl's research, and vice versa, because they had more in common than their interest in fluid dynamics. When Prandtl acknowledged in 1909 that theory would benefit from Ahlborn's results he probably hoped that they could provide visual evidence for his boundary layer theory. For example it was obvious from Ahlborn's slides presented to the Göttingen Physical Society that "at the surface of the plates a fluid layer is adhered"—such was the no-slip condition of the boundary layer theory confirmed experimentally. But what Prandtl regarded as a boundary layer was different from Ahlborn's conception. According to Ahlborn the thickness of the layer was primarily dependent on the roughness of the body surface from which the layer becomes rolled up into a sequence of vortices. He compared it to a ball bearing: "like frictional rolls between the solid surfaces and the surrounding still water" [74, p. 204]. Ahlborn also had a special name for it; he called it the "balanus layer," from the Latin word "balanus," meaning barnacle [77, pp. 40–44]. Barnacles grow on the hulls of ships; it increases the roughness, and therefore also the surface resistance. Ahlborn occasionally remarked that there was a similarity between his "balanus layer" and Helmholtz's vortex sheets, just as Prandtl had indicated in his Heidelberg Congress paper that vortex sheets could result from his boundary layer. These observations could have sparked a debate about their mutual conceptions, but Ahlborn never developed his views into a theory, and Prandtl's theory offered no interpretation of what Ahlborn saw in his photographs. It was clear that skin friction originated from a thin layer, which was recognizable as a bright layer because the spores of lycopodium reflected the flash light stronger there than further away from the surface of the body, but whether the increase of brightness was due to slowly moving spores (as one would assume in a laminar boundary layer) or to vortical motion (which could have been interpreted as frictional rolls) was impossible to decide. Ahlborn believed that "tender cycloidal serpentines" were evidence supporting his view that the transitional space to the free stream at some distance from the surface was "filled with a long chain of vortices," but this hypothesis was hard to verify [78, Figs. 5 and 6].

In the 1920s Ahlborn and Prandtl became engaged in sometimes harsh scientific quarrels (see Chapter 6). Despite their common interest in the nature of the boundary layer and the mechanism of skin friction, their mentalities as researchers had little in common. "Per experimentum et inductionem omnia," Ahlborn's leitmotif, was not Prandtl's, as will become apparent by a closer inspection of his Göttingen research program.

## 2.3
### The First Doctoral Dissertations on Boundary Layers

Between 1905 and 1914, Prandtl supervised 17 doctoral dissertations. The themes of the dissertations ranged from elasticity theory to gas dynamics: 7 may be categorized as hydro- or aerodynamics, and three among these addressed the new issue of boundary layers.[1] This pattern illustrates that Prandtl's research interests were oriented towards the entire field of applied mechanics when he came to Göttingen University in 1904 as associate professor of technical physics. In 1905, the new institute for applied mathematics and mechanics was opened under Prandtl's and Runge's directorship. In 1907 Klein and Prandtl conceived plans for a model airship testing facility; henceforth, aerodynamics was put on Prandtl's agenda as a major focus of research. In 1909 he started to lecture regularly on aviation sciences. Two years later Prandtl came forward with a plan to found a "Kaiser Wilhelm Institute for Aero- and Hydrodynamics" [67].

---

**1)** A list of doctoral dissertations supervised by Prandtl is presented in [79, vol. 3, pp. 1612–1617].

In view of Prandtl's ambitious plans, boundary layer theory was just one out of many research problems before the First World War. The first doctoral dissertation to emerge out of Prandtl's supervision was a careful investigation of the simplest case: deriving the velocity profile in the boundary layer along a flat plate, and hence the resistance of the plate. From Prandtl's unpublished notes it is apparent that he had solved the problem already at the time of the Heidelberg Congress in a rather crude approximation. It was left to his first boundary layer doctoral student, Heinrich Blasius (1883–1970), to arrive at a more satisfying development of this problem. Blasius had studied mathematics and physics in Marburg and Göttingen, where he became particularly interested in applied mathematics. He tended to theory, but had "certainly no intention to focus on pure physics," explained Blasius when Prandtl offered him financial support and temporary employment in his institute for some time after his dissertation [80]. In accord with his theoretical orientation, the scope of the doctoral work was limited to the mathematical development of the boundary layer differential equations without additional experiments to check the results. Blasius first considered the case of a thin flat plate submerged in a flow parallel to the plate's surface; he could simplify the problem by similarity considerations so that it was reduced to solving an ordinary differential equation, which Blasius achieved by a sophisticated power series expansion. His result largely confirmed Prandtl's velocity profile presented at the Heidelberg Congress. He was able to improve Prandtl's formula for the resistance of the plate by slightly changing a numerical factor. Although these results made Blasius a name in the history of fluid dynamics (via what became known as the Blasius equation), most of these results were obtained by Prandtl earlier in his unpublished notes. Probably for this reason, Prandtl did not consider them sufficient for the doctoral degree, and Blasius had to add a second part to his dissertation, in which he studied the problem of boundary layer separation from the surface of a circular cylinder and with the time-dependent formation of a boundary layer at the start of flow [81].

In another doctoral work, the boundary layer at the surface of rotational bodies was made the subject of a detailed investigation [82]. After this theoretical work was complete, it was time for a "quantitative experimental test of Prandtl's ansatz," as a third dissertation on boundary layers was introduced in 1911. This work focused on pressure measurements along the circumference of a cylinder in a uniform flow of water, because the pressure distribution allowed one to determine the velocity profile in the boundary layer and the site where it became detached from the surface. For this purpose, a water canal, depicted in Fig. 2.6 was constructed, in which the flow velocity could be varied both by the power of the pump used to circulate the water and a set of exchangeable cartridges with mouths of different sizes. The pressure along the circumference of the cylinder immersed into this flow was measured

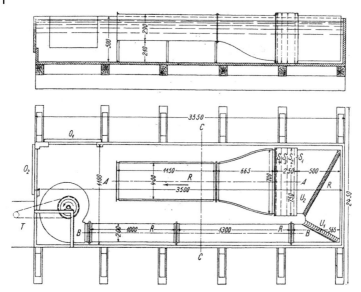

**Fig. 2.6** Vertical section and plan view of a Göttingen water canal used in 1911 for testing boundary layer theory.

through holes drilled through the surface of the cylinder using a novel device for measuring water pressure. This water canal was a hydrodynamically precise instrument. It had little in common with Prandtl's first water canal and its hand-driven paddle-wheel, except that the water circulated, unlike Ahlborn's canal, in which the object was moved through still water [83].

In order to compare the experimental values obtained from this device with theory the boundary layer equations were solved numerically (using a method conceived by Wilhelm Kutta in 1901 [84]); the result was "a quantitatively very satisfactory agreement of observation with calculation" [83, p. 410].

This doctoral work also became important for Theodore von Kármán's (1881–1963) first contribution to fluid mechanics. Kármán had come to Göttingen from Hungary in 1906 in order to complete his doctoral studies under Prandtl's supervision. Like Prandtl himself, Kármán had first focused on elasticity theory. In 1910 he presented for his habilitation as Privatdozent (this is the qualification to teach in German universities) a treatise on plastic deformation. A year later he collaborated with the theoretical physicist Max Born in a pioneering paper on the quantum theory of the specific heat of solids. Before 1911 Kármán had paid little attention to hydrodynamical problems. The occasion to familiarize himself with hydrodynamics came during a discussion between Kármán and Prandtl's doctoral student who built the water canal to test the boundary layer theory. Detached vortices from the cylinder "caused an irregular oscillation of the entire wake," described the student in his dissertation [83, p. 372]. It proved difficult, therefore, to measure the pressure around the cylinder with the desired precision.

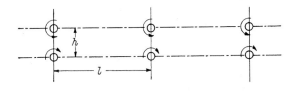

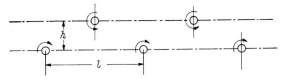

**Fig. 2.7** Only rows of mutually displaced vortices (lower sketch) proved as a stable configuration; the upper parallel configuration is unstable. Kármán's result for $h/l = 0.28$ agreed well with observations.

Kármán became curious and tried to find out the cause the oscillating wake [85]. He compared two arrangements of detached vortices – one with pairs of vortices trailing away after being simultaneously shed from two opposing sites of the cylinder, and one in which the vortices were staggered (see Fig. 2.7). Only the latter turned out to result in a stable configuration – a Kármán vortex street, as it was later named. Kármán's stability analysis also accounted for the geometry of this vortex street, i.e., the ratio $h/l$, where $h$ is the distance between the two rows of vortices, and $l$ is the distance between vortices in a row. Because the trailing vortices carried momentum away, it was also possible to account for the resistance of the cylinder in the flow that resulted from this vortex-shedding. Instead of the Kirchhoff–Rayleigh theory of fluid resistance (see Chapter 1), which predicts the unrealistic result of dead-water in the wake of an obstacle, Kármán's theory regarded the vortex-shedding and the concomitant transport of momentum as a major mechanism for the resistance [86].

The theory was immediately subjected to experimental tests. Photographs obtained in a water canal according to Ahlborn's method (moving the object through still water with lycopodium sprayed on the water surface) confirmed Kármán's results [87].

## 2.4
## Airship Research

In addition to these experiments in Prandtl's institute at Göttingen University, there began in 1908 aerodynamic laboratory work at the new model testing facility of the motorized airship study society, the "Modellversuchsanstalt," abbreviated as MVA. The central experimental device of this laboratory was

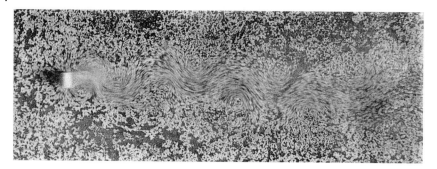

**Fig. 2.8** Kármán vortex street photographed using Ahlborn's method.

a wind tunnel, as depicted in Fig. 2.9. In contrast to the earliest wind tunnels elsewhere, which sucked air in at one end and blew it out at the other, Prandtl had conceived the Göttingen tunnel like his water canal with a closed circuit. Experiments in this new facility had "to serve practice immediately," as Prandtl explained at an aviation conference in 1911 [88]. In contrast to the academic milieu at Prandtl's university institute, where the motive to undertake an experiment was embedded in a scientific investigation and most often carried out as part of a doctoral dissertation, the research goals at the MVA were entirely dictated by external contracts that usually addressed specific aerodynamics problems of aviation. The first flying devices – cigar-shaped airships and strut-and-wire bird-like airplanes—were designed with little aerodynamic underpinning. Laboratories like Prandtl's MVA played a crucial role in bringing aviation out of its initial trial-and-error phase.

Before the First World War, a large part of the aerodynamic experiments at the MVA was concerned with balloons and airships. Research contracts came from such famous airship builders as August von Parseval (1861–1942) and Ferdinand von Zeppelin (1838–1917) and were oriented towards measurements of the forces on airship models. The initial aim of the facility was to perform "measurements of air resistance on models of sufficiently large models of airships for determining the most favorable shape of the balloon" [89]. Prandtl and Georg Fuhrmann (1883–1914), who was the MVA's first employee after finishing his engineering studies at the Technical University Hanover, pondered carefully what equipment was most appropriate to undertake such experiments. The simplest device for aerodynamic tests, a whirling arm, was not considered because the air would still be in motion from preceding revolutions. The first tests on airship models were performed with a sucking tube, a straight pipe with a diameter of 30 cm through which air was ventilated by a fan. Originally the wind tunnel was designed as a larger version of the sucking tube, but Prandtl came to the conclusion that a closed-circuit tunnel was more advantageous: "As experiments on a small scale have shown." Prandtl

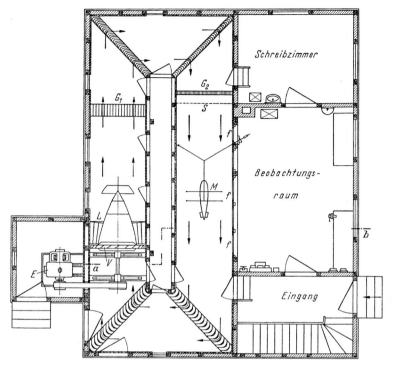

**Fig. 2.9** Plan view of the MVA with its wind tunnel. (Schreibzimmer = office; Beobachtungsraum = observation room; Eingang = entrance)

argued against an open wind tunnel of the sucking tube–type, since "even winds with a very small velocity, which are almost always present, cause considerable fluctuations of the airstream." Furthermore, instruments for measuring pressures and air forces had to be designed. In order to determine the velocity of the air stream across a testing section Prandtl developed a precise instrument, which finally became known as "Prandtl'sches Staurohr," a special tube for measuring the pressure difference at the stagnation point and in the free stream [90], [67, pp. 39–50].

As soon as the wind tunnel was operational, Fuhrmann analyzed various airship models; the results, some shown in Fig. 2.10, were published in a new journal, the *Zeitschrift für Flugtechnik und Motorluftschiffahrt*, in a special feature under the headline "Communications from the Göttingen model testing establishment" [91]. In addition, Prandtl reported on the progress of airship model testing in the annual reports of the "Motorluftschiff-Studiengesellschaft," the funding society of this establishment. Already the first trial experiments with the sucking tube had shown that the cigar-shaped balloons with a blunt stern – then a common shape of airships – was less favorable than balloons with the same diameter but a pointed stern. This result was verified with de-

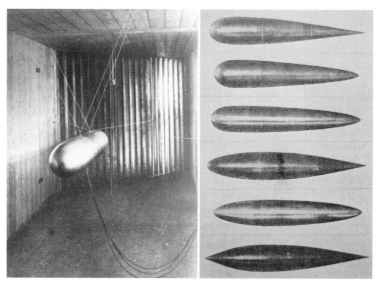

**Fig. 2.10** *Left*: test section of the wind tunnel with an airship model. *Right*: Fuhrmann's airship models subject to comparative wind tunnel tests; the model with minimal drag was the third from below.

tailed wind tunnel tests on a variety of airship shapes. "It turned out," Prandtl wrote in the final report of this test series, "that a pointed stern is much more important than a pointed bow tip" [92].

With these results airship design changed considerably – the contrast can be seen in Fig. 2.11. Airships built after the First World War more or less looked the same – like the optimal shape tested in Prandtl's wind tunnel. This success motivated Prandtl and his assistant at the MVA for more fundamental investigations about fluid resistance. Fuhrmann extended his experimental results in a theoretical analysis of the behavior of rotational bodies in an ideal fluid. Based on a method developed by Scottish engineer William John Rankine (1820–1872), he analyzed flows composed from a suitably arranged distribution of sources and sinks, such that the resulting flow would come close to the one around an airship-like rotational body. He was able to show that the experimentally determined pressures along the surface of such bodies agrees quite well with those determined theoretically from potential theory, except at the rear of the body where the theoretical and the experimental values deviated from another for all models in the same manner [93].

This result confirmed the qualitative conclusions from Prandtl's boundary layer theory: due to skin friction, the flow of air is subject to increasing pressure as it approaches the stern of the model, where it becomes detached from the surface; only up to that point was agreement between ideal fluid (i.e., potential) theory and experiment to be expected.

**Fig. 2.11** Airships before and after research on minimal air resistance (Source: Deutsches Museum, Munich).

## 2.5
## The Discovery of the Turbulent Boundary Layer

There were, however, unexpected discrepancies between theory and experiment. Otto Föppl (1885–1963), a researcher at the MVA from 1909 to 1911 and Prandtl's brother-in-law, performed experiments on air resistance and lift on plates inclined at various angles of attack against the air stream and on other test bodies such as disks and spheres. Similar tests had been undertaken in the laboratory of Gustave Eiffel (1832–1921) in Paris, one of the few other aerodynamic research establishments in the world equipped with a powerful wind tunnel and sophisticated measuring techniques [94]. In 1912, Föppl published his results and compared them to Eiffel's data. He found "generally a very good agreement," except for the drag of spheres where "apparently a mistake was made" in Eiffel's laboratory, as Föppl supposed. The discrepancy was so blatant that he thought Eiffel or his research collaborator had omitted a factor of 2 in the final evaluation [95].

When Eiffel learned about his Göttingen rivals' suppositions, he "became very angry," as Kármán recalled [14, p. 87]. Eiffel was renowned as a careful observer who would not rush into the publication of unconfirmed data. Furthermore, these experiments were not his first measurements of aerodynamic drag. Before his first wind tunnel was established in 1909, he had performed free fall experiments from a platform of the Eiffel tower at a height of about hundred meters and recorded the drag of spheres as a function of fall velocity. In 1911 he moved his laboratory to Auteuil where he installed a larger wind tunnel. The terminal velocities reached with his fall experiments from the Eiffel tower were as high as 40 m/s – much faster than the velocities of air flow in wind tunnels at that time. Eiffel's 1909 wind tunnel attained 20 m/s after the insertion of a nozzle, which narrowed the diameter of the cross-section from 3 to 1.5 m. In his new wind tunnel at Auteuil, the air speed reached a maximum of 30 m/s, compared to only 10 m/s in the smaller Göttingen wind tunnel. Provoked by the Göttingen data, Eiffel performed a new test series to measure the drag coefficient with systematically varied air speeds (from 2 to 30 m/s) and sphere diameters (16 cm, 25 cm, 33 cm)—and discovered a new phenomenon: "At speeds below a critical velocity, the coefficients do not much differ from those obtained by the Göttingen laboratory. If one did not find there the value I had published, this is simply because they could not measure at velocities above 10 m/s. This case shows clearly that it is necessary to perform experiments not only in a stream of air with a large diameter but also with sufficiently high velocity. Only at these higher velocities are the new phases of the phenomenon are revealed" [96].

In other words, the alleged slip of Eiffel's measurements turned out to be a new aerodynamic phenomenon, a phenomenon that could not be observed in the Göttingen wind tunnel because its speed was not high enough. Prandtl immediately responded to this deficiency and made plans to insert a nozzle into the air stream [97]. He visited Eiffel's laboratory and was presented with detailed design plans of Eiffel's new wind tunnel [98]. For Prandtl's colleague Carl Runge, Eiffel's experiments prompted an appeal to the Göttingen Association, the funding organization for applied sciences at Göttingen university, to demand more attention for aerodynamics. One should be "prepared for surprises," he said, alluding to the experiments on the drag of spheres presented in a paper on the importance of aerodynamic model testing in wind tunnels. Eiffel's discovery provided evidence "that the entire mode of air motion changes at a critical velocity" [99].

After inserting the nozzle according to Eiffel's scheme, the critical range of velocities for observing the new phenomenon became accessible in the Göttingen wind tunnel as well. Prandtl asked Carl Wieselsberger (1887–1941), who was employed in September 1912 as a new collaborator in the MVA, to perform these measurements. The results were published in the proceedings of the Göttingen Academy of Science [100] and in the regular series of communications of the MVA [101], which indicates that Prandtl regarded them as both of academic and of technological interest. The phenomenon of an abrupt change of drag at a critical velocity had been observed in the meantime in other laboratories as well; for example, in a military testing facility in Rome where drag experiments were made with spheres in a water canal. Prandtl and Wieselsberger, therefore, were not only interested in verifying the phenomenon in the improved Göttingen wind tunnel but also to identify its physical cause. Eiffel had abstained from explaining the sudden change of the drag coefficient and kept his publication descriptive. It was Runge who first introduced the notion of turbulence when he mentioned Eiffel's discovery in his address before the Göttingen Association: "At small velocities there is a cone of turbulent air behind the sphere; above a critical velocity the turbulence disappears almost completely" [99].

It was obvious that turbulence was involved in this phenomenon. Reynolds had already shown that in pipe flow, a change from laminar to turbulent flow occurs when a critical velocity is exceeded (see Chapter 1), but it seemed unlikely that the mode of motion would change, as Runge suggested, from turbulent to laminar with increasing velocity. Prandtl offered another explanation: he assumed that the transition happened as in Reynolds's case of pipe flow – from laminar to turbulent – but within the boundary layer rather than in the free fluid. He arrived at this explanation because he and Wieselsberger interpreted the data as a function of the Reynolds number, $UD/v$ (where $U$ is the flow velocity, $D$ is the sphere diameter, and $v$ is the kinematic viscosity), rather than as a function of the velocity as Eiffel did. If plotted in this manner, it turned out that the different curves for the drag coefficient of spheres with different diameters coincided. But the transition happened at a much higher Reynolds number (about 200,000) than the transition observed by Reynolds in pipe flow. Because the Reynolds number is a measure for the ratio of inertial to frictional forces, the high Reynolds number indicated that in the free flow around the sphere inertial forces far exceeded frictional forces, such that a transition in this flow regime was not explicable. It was this observation that told Prandtl that the phenomenon had its origin within the boundary layer rather than the free fluid: "This ratio [of inertial to frictional forces] holds for the 'free fluid' only," he argued, "but not for this usually thin layer in the closest vicinity of the body surface, in which the free flow velocity is changed to the velocity at the surface" [100].

In other words, within the boundary layer were inertial and frictional forces of the same order of magnitude; only there did it make sense to locate the cause for the transition from laminar to turbulent flow. But that interpretation did not yet explain why the drag of the sphere would decrease at that transition so dramatically. Prandtl and Wieselsberger explained this in terms of boundary layer separation: if the boundary layer becomes "vortical" before it separates from the surface, it sweeps away more fluid in the wake than if it is laminar; the boundary layer stays longer attached to the surface and "this results in a considerably smaller system of vortices and therefore also a smaller drag."

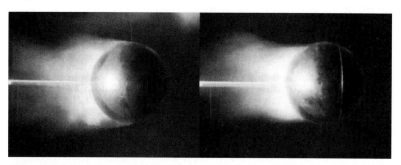

**Fig. 2.12** A trip wire around a sphere (right image) makes the boundary layer turbulent, resulting in a smaller vortical wake and therefore a decrease of the drag.

The explanation that a transition from laminar to turbulent flow results in a smaller drag seemed paradoxical. Therefore, "a true experimentum crucis seemed desirable," as Wieselsberger introduced the now-famous trip wire experiment. He wound a thin wire (1 mm) around a sphere (28 cm diameter) so that it formed a circular threshold shortly before the equator (see Fig. 2.12). Due to this obstacle, the boundary layer would become turbulent, and the site of boundary layer separation would move further to the rear than before, resulting in a reduction of drag. This was indeed the case. Furthermore, "we made the vortical wake behind the sphere visible by introducing smoke and photographed it" [101]. The "crisis caused by Eiffel's discovery," Prandtl concluded, "is essentially repaired" [100].

## 2.6
### The Beginnings of Airfoil Theory

Another series of early wind tunnel experiments at the MVA was dedicated to wings. As early as 1911, Prandtl had put "the determination of lift, drag, and pressure center of airfoils with various forms and profiles" on the agenda [88], [67, pp. 188–193]. This experimental program was accompanied by theoretical efforts to cope with the airflow around wings, which involved two problems: first, the calculation of the two-dimensional flow in a cross-section of the wing; and second, the three-dimensional flow including the phenomena at the wingtips and the effect of the wing's plan view. By 1910, Wilhelm Martin Kutta (1867–1944) and Nikolai Joukowsky (1847–1921) had offered mathematical approaches to address the first problem. The two-dimensional flow around a wing profile could be regarded as a superposition of a uniform and a circulatory flow. Such flows were accessible by a mathematical method known as conformal mapping. The lifting force $L$ per unit span of an airfoil was calculated by Kutta und Joukowsky as $L = \rho V \Gamma$, where $\rho$ is the density of air, $V$ is the uniform flow velocity, i.e., the speed of the airplane in free flight, and $\Gamma$ is the so-called circulation, a mathematical quantity that accounts for the circulatory part of air flow around the wing. Aerodynamic lift, according to this circulation theory, was entirely due to the circulatory part of the flow [102, 103].

The Kutta–Joukowsky theory also offered an approach to cope with the second problem of the three-dimensional flow. If it is the circulatory part of the flow that contributes to the lift, one can imagine that a vortex around the wing accounts for the lift, and, for theoretical purposes, replace the wing by a vortex line. According to Helmholtz's laws on vortex motion, however, a vortex line in an ideal fluid has to be closed (or start and end at the walls of the container). Therefore, Prandtl imagined that the vortex line is bent backwards at the wingtips to form a U-shaped horseshoe vortex, as shown in Fig. 2.13, closed far behind the wing to form a ring. The front part of this vortex, which travels with the wing, was described as a bound vortex, the rear part left behind was described as the starting vortex, and the two connecting parts stretching backwards were described as wingtip vortices. The last of these is a consequence of the pressure difference between the upper and lower side of the wing, which gives rise to pigtail-like downward and upward flows from the upper and lower sides.

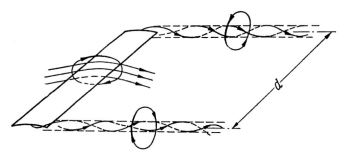

**Fig. 2.13** Vortical flow of air around a wing as visualized by Prandtl in 1913.

Prandtl presented this concept in its simplest form for the first time in a lecture in the summer of 1909 [67, p. 190]. He delineated a rectangular area behind the wing by a single vortex thread composed of three straight pieces: one for the bound vortex parallel to the wing and two perpendicular pieces, which extended backwards to infinity from the wingtips. The lift of the wing came from a "carrying vortex thread," and the air within the area delineated by it is swept downward (see Fig. 2.14). In 1911 Prandtl presented this formulation: "According to the principle of action and reaction the lift created by the wing is necessarily connected with a downwash of air behind the wing (...) It turns out that the downwash is caused by a pair of vortices whose vortex threads originate at the wingtips. Their distance is equal to the span of the wing, their strength equal to the circulation of the flow around the wing" [88, p. 34–35].

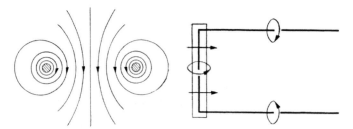

**Fig. 2.14** Cross-section and plan view of Prandtl's "carrying vortex thread".

With this theory, lift calculation was reduced to accounting for the strength of the vortex thread. Another feature of this concept concerned the vortical motion at the wingtips: it gives rise to a resistance unrelated to viscosity—the carrying vortex thread concept is based on ideal fluid theory, i.e., no viscous terms are involved. The first results were published in 1911 [104]. The concept of carrying vortex threads also allowed one to estimate and compare the lift and resistance of various arrangements of wings, such as in biplanes. Albert Betz (1885–1961), who became an employee in the MVA in 1911, demonstrated that biplanes were more advantageous than monoplanes if high lift was required at very low velocities; for higher velocities, however, monoplanes yielded a better lift-to-resistance ratio. Wind tunnel experiments confirmed these theoretical conclusions, and the Göttingen aerodynamicists felt confident that the assumptions upon which airfoil theory was based were "basically correct" [105].

Despite such confidence Prandtl did not consider airfoil theory ready for publication. In 1913 he mentioned the basic idea together with a drawing of the horseshoe vortex (see Fig. 2.13) in a review article on "fluid motion" [106], but he did not further elaborate the theory. In the following year, Wieselsberger published a paper on the V-shaped configuration of the flight of migrant birds based on Prandtl's wing theory, which "already went a long way in several investigations." He argued that outside the downwash area of the horseshoe vortex the air is swirled upwards along the wingtip vortices, so that a bird flying on the left and on the right behind a leading bird experiences added lift. Extending this argument for additional birds, a V-shaped formation resulted as the configuration with the maximal lift for the entire group of birds. Wieselberger's calculation also revealed to some extent how calculations according to Prandtl's airfoil theory were done: they were analogous to calculations in electrodynamics, where the Biot–Savart law allowed one to calculate the strength of the magnetic field in the vicinity of an electric current; the same formalism was applied in airfoil theory to calculate the motion of air in the vicinity of the horseshoe-shaped vortex threads [108]. In the same issue Betz published a formula for the optimal resistance-to-lift-ratio of a wing without further derivation, hinting at a "soon to be published" theoretical investigation by Prandtl [107].

But the airfoil theory was presented only in 1918 in the proceedings of the Göttingen Academy of Science and in subsequent dissertations. With the hindsight afforded by these publications, it is obvious that there was still a long way to go before the concept was shaped into a comprehensive theory. When it finally arrived at this stage, it was more the result of a collaborative effort than a single stroke of genius. Prandtl's airfoil theory, as it is often called, involves more than the concept of the carrying horseshoe vortex. Several problems had to be solved before this concept could be molded into a coherent theoretical framework. A major problem dealt with the distribution of lift over the span of the wing, and how it approached the zero-lift condition beyond the wingtips. Prandtl's doctoral student Ernst Pohlhausen, for example, achieved a breakthrough when he came up "by the end of 1913," as Prandtl acknowledged in 1918, with the "very remarkable result" that in order to obtain a minimal resistance-to-lift ratio, one has to assume a lift distribution over the span "according to half of an ellipse" [109, pp. 342–343]. Other problems were solved in future dissertations (see Chapter 3).

With regard to the debate of science versus technology, the emergence of boundary layer and airfoil theory during the decade before the First World War in Göttingen cannot be sorted along the traditional divide into academic versus technological research. Fluid dynamics, as studied in Prandtl's school, proceeded as traditional academic research in a university institute *and* as contractual research in a separate aerodynamic establishment aimed at technological ends. Themes for doctoral dissertations arose in both contexts. Results from this research could appear as science or technology, depending on opportunities and circumstances beyond the scope of the research itself. Airfoil theory, for example, became a target of opportunity for aspiring mathematicians as well as for theoretically minded engineers. Ernst Pohlhausen, for example, was "according to his study first of all an applied mathematician," Prandtl wrote in a recommendation for his doctoral student, but he was also "pervaded with the way of thinking of an engineer" [110]. Such dual talents were not unusual in Prandtl's school, as we will see shortly.

# 3
# Aviation and the Rise of Aerodynamics in the First World War

The role of science in war has been expressed as follows: the First World War has been described as a "chemists' war," while the Second World War was dubbed a "physicists' war." The former label alludes to poison gas—the latter to radar and the atomic bomb as science-based weapons. But such labels distort the role of scientific technologies in these wars. They narrow the focus to a few spectacular events and do not address, for example, the new modes of organizing science and technology for the purposes of war. With regard to other war technologies like shells, submarines, ships, and airplanes, First World War deserves greater interest. Due to these new technologies the "Great War," as it was called then, could literally be carried beyond frontiers. With new organizations in the war-waging countries aimed at the mobilization of all available scientific and technological resources for the purposes of war, the First World War was characterized also as paving the way for an "institutional modernization" and a "hinge-phase of the modern western societies for their path to modernity" [111, p. 99].

Among the various scientific war technologies, those involved with aeronautics became particularly important. The airplane embodies like no other device the potential of all-out destruction by science-based technology. However, even if we restrict ourselves to aviation we are concerned with no single technology but rather with a host of specialties: beyond those involved with the airplane itself, such as stability, aerodynamics, or strength of materials, disciplines like meteorology or radio science also play an important role. In this chapter, the focus is on aerodynamics only, a specialty that has been called the "fundamental engineering science of airplane technology" [112].

Is it only by coincidence that Prandtl's airfoil theory was published in 1918, at the end of the First World War? Was it kept secret for the duration of the war so that only German airplane manufacturers would benefit from its results? Prandtl's model testing facility is a primary example for the mobilization of scientific resources for the war; it was considerably expanded during the war and funded by the War Ministry. What was the role of science and technology

*The Dawn of Fluid Dynamics: A Discipline between Science and Technology.* Michael Eckert
Copyright © 2006 WILEY-VCH Verlag GmbH & Co. KGaA, Weinheim
ISBN: 3-527-40513-5

when airplanes became weapons of war, and what role did Prandtl and his Göttingen facility play when his research was at the threshold of coming up with the fundamental engineering science of airplane technology?

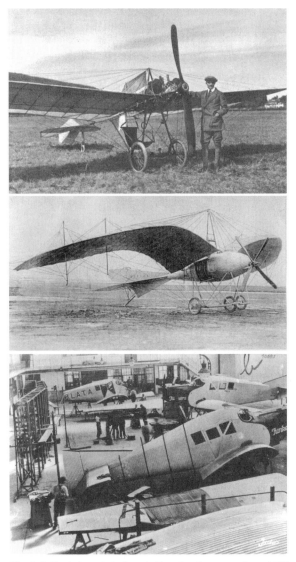

**Fig. 3.1** Evolution of airplanes. *Top*: Grade mono-plane 1909. *Middle*: Garuda seagull 1913/14. *Bottom*: Junkers F-13 1919 (Source: Deutsches Museum, Munich).

In the First World War, the airplane became a weapon. A comparison of aircraft before, during, and after the war shows how much the airplane changed in the course of only a few years (see Fig. 3.1). Before 1914, airplanes were frequently based on the model of the Wright Flyer, with control surfaces in front of the wings (canard-type), or with wings shaped like those of birds (pigeon-type). The fuselage and the wings were made from plywood covered with fabric and held together by string such that the entire structure had the least possible weight, which was regarded as the most important requirement for the flight "heavier than air" during the birdman era, as those early years in the history of aviation have been called [113, 114]. The use of self-supporting cantilevered wings rather than thin airfoils, held in place by struts and wires, was a major innovation in airplane design. It resulted from the availability of strong motors and the insight that with increasing speed it was more important to reduce the air resistance than the weight. If the drag due to struts and wires was avoided, more weight could be lifted. The evolution of airplanes within a decade from pigeon- or canard-like shapes to full-metal constructions like the Junkers F-13 clearly illustrates this message. What was the role of science in this change? How did the war affect the relationship between airplane design and aeronautical research?

## 3.1
### A Symbiotic Relationship

Before the First World War, airplanes were produced in the workshops of individuals or small companies. In 1914 the annual production of airplanes was in the hundreds. In 1918 wartime delivery of airplanes amounted to tens of thousands [115]. The mobilization of the resources in manpower and material necessary for mass production on such a scale resulted from the conviction that the airplane can be used as a versatile weapon of war. Each intended use – aerial reconnaissance, bombing, air fights, guidance for artillery, or naval applications – required special designs. From this perspective, the First World War has been called a "proving ground of aerial war" [116].

In many countries arrangements were made in order to adjust the development of new airplanes to the needs of the military [117]. In Germany a special branch of the army, the Inspektion der Fliegertruppen (Idflieg), was founded for this purpose. In 1915 it started to operate a test facility at the site of the Deutsche Versuchsanstalt für Luftfahrt (DVL) in Berlin-Adlershof, the "Prüfanstalt und Werft der Fliegertruppe," which was combined two years later with the Flugzeugmeisterei (FLZ), previously a civilian institution of the DVL, to a central airplane testing facility of the army. A similar institution was established for the navy at Warnemünde on the coast of the Baltic Sea, the Seeflugzeug-Versuchskommando (SVK) [118]. The military performed a crucial role at these testing facilities: they conveyed the specific requirements to the aircraft manufacturers who competed for contracts for the development of new airplanes. In the course of this process, the testing facilities became a clearing house for technical problems which called for solutions beyond the capabilities of the individual manufacturers. To this end, in 1916, the FLZ founded a Wissenschaftliche Auskunftei für Flugwesen (WAF) and entrusted it with the task of editing a series of secret technical reports, the "Technische Berichte" (TB). As was expressed in the preamble of these reports, the WAF intended to create a marketplace between the military, the nascent aircraft industry, and the sciences involved with aeronautical research [119].

These measures aimed at the mobilization of science and technology when the war reached a stalemate. At Berlin-Adlershof, the experimental facilities of the DVL were directly placed at the military's disposal. But science also mobilized itself for war purposes. In Göttingen Prandtl whetted the military's appetite with the potential of the Modellversuchsanstalt (MVA) for war-related research. The military, airplane manufacturers, and aeronautical researchers developed in this situation what has aptly been called a "symbiotic relationship" [120, pp. 89–108]

The concerted action of such diverse groups did not happen at once. At the beginning of the war Prandtl's laboratory had experienced a setback. Both assistants at the MVA, Betz and Wieselsberger, went to war as volunteers; his doctoral students and other personnel from his institute for applied mechanics at Göttingen University were called to arms. Prandtl had offered himself for service in a flight battalion, but the military had no use for the almost 40-year-old professor. Expecting a short war, his staff left him at the orphaned academic post in Göttingen. His ambitious plans to expand his research facilities, which had come close to realization shortly before the war, had "receded into the distance again," Prandtl wrote shortly after the outbreak of war in August 1914 [67, p. 121].

Prandtl had planned a large institute for aerodynamics and hydrodynamics since 1911, after the foundation of the Kaiser-Wilhelm-Society for the Advancement of Sciences (Kaiser-Wilhelm-Gesellschaft zur Förderung der Wissenschaften, or KWG). Together with the influential sponsors of applied research at Göttingen University, he raised great hopes of establishing a Kaiser Wilhelm Institute dedicated to the "science of air and fluid motion (with regard to application for aviation)." The KWG had already agreed to this plan but the outbreak of war prevented its realization. During the second year of war, however, Prandtl's hopes rose again when both the War Ministry and the Naval Ministry declared it expedient to establish a large research institute "in the interest of the rapidly developing aviation." A month later, in June 1915, the Berlin Ministry of Culture reached an agreement that the army and navy administration was to place at Prandtl's disposal an amount of money from the war funds to begin construction. Both administrations urged Prandtl to proceed with utmost speed "so that the start-up can still happen during the war" [121].

Prandtl and his Göttingen colleagues acted swiftly. In the fall of 1915, the foundation was laid; in the winter of 1916/17 the new institute was ready for operation. It was not, however, what Prandtl had originally conceived as a "Kaiser-Wilhelm-Institut für Aerodynamik und Hydrodynamik"—this vision had to wait another ten years to be realized. Since the money for the institute came from the military, it was earmarked for applied research for aviation rather than for basic research. Prandtl therefore proposed dropping Hydrodynamik from its name and calling it "Modellversuchsanstalt für Luftfahrt;" but in order to avoid confusion with the Deutschen Versuchsanstalt für Luftfahrt (DVL) in Berlin-Adlershof, it was finally named "Modellversuchsanstalt für Aerodynamik" (MVA). After the war, it was renamed "Aerodynamische Versuchsanstalt" (AVA). The small precursor institute, founded in 1908 as the "Motorluftschiffmodell-Versuchsanstalt" and also abbreviated as MVA, was dismantled. Since operation started in the new MVA in the winter of 1916/17 "the establishment was busy to fulfill the needs of the Army Administration and the airplane factories by constantly expanding its personnel," Prandtl reported after the war. The number of personnel reached 50 in 1918. Prandtl was particularly proud of the new wind tunnel, which became operational in March 1917. As in the first Göttingen wind tunnel (which was modernized and kept in use in an attached building), the air flowed in a closed circuit – a diagram of the building plans is seen in Fig. 3.2. Compared to the 30 horsepower fan of the first wind tunnel, however, the new wind tunnel had a motor ten times more powerful, which blew the air at a maximal speed of 200 km/h through a test area of circular cross-section with a diameter of 2 m. Special guide vanes deflected the air in the bends of the tunnel. The entire installation was designed such that the air flow in the test section was as homogeneous as possible. At the time of its inauguration and still for several years after the war, this wind tunnel was the most advanced aerodynamic test facility in the world [67, 122].

The new wind tunnel could have been used for a variety of aerodynamic tests for airplane manufacturers, but Prandtl offered to contribute to the war effort with more aerodynamic research. In September 1918 he proposed that his institute become engaged in gasdynamic research at supersonic velocities for ballistic applications besides the low-speed aerodynamics for aviation. For this purpose, he applied for funds to build a supersonic wind tunnel. With regard to the future Kaiser-Wilhelm Institute, "such a facility for experiments at supersonic speed has already been planned and the project has been elaborated to some extent," he wrote the War Ministry. However, ballistics was a new research area for Prandtl. Carl Cranz (1858–1945), a professor at the Technical University in Berlin-Charlottenburg and author of a famous textbook on ballistics, proposed at the same time to build "an aerodynamic laboratory for the purposes of the artillery" in Berlin. In an attempt to calm feel-

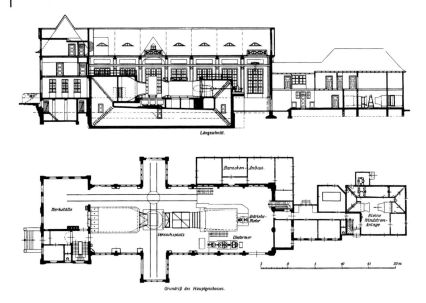

**Fig. 3.2** Cross-section and plan view of the new MVA, from [67, p. 150].

ings of rivalry, Prandtl suggested that his proposal should not be regarded as a competition with Cranz but rather as division of labor, such as that between the DLR in Berlin and his MVA in Göttingen, which fruitfully complemented one another: "The division of labor should be accomplished such that the Berlin establishment serves the immediate needs and daily problems of the artillery, while the Göttingen facility focuses on the general laws of air resistance at high velocities without a too narrow involvement with artillery problems" [123].

Prandtl tried until the very last days of the war to persuade the War Ministry of his plans. On 29 October 1918, he submitted a detailed proposal for a "test facility for measurement of air resistance at high velocities" to the Artillerie-Prüfungs-Kommission (APK), the Army's expert unit for ballistics: by evacuating a $40 \, \text{m}^3$ vessel, a pressure difference should be created in an attached tube of 20 cm × 20 cm cross-section so that when a valve was opened, air would be sucked through this tube into the vessel at supersonic speed. Using this method tests could be performed within a time span of a few seconds. The test results would be recorded photographically using the Schlieren method, a method introduced in experimental ballistics for visualizing shock waves. At the end of his five-page proposal, Prandtl admitted a grain of doubt whether the project could be realized in view of the war situation at the time, but with regard to the "coming scarce times," facilities like the one he proposed would merit particular support because they are cheaper than shooting tests [124].

The end of the war prevented the realization of these plans until they were renewed with less emphasis on their use in ballistics a decade later (see Chapter 7), but the aviation-related facilities alone sufficed to turn the MVA into a prominent aerodynamic research center. "I believe that this establishment, as far as the news from the enemy countries are pertinent, at least until recently, was the biggest and most powerful of its kind," Prandtl wrote after the war. And he knew whom he had to thank for its creation: "We owe it, of course, to the generosity of our military administration" [122].

## 3.2
## War Contracts

It will not come as a surprise that the new facilities were largely used for their intended purpose. To what extent Prandtl's research was oriented toward war-related goals became evident with a confidential survey that the chairman of the German Mathematical Association (Deutsche Mathematiker-Vereinigung, DMV) had administered to the directors of institutes for mathematics and allied disciplines: "Is your institute or seminar working for purposes of the Army, Navy, or Aeronautics," was one question, "and if so, in what direction." Prandtl responded that his university institute "was practically completely absorbed into the MVA which works at the moment exclusively for the interests of the Army (aerodynamic measurements, mainly on models of airplanes, parts of airplanes, etc., calibration of instruments for measuring the air speed)." Another question asked how many mathematicians, from students to professors, were engaged in such work. Prandtl answered that "about 10 students of mathematics are employed in performing test work and extensive numerical and graphical elaboration of test results." He added that this work was largely routine, akin to reading the scales of balances" or calculations using a slide rule. "A few advanced mathematicians deal with more difficult calculations (evaluation of integrals, etc. for hydrodynamic problems, etc.)" [125].

What Prandtl was referring to were wind tunnel measurements performed on behalf of the army's FLZ or directly under contract for one airplane manufacturer or another. Occasionally the MVA also received contracts from the navy's SVK to perform tests of sea planes. As a rule, the results of such measurements were transmitted to the WAF, the Berlin clearing house for the testing of new military airplanes, which printed them in its secret TB-series and subsequently passed them along to the manufacturers of war planes and others involved with the aeronautical war effort. If a manufacturer wished to have his design tested, he had to disclose it to the WAF and agree that the test result would be printed anonymously in the TB. From the perspective of an individual airplane manufacturer, this procedure involved a balance of advantages and disadvantages: in order to participate in the pool of collective aerodynamic test results, he had to communicate know-how that was otherwise jealously guarded as the private property of a firm. But in order to attract remunerative contracts from the military for the production of new airplanes manufacturers had not much choice other than to comply with the rules of the FLZ, and almost all firms participated in this knowledge pool according to the rules of the WAF, which insisted in its authority on all matters of aeronautical innovations: "The WAF thinks it important that firms do not possess innovations which are not yet known to the Military Administration or the Naval Office," the chief executive officer of the FLZ explained to Prandtl when they debated how knowledge between the manufacturers, the military, and the research institutes should be shared. In June 1917 he sent Prandtl a list with the names of 24 firms that complied with this rule [126].

However, the symbiotic relationship among the airplane manufacturers, the military, and the aeronautical research facilities, as established by the WAF, was not free from tensions. Prandtl was critical of the fact that the WAF forced the manufacturers into a collaboration with his laboratory by imposing conditions on them, which "to some extent from the perspective of their business interests have to be designated as unusual." He forwarded to the military the complaint of some manufacturers "that the FLZ had to be informed at any time about tests made on contract for a firm." Some firms declined to collaborate with the WAF; others refrained to issue important tests [127]. The Pfalz Flugzeugwerke, for example, asked Prandtl to keep the contractual test results secret because they were concerned that competitors could become aware of their own design secrets; in this case, the chief executive officer of the FLZ insisted that the firm comply with the prescribed conditions – otherwise, it would lose the advantages of participating in the WAF-pool such as being provided with the TB. The firm's concern was calmed by the assurance that the communication of the test results in the TB happens in such a way "that the name of the Pfalz-Flugzeugwerke is kept out, so that spying into the respective secrets of firms will not be possible" [128]. Another manufacturer complained to Prandtl: "We are constructing airplanes but lack various results from measurements in an artificial stream of air for our designs. Because we did not join the association of airplane manufacturers and also do not intend to become a member, the material of your testing facility is not available to us." Although the firm only asked for "material as far as it is free" Prandtl had to respond that the only viable way to obtain such material was via the FLZ, whose "TB are handed out to airplane manufacturers provided certain commitments of secrecy are obeyed" [129].

Prandtl was not happy with his role as mediator between the military and the airplane manufacturers. Furthermore, he complained about a lack of feedback: "We are left ignorant about what the firms accomplish on the basis of our results," he once argued. "It would further our work if we were instructed, be it by the firms or the FLZ, about the respective aerodynamic progress in such a way that we learn something about whether such progress was achieved because of or despite our work" [127]. Wind-tunnel tests had not yet progressed so far that they could be regarded as a routine measurement technique. No standards had been agreed upon between airplane manufacturers, the military, and the research institutes. Prandtl therefore argued that a lot of technical detail needed clarification, but the chief executive officer of the FLZ trusted Prandtl's authority and flattered him: "We all here in Adlershof may regard us with our knowledge of aerodynamics as your pupils," and they took the units, diagrams, and other details of wind tunnel measurements used in Göttingen as binding for the communication in the FLZ's TB [130].

**Fig. 3.3** Struts and wires on a 1916 warplane (Rex D16) (Source: Deutsches Museum, Munich).

An example illustrates how war contracts that originated in the course of specific design problems resulted in new knowledge made available to the airplane manufacturers who had joined the WAF-pool. In June 1917 the TB contained a report about the air resistance of struts [131], [67, p. 172]. Before self-supporting cantilevered wings became the rule in airplane construction struts and wires were used to attach the wings to the fuselage or, such as in bi- and triplanes, to one another (see Fig. 3.3). The drag caused by the struts was soon recognized as an obstacle to reaching higher velocities. In 1915 the Reich's Naval Office (Reichs-Marine-Amt) issued a first contract to the MVA to measure the resistance of struts with different shapes. During the following two years, at least five airplane manufacturers (Flugzeugbau Friedrichshafen, Albatroswerke, Flugzeugwerft Staaken, Luftschiffbau Zeppelin, AEG) addressed requests for wind tunnel tests for the specific shapes of struts they intended to use in their design of new airplanes. The first measurements were still performed with the small wind tunnel of the old MVA, but in the spring of 1917, the new wind tunnel became operational "and then we can measure the struts in a stream of air with 2 m diameter at velocities up to 60 m/s," as the MVA informed a contractor [132]. The MVA coordinated its test series on struts with the FLZ and the individual contractors. For example, it was arranged between Prandtl and the FLZ that the manufacturers had to accept a reference velocity of 40 m/s (144 km/h) as a characteristic velocity for which the drag data were presented [130], which was a normal flight velocity of airplanes at that time. Accordingly the MVA, normalized its data and made it anonymous for practical uses.

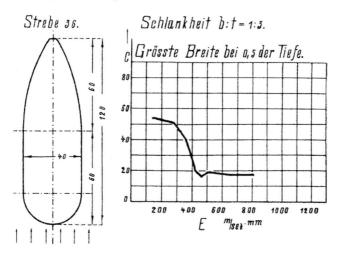

**Fig. 3.4** The drag of a strut with a streamlined profile [131].

The profiles of struts were numbered, and the results of the drag measurements were presented in associated diagrams (see Fig. 3.4). The drag coefficient was not displayed as a function of the Reynolds number but as a function of the more handy "Kennwert," the product of the width of the strut and the velocity. The sudden drop of the drag coefficient due to the transition from laminar to turbulent flow in the boundary layer at a certain value of the Kennwert, which had been discovered shortly before the war (see Chapter 2), was found to occur with most struts and called for special attention because of its counterintuitive consequences: "In particular, a reduction of speed, for example when the plane changes from horizontal flight to climb, causes a sudden increase of the resistance coefficient," the report explained. It therefore suggested that manufacturers choose strut profiles which are beyond the critical transition at normal flight velocities. Because of the dependence on the Kennwert, i.e., the product of strut width and velocity, rather than the velocity alone, a profile which was good for one application could be bad for another. Thick struts required different profiles than thin wires [131, pp. 88–89].

Already, these first reports indicated that wind tunnel tests at the MVA provided important lessons for the design of airplanes in the First World War. Although slim strut profiles, for example, caused less drag than thicker struts, they were regarded as less advantageous because a change of the angle of attack altered the resistance considerably: the drag coefficient of a strut with a slender profile increased by a factor of four if the angle of attack changed from zero to nine degrees, while the drag coefficient of a thicker profile only doubled [133]. Striking results were also reported from wind tunnel measurements of wires with circular and oval cross-sections: the drag coefficient of 3-mm-thick wires with an oval profile was only one third of that from a 2.8-mm-thick wire with circular cross-section [134].

Measurements of air resistance of struts and wires were just one example out of a variety of aerodynamic tests. It is difficult to summarize the MVA's war contracts within a coherent scheme. Most contracts involved wind tunnel tests, i.e., the measurement of air forces – lift and drag—on parts of airplanes or scale models. Based on a list of 24 reports communicated by the MVA to the FLZ for publication in the TB between June 1917 and August 1918, we may estimate that about one third of the war contracts dealt with the drag of various parts of an airplane; roughly another third addressed the aerodynamic behavior of control surfaces and the combined behavior of the fuselage and propeller; the rest, more than one third of all wind tunnel measurements, was concerned with the wing [67, pp. 170–172].

## 3.3
## Göttingen Profiles

A series of reports about wind tunnel tests at the MVA was introduced in 1917: "Airfoil investigations were performed as issued on external contracts and in a systematic manner at the establishment's own account." To undertake more than just the contractual measurements implied a strong motivation, because there was no lack of contracts from airplane manufacturers, and each test of a specific airfoil required a laborious construction of a scale model according to the manufacturer's design. But the designs of the various manufacturers were incoherent and it was "difficult to discern the crucial and important aspects from such arbitrarily amassing material." Although from the perspective of an individual firm the measurements according to the respective contract was certainly useful, the reporter at the MVA argued that for the aviation industry as a whole, the benefit would be greater if tests were done more systematically [135].

As was done with the drag measurements on struts, the various profiles of wings subjected to wind tunnel tests were numbered and sorted by geometrical criteria independent of the respective contractors. Easily measurable quantities of a profile were taken as its characteristics: the curvature ("Wölbung" $W$, taken as the maximal distance to the upper surface from the baseline) and maximal thickness ("Dicke" $D$), each expressed as a percentage of the profile's depth ("Tiefe" $t$). The height ($H$) at 2/3 of the profile's depth was also chosen as a parameter. Profiles classified in this manner were displayed in a Cartesian coordinate system as a graphical inventory with the $x$- and $y$-axes representing the thickness and curvature, respectively (see Fig. 3.5). Each profile was represented as a numbered disk at a location corresponding to its respective $D$- and $W$-coordinates, with the disk itself coded according to the quantity $H$. Profiles with little difference in curvature and thickness were recognizable immediately as closely neighboring points in this graphical inventory [67, p. 178].

All wing models were manufactured with a rectangular planform with a span of 100 cm and a depth of 20 cm, the "normal wing model" size. (For measurements in the old wind tunnel 72 cm span and 12 cm depth had been chosen.) Originally, model wings were made from two iron metal sheets folded such that they corresponded to the upper and lower sides of the prescribed profile. Later, the models were produced in a more sophisticated routine procedure: a special machine milled gypsum into a mold according to the shape of a profile, around which a single piece of sheet metal could be folded. The model wing was suspended in the wind tunnel in such a way that the vertical and horizontal forces could be measured at each angle of attack. From these measurements the coefficients of drag ($C_w$) and lift ($C_a$) were deduced for a set of angles of attack and displayed in so-called polar diagrams. (The principle of this mode of representation goes back to Lilienthal – see Chapter 1. A description of the measurement techniques in the Göttingen wind tunnel is given in [122]).

By the end of the war, the list of "Göttingen profiles," as these measurements later became widely known among aeronautical engineers, comprised 346 different profiles. The majority of profiles was slim like most of the real wing profiles at that time, but there was also a series of "extremely thick" profiles, as was explicitly emphasized in a TB communication in August 1918 [136, p. 450]. The results clearly demonstrated that in terms of lift-to-drag ratios, thick profiles with appropriate curvatures were more favorable than slim profiles. The choice of thick profiles enabled the use of cantilevered wings so that struts and wires – a major source of drag – became superfluous.

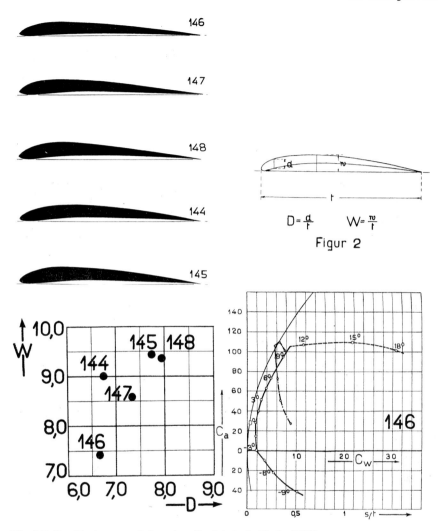

**Fig. 3.5** Graphical representation of profile data tested in the MVA's wind tunnel.

Airplanes with thick cantilevered wings were heavier than the thin-winged airplanes made from plywood and fabric. But with the availability of strong motors, the weight was no longer the primary problem if the drag did not prevent the required higher speeds, where the lift would compensate for the increased weight, from being reached. This was not a novel discovery in 1918. Hugo Junkers (1859–1935) had noticed this phenomenon almost ten years earlier. Junkers was the owner of a factory for heat boilers a professor of heat technology (Wärmetechnik) in the faculty of machine engineering (Maschinenwesen) at the Technical University in Aachen. When Hans Reissner (1874–

1967), professor of mechanics at the same university, infected him in 1908 with his enthusiasm for aviation, Junkers and Reissner founded a joint venture to build airplanes and to study the principles of airplane construction. Junkers also performed aerodynamic measurements with a wind tunnel in his private laboratory. In December 1909, he formulated his ideas in a "Patent concerning a bodily design of airfoils" in which he argued that "for the purpose of reducing the drag as much as possible," the entire weight should be stored inside the wing (see Fig. 3.6). Although the patent of the "thick wing," as it later became known, was ahead of its time, Junkers's message that "the crucial part of the problem of flight is not concerned with weight but with the technology of fluid dynamics" hit the mark [137, p. 114]. The data sheets of thick profiles, like No. 290 from the Göttingen profiles, amply confirmed this insight.

But to what extent could the data from wind tunnel tests on model wings be trusted as pertinent to real wings under free flight conditions? Measurements in the first Göttingen wind tunnel were made at an airspeed of only 9 m/s. Air speed and model size enter into the "Kennzahl" or the Reynolds number, and the law of dynamic similarity predicts that the aerodynamic forces are the same if the Reynolds number is the same in small scale model tests and in free flight with real wings. But having the same Reynolds number alone was not sufficient. Because of the "uncertainty of the transferability of model test to the large scale it would be of little use if we aim for higher accuracy," the MVA's engineer in charge of model tests once admitted in a report [135, p. 135]. In order to check the transferability between different wing sizes and flow regimes, five different profiles were measured in the large wind tunnel at three different air speeds (10, 25, and 40 m/s), each one at three different rectangular wing sizes (72 cm × 12 cm, 100 cm × 20 cm, 150 cm × 60 cm). Furthermore, the texture of the wing surface was varied from smooth to rough (using varnished fabric, plywood, and gypsum). The law of similarity was found to hold quite well in a range of high Kennzahlen from about 600 to 30,000 m/s mm (corresponding to Reynolds numbers from about 42,000 to 2,100,000) for small angles of attack and smooth surfaces. These measurements were taken as evidence that it was indeed justified to apply the data obtained in model tests to real flight conditions [138].

This conclusion, however, relied on further theoretical assumptions. A major problem concerned the influence of the wingspan. Prandtl's airfoil theory was still not published, but was used within the MVA in order to compare different airfoil measurements. For example, the theory provided a transformation formula for the drag-to-lift ratio of an airfoil of a given span if this ratio was known for a different span [139]. According to this formula, it was possible to derive the required aerodynamic data of wings that had a different planform than the normal model wings used in the wind tunnel tests. In order to check the validity of this formula, special tests were performed in

**Fig. 3.6** A cantilevered wing as tested in October 1915 for strength in Junkers's factory; such wings were used in the first all-metal airplanes such as the Junkers Eiseneindecker J-2 constructed in the summer of 1916 (Source: Deutsches Museum, Munich).

which the wing was successively shortened. Each measurement resulted in a slightly different polar curve, but the transformation formula was able to account for the differences such that one could make the curves coincide (see Fig. 3.7) [140].

Another uncertainty concerned the planform of the wing, particularly the shape at the wingtips. A war contract from the Bayerische Flugzeugwerke offered the researchers the opportunity to compare wings of rectangular plan-

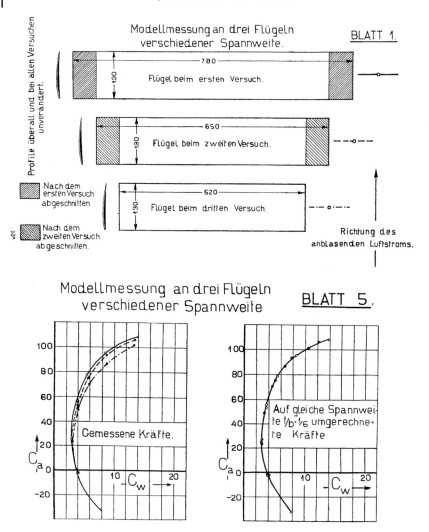

**Fig. 3.7** Comparison of wings with different span.

form with wings of different shapes. The measurements showed that it made little difference how the wings were shaped at the tips provided their planforms had the same surface area [141]. Other investigations addressed the combined effect of multiple wings. The "aerodynamics of carrying organs of airplanes," as one series of measurements was titled, was less academic than this clumsy title suggests; it referred directly to biplanes and triplanes, fighter planes of extreme maneuverability, such as the red triplane flown by the legendary Red Baron, Manfred von Richthofen. Related to these were tests of staggered wings. The data of such investigations were used to validate formulae derived from airfoil theory. If this theory could be trusted, the measurement of a single simple model wing would suffice "to derive all required data for a complicated carrying organ," as Prandtl's research collaborator argued [142, p. 188].

## 3.4
## Max Munk and the Foundation of Airfoil Theory

The MVA contributed a total of 24 communications to the Technische Berichte. Among these, 10 were authored or coauthored by Max Munk (1890–1986), Prandtl's assistant from 1916 to 1918. Munk and Betz were Prandtl's closest research collaborators in the development of airfoil theory. Both wrote doctoral dissertations on fundamental parts of airfoil theory—Munk during and Betz shortly after the war [143, 144]. At the same time, Munk was responsible for the wind tunnel measurements issued by war contracts. The beginnings of Munk's career at the MVA, therefore, provides us with first-hand insights into the emergence of airfoil theory during the First World War. From Prandtl's first publications in the proceedings of the Göttingen Academy of Science, the theory appeared largely as a mere mathematical accomplishment [109], but with the focus on Munk, it becomes obvious to what extent this theory was related to the war work at the MVA.

Munk did not study in Göttingen. Nevertheless he belongs – next to Betz, Blasius, Fuhrmann, Kármán and Wieselsberger (see Chapter 2)—to Prandtl's most prominent early disciples. He was still a student at the Technical University in Hanover in 1912 when he addressed a letter to Prandtl in response to a job offer in Prandtl's institute: He had passed the bachelor's examination (Vorexamen) in mathematics and mechanics with the best grade and expected to finish his studies in 1914, he wrote to Prandtl, in an attempt to be employed as Prandtl's assistant at that time [145]. Prandtl was impressed by the ambitious student, and Munk felt encouraged to renew his application after receiving his master's degree in engineering (Dipl. Ing.) in 1915. He was exempted from military service, Munk wrote to Prandtl, but he would consider employment as Prandtl's assistant for the duration of several years only if allowed to pursue a doctoral degree during this time period [146]. Prandtl could not offer an ordinary assistantship, but because his regular assistants had been called to arms, he could employ Munk as a stand-in for the duration of the war. On 1 April 1915, Munk was hired with a one-year contract as an assistant-in-aid (Hilfsassistent) at the MVA [147]. When the war dragged on, his contract was prolonged. Altogether Munk's Göttingen sojourn lasted until the spring of 1918. For the final months of the war, he was employed by the navy's Seeflugzeug-Versuchskommando (SVK) in Warnemünde on the coast of the Baltic Sea, but he kept in close contact with Prandtl, and his entire wartime activity as a test facility engineer left numerous traces in the MVA's archival records.

In view of Munk's year-long experience with wind tunnel tests during the war, it is understandable that he became eager to exploit his expert knowledge for the benefit of his own career. The "only disappointment I experienced in Göttingen," he wrote to Prandtl after his move to the Baltic coast, was that he could not realize his intent to accomplish this sojourn with a doctoral dissertation. He felt inferior to his colleagues in the navy testing facility who usually had a doctoral degree in addition to their engineering diplomas. Munk had presented a collection of some results from his Göttingen wind tunnel tests concerning the aerodynamics of carrying organs of airplanes as a dissertation to his Hanover professor but was left without response for a while. Prandtl consoled him that "such affairs tend to advance slowly in Hanover" and offered to accept a polished version of Munk's work as a doctoral dissertation in Göttingen. However, upon reading the dissertation, Prandtl criticized that the theory was presented in such terse manner that it was barely understandable. Munk was "overjoyed" about Prandtl's offer nonetheless, as he wrote to Prandtl in the course of the ensuing correspondence. The "reproach of terseness" did not surprise him: "To be honest: I thought you and a few others will understand it, others will only read the final results" [148].

The dispute about theoretical explanations versus practical data finally resulted in two dissertations, one for Hanover [149] and the other for Göttingen [150]. Prandtl had to defend Munk against the reproach from the Hanover professor that "you submitted a dissertation of minor quality at Hanover while the finer work was reserved for Göttingen." Munk found it deplorable that his former professor in Hanover was unable "to judge the case by himself." With regard to the Göttingen dissertation, Prandtl suggested that Munk render it more understandable by adding explanatory remarks and to replace the clumsy title "isoperimetric problems" by something more down-to-earth: "Why don't you simply name it: On wings of minimal drag?" [151]. Munk, however, kept the original title, presumably because he was eager to make it appealing to mathematicians ("isoperimetric" means "of equal perimeter" and addresses the specialty of variational calculus).

Munk's Göttingen dissertation – completed in May 1918—contained the foundation of what became known shortly afterwards as Prandtl's airfoil theory. It explained such fundamental phenomena as the induced drag, a new form of air resistance due to the formation of vortical air motion, which inevitably arises with wings of finite span. Although the induced drag and other aspects of airfoil theory were known to Prandtl and his circle, they had not been treated before on a rigorous mathematical basis. As Prandtl explained in his report about Munk's dissertation, these problems "sprang up from the general circle of ideas familiar to the collaborators at the MVA, but the ideas and mathematical methods which gave rise to their solution are the sole intellectual property of Mr. Munk." It addressed "with enjoyable generality the task to derive for a given geometrical arrangement of wings the conditions under which the drag becomes minimal at a given speed and total lift" [152]. Munk passed the oral doctoral examination on 17 June 1918. The printed version of his dissertation appeared a year later; it was only 31 pages long and consisted mainly of mathematical proofs. One of the theorems in Munk's dissertation stated, for example, that the induced drag of a wing is minimal if the downwash velocity is the same at all positions along the span. Munk also proved that the total induced drag of a parallel arrangement of wings (such as in a bi- or triplane) does not depend on the displacement of the wings in the direction of flight (stagger theorem). A most important theorem derived the then unproven conclusion that the induced drag is minimal if the lift per length of span is distributed according to half of an ellipse. For this case, the formula was derived by which wings of different span could be compared; this was the same formula "which was found earlier by Prandtl in a different manner" and which occurred in several publications of the MVA since 1914 without derivation.

Munk's role in the development of airfoil theory did not become widely known. His Göttingen dissertation was quoted in subsequent publications mainly with regard to the proofs of the theorems used in airfoil theory. When Prandtl presented his "Airfoil Theory I" in July 1918 to the Göttingen Academy of Science, he referred to Munk's "forthcoming" dissertation because it contained "an important extension of the range of applications of the theory" [109, p. 322]. Half a year later, in "Airfoil Theory II," he pointed to the stagger theorem as an example for such an extension. He praised Munk's "meritorious dissertation" for its generality but at the same time criticized "Munk's derivation based on classical variational calculus" and presented the same conclusions with simpler derivations [109, pp. 350 and 353].

For readers of Prandtl's two-part airfoil theory who were not familiar with Munk's work at Göttingen such praise must have created an impression – surely against Prandtl's own intent – that Munk was merely the theorist who added some mathematical rigor to an already well established theory. Munk's Göttingen doctoral degree fits with this image as much as Munk's own self-portrayal appended to his dissertation, in which he emphasized the employment as Prandtl's assistant "at Göttingen University," where he could focus on "mathematical and physical studies." Nowhere does he mention how he spent most of his time in Göttingen: with wind tunnel measurements at the MVA issued by war contracts. An entirely different portrayal is presented to readers of Munk's Hanover dissertation which earned him a doctorate of engineering, and which consisted almost entirely of tables and polar diagrams of wings. In the attached curriculum vitae to this dissertation, Munk portrayed himself as a practical engineer who spent his three-year sojourn at the MVA "mainly with aerodynamics and the associated laboratory experiments." He introduced this dissertation with the remark that he did "not aim at a comprehensive intellectual assessment of the displayed data but rather at the communication of the obtained results for the purpose of practical application" [149, p. 3].

## 3.5
### Theory and Practice in Airplane Design

With his two doctoral dissertations and degrees, Munk embodied the juxtaposition of theory and practice often observed with aerodynamics since the First World War. Depending on the preferences of a textbook writer or the envisioned audience, the mathematical, physical, or technical aspects may prevail. So far we focused on Prandtl and his Göttingen circle who laid the foundations of modern airfoil theory. How was this theory received by those who would use the new knowledge for practical ends – the design of airplanes? The images of airplanes before and after the war clearly reveal to what extent aviation technology leaped forward within those few years. Were these advances due to the simultaneous advances of aerodynamic theory?

Among the many manufacturers of war planes, Anthony Fokker, the "Flying Dutchman," as he called himself, was one of the most successful ones. As an adventurous pilot and designer, Fokker perceived the war as an opportunity to manufacture airplanes on a much larger scale than during the years before. "My own factory built about 4,300 airplanes during the war," he recalled in his autobiography, "altogether roughly 7,600 Fokker machines were produced." He titled the chapter about the expansion of airplane production during the First World War "I become an industry" [153]. Fokker airplanes amply illustrate the transformation of fragile thin-airfoil objects into cantilevered-wing airplanes. Fokker's triplane Dr. 1 (see Fig. 3.8) was regarded as evidence for the rapid transfer of new aerodynamic know-how from the research laboratory into industrial practice because the wing profile of the Dr. 1 was listed as No. 298 in the Göttingen profile catalog. "That revolutionary discovery was immediately picked up by the famous designer Anthony Fokker, who incorporated the 13-percent-thick Göttingen 298 profile in his new Fokker Dr. 1," a text on the history of aerodynamics remarked [42, p. 309]. But the fact that the wing of the Dr. 1 figures in the list of Göttingen profiles does not justify this conclusion. In fact, as Fokker's chief engineer Reinhold Platz recalled: "This aerofoil was not tested aerodynamically in a wind-tunnel or in any other way." Platz had come in 1913 to the Schwerin factory of the Flying

**Fig. 3.8** Fokker's triplane Dr. 1 and its wing section (Source: Deutsches Museum, Munich).

Dutchman; by training, he was a welder and had no engineering degree. His guiding design principle was simplicity. The Dr. 1 wing was a simplified version of a more slender profile of a precursor test plane whose lower side was curved. By making the lower side straight, the profile became thicker, and the wing became easier to build. "Later on, unknown to Platz, the FLZ subjected it to wind-tunnel tests at Goettingen as the Goettingen 298 aerofoil," a historical account of Fokker's airplane construction (based on Platz's recollections) explained how this profile entered the Göttingen profiles [154, p. 225]. That thicker cantilevered wings were given preference over thin wings with wires and struts also resulted not only from aerodynamic considerations: "If struts and wires were shot away these [thin] wings simply collapsed, while the cantilevered wings suffered almost no damage" [153, p. 228].

Other manufacturers also based their design of new airplanes primarily on practical experience with precursor planes – modified according to the expectations of their military clients – rather than upon theoretical knowledge from recent aerodynamic research. "A conference about a new design seldom took longer than two hours," recalled airplane designer Ernst Heinkel about the meetings with the navy at which his firm, the Hansa Brandenburg, hoped to acquire new contracts for the design of seaplanes. As a result of these conferences, the military requirements concerning armament, duration of flight, and weight were specified. "When I left I already had the first draft sketches ready. A few pieces of paper and a pencil were all I needed. I sketched the then quite famous seaplane 'Hansa Brandenburg W 12' on the backside of a beer mat." The new aircraft, a biplane, was built within "barely eight weeks." The first flight was a dangerous adventure for the test pilot; changes such as the use of shorter struts were made overnight before the decisive test flight at the SVK, but in such a way that nobody noticed the modifications, remarked Heinkel. "Estimating roughly in the hopes of obtaining correct results (Richtig über den Daumen zu peilen) was then the secret of all designers" [155, pp. 79 and 89].

Even in the factory of Hugo Junkers, who was proud of the scientific principles behind his industrial success, airplane design was a matter of practice rather than theory. "There was not much drawing when a new airplane was designed," reported an engineer about the beginnings of the all-metal airplane construction in Junkers's Dessau factory. "Little sketches were all the workshop needed. The larger parts were immediately drawn upon the sheet metal and cut out by foreman Seifert" [156]. Rumors were spread that Junkers did not come forth to fulfill the needs of aviation in the war because he regarded aircraft design primarily as an opportunity for interesting research; such reproaches, however, were unfounded and should be interpreted as the result of the harsh competition among airplane manufacturers. Junkers was the only airplane manufacturer who possessed his own wind tunnel. As the first engineer in charge of the operation of the wind tunnel, Philipp von Doepp recalled, "Hugo Junkers was very keen to portray his technical activity as 'research,'" but Junkers had an idiosyncratic notion of research: "Research activity was not meant to serve for the extension of our knowledge but to achieve technological progress" [157]. Junkers himself once admitted in his diary: "It was alleged that I am a theorist, and because I am a technologist (Techniker) and no artist or scientist this means that I allegedly do not regard sufficiently the economic aspect. Nothing could be more wrong about the motivations and principles of my activities. The direct opposite is the case: I am convinced that each activity, also in science and art, has to be guided by economic principles if it is supposed to be successful." Junkers reflected in this manner about his own mentality on the eve of negotiations with the Army Administration for the acquisition of war contracts. Under the headline "Aviatik," he formulated his main objective: "It is crucial to arrive quickly at an agreement" [158].

With regard to airfoil theory, Doepp recalled that "it was characteristic for Junkers that he was always very reserved when confronted with theoretical results." Although the theory was not yet well enough known that he could have made it the subject of closer scrutiny, he would hardly have found this worth the effort: "Theoretical deliberations spun too far appeared to him uncertain, particularly if they dealt with involved mathematical investigations" [157, p. 10]. Also telling is Prandtl's recollection that Junkers never paid a visit to the MVA, although Prandtl visited Junkers's laboratories several times. Prandtl even "crawled into the first wind tunnel" during a visit to Junkers's institute in Aachen, and he investigated the Dessau wind tunnel built after the Göttingen model with a closed circuit. But unlike the Göttingen wind tunnel measurements, the tests in Dessau did not serve theoretical ends. Their purpose was to obtain data for the design of airplanes with optimal lift-to-drag ratios. Without the theoretical underpinning of airfoil theory a considerable number of tests was required because the polar diagrams of model wings of different spans could not be compared to one another. When Junkers learned in April 1918 that the Göttingen airfoil theory provided such transformation formulae, he was "very surprised" and exclaimed: "Had we known them earlier we could have spared all our test runs" [159].

A glance into Munk's report in the Technische Berichte could have offered Junkers this insight one year earlier. That even Junkers, the "theorist" among the airplane manufacturers and himself a former professor at a technical university, was unaware of these fundamental results suggests that outside Prandtl's Göttingen circle, progress in theoretical aerodynamics was largely ignored by airplane designers during the war. Only against this background does the zeal with which Munk tried to "win friends" for the formulae of airfoil theory become understandable: he introduced his report on "span and air resistance" with the remark that "Prandtl's wing formulae" were not received with the friendliness as they deserve "because they are based on theoretical foundations." He found this "very deplorable, for the formulae contain more and accomplish better things than the practitioners are willing to believe they are capable" [160]. The lack of interest on the side of airplane manufacturers also explains why the Göttingen aerodynamicists communicated only as much detail about the airfoil theory as was required to apply the formulae. Once in May 1918, an engineer from the Deutsche Flugzeugwerke G.m.b.H. in Leipzig wondered why certain results were presented "without any reference to a publication." He was particularly interested to learn "whether the downwash of air behind the wings of a mono- or biplane already has been assessed computationally." Apparently the Leipzig airplane manufacturer did not know that this issue was already a recurring theme among the Göttingen theorists for some years. Prandtl responded, "The theory of the monoplane for which you are asking has not been published so far in print; it was presented only in lectures and seminars" [161].

When Prandtl published the theory two months later, he did not choose the airplane designers for his audience but the academic circle of his Göttingen colleagues: he presented it to the mathematical physics class of the Göttingen Academy of Science [109]. The communications of the Academy, where the airfoil theory appeared in print the following year, certainly did not belong to the literature of which an aeronautical engineer or aircraft manufacturer would likely take note. Practitioners learned the new theory only later after it was published together with systematic data of wind tunnel measurements in the postwar communications of the Aerodynamische Versuchsanstalt (AVA), as the MVA was renamed in 1920 [162]. The new theory was also presented together with the wartime wind-tunnel data from the TB in 1922 in the form of a textbook by former members of the FLZ. They introduced their treatise with the explicit remark that the practical importance of Prandtl's airfoil theory is "beyond all doubts" and an "ever more intimate coalescence of these theoretical physical considerations with the fundamental ideas of airplane design must be expected in the future" [163, p. III].

What lessons may we draw from these events about the relationship between theory and practice in aerodynamics during the First World War? Was it truly symbiotic, as was suggested in the beginning of this chapter? Yes, if we address the mutual benefit of science, technology, and the military in terms of their institutions: the new MVA was motivated by and operated for the interests of military aviation; Prandtl, who embodies the science of aerodynamics in this three-sided relationship, was presented with the world's most advanced wind tunnel; the airplane manufacturers acquired remunerative contracts by participating in the triangle; the military was provided with an ever increasing number of airplanes for every kind of war purpose. On the other hand, the relationship was not symbiotic, as far as the mutual transfer of knowledge was concerned. Practice did not take note of airfoil theory before the 1920s, while theory was presented with a host of practical experience as a consequence of war contracts, and theoretical results could be tested. Uncertainties that prevented the publication of airfoil theory before 1914 were resolved, which gave way to a growing confidence that the theory was founded on solid ground. To speak of "successes of Prandtl's airfoil theory" (such as in the 1922 textbook) did *not* mean that airfoil theory was successfully applied in airplane design but rather that certain theoretical results (such as the influence of the span on the polar diagram) had been confirmed by wind tunnel measurements. The war was a phase of maturation for aerodynamic theory. Only when it stood the test of hundreds of wind tunnel measurements was it clear that airfoil theory would play an important role in future airplane design. However, this was still many years in the future (see chapter 9).

Although theory did not play such a crucial role in the First World War, the performance of airplanes improved considerably in the course of the few years starting in the second decade of the twentieth century. If not by the application of theory, how can this change be otherwise explained? The competition and rivalry among more than two dozen airplane manufacturers suggests another cause. Driven by ever-changing and growing military demands, a great variety of airplane types was designed among which the military, with its FLZ and supported by aerodynamic testing in scientific institutions such as Prandtl's MVA, selected those designs that best met their requirements. Similar variation-selection processes have been observed to advance technological development in other cases [164]. Between 1910 and 1914, for example, 25 airplane types were developed in Fokker's factory; from 1914 to 1918, this number rose to almost 100 [154, Appendix 1]. From one type of aircraft to another, there were often only minor differences, but across the entire spectrum of airplane manufacturers, there was an enormous variety from which only a few were selected for mass production. Only those that not only survived the tests at the army or navy test facilities but also surpassed their competitors were awarded a contract. Technical improvements were achieved largely within a balanced interplay of keeping proven designs and adding a few new ones by trial and error. A number of factors played a role, including the choice of materials, stability, motor and propeller technology, and armament, to name only a few considerations an airplane designer had to keep in mind. Among these considerations, aerodynamic theory was not the most important component, and certainly not yet a factor influencing selection in this evolutionary process of airplane design.

# 4
# The Internationalization of Fluid Mechanics in the 1920s

The interaction of science and technology in the First World War, as the case of airfoil theory indicated, served the advancement of theory more than the advancement of practice. Nevertheless the same case made evident that scientific investigations of fluid phenomena would become increasingly important. The situation at the outset of this process, however, was very different in different countries. In the USA, the theoretical progress achieved in Germany during the war caused an "aerodynamic culture shock" in the early 1920s [42, p. 292]. In Great Britain, Prandtl's airfoil theory was received hesitantly and accepted only after years of critical investigations. In Germany, further aeronautical progress was faced with harsh economic and political conditions. The Versailles Peace Treaty prohibited the construction of military airplanes, and no research contracts were issued to the Göttingen research establishments from the military as they were during the war. Civil contracts also seemed unlikely because the aviation industry as a whole faced an uncertain future under the restrictions imposed by the Allies. Although their primary goal was to prevent German rearmament, there was also the economical motivation "to effectually cripple the German aircraft industry as a competitor in the markets of the world," as was formulated in a U.S. Military Intelligence Report [165]. As a consequence, the "symbiotic relationship" in which Prandtl's MVA flourished during the war broke down, and even the mere existence of this institution seemed endangered [67]. Furthermore, German science was boycotted by a newly founded International Research Council (IRC) which, for example, excluded German scientists from participating in international conferences [166].

It seems paradoxical that under such adverse conditions the Göttingen research in fluid mechanics not only continued operations but was further expanding. The MVA's wind tunnel and Prandtl's airfoil theory in particular attracted the curiosity of American aeronautical researchers, and politically motivated boycott measures were brushed aside. Within few years after the war, Prandtl arose to international fame, and his Göttingen research facilities attracted visitors from all over the world.

*The Dawn of Fluid Dynamics: A Discipline between Science and Technology.* Michael Eckert
Copyright © 2006 WILEY-VCH Verlag GmbH & Co. KGaA, Weinheim
ISBN: 3-527-40513-5

## 4.1
## American Emissaries at Prandtl's Institute

Before the First World War, the establishment of aerodynamic research establishments in Europe, such as Gustave Eiffel's laboratory in Paris or Prandtl's Göttingen institute, were already attracting the curiosity of American observers [167, 168]. In 1913, for example, Jerome C. Hunsaker (1886–1984), a naval officer who had been trained as an engineer at MIT, together with Albert F. Zahm, a professor of physics at the Catholic University of America in Washington, DC, visited European aeronautical research laboratories. In America, the country of the WrightWright brothers, it was regarded as a shame ten years after the first motorized flight to have fallen behind European countries like France, Great Britain, and Germany. "In aeronautics, we in America are still in a transition stage," Hunsaker concluded in his survey of European research facilities. His report included a clear lesson: "We are at the point where the inventor can lead us but little further, and it is to the physicist and the engineer that we must look for perfection of air craft and the development of a new industry growing out of their manufacture and operation" [169, p. 31]. The foundation of a central organization for flight research in 1915, the National Advisory Committee for Aeronautics (NACA), was motivated, among other reasons, by the desire to catch up with Europe in this new technology [170].

The urgency to catch up became more obvious in 1917, when the USA entered the First World War and were confronted with expectations of its European allies to send thousands of war planes to the European war theater [171]. Another impetus came from the National Research Council (NRC), which was founded in 1915 in order to explore what science could contribute to the war effort. "War should mean research," was the NRC's motto. War also meant scientific intelligence. Following a request from the War Department, a military committee involving the Director of Naval Intelligence and the Chief of the Military Intelligence Section was formed under the guidance of the NRC. Among its first activities was the establishment of a Research Information Committee, with offices in Washington, London, and Paris, for the purpose of gathering scientific and technical information on war-related problems. To meet these goals, scientists were assigned to the American embassies in Rome, Paris, and London as "scientific attachés." Their major task was "the securing, classifying, and dissemination of scientific, technical, and industrial research information, especially relating to war problems, and the interchange of such information between the allies in Europe and the United States" [172, p. 41]. In particular, they were to acquire information on European developments in submarine detection, chemical warfare, trench warfare, and aeronautics [173]. A review issued by the Research Information Committee after the war reported that the European offices sent 1101 special reports to Washington from "practically every field of activity on war problems"—among these, 384 were on aeronautics. Much of this information was "of value for peace purposes as well as for those of war" [174, p. 45].

With aeronautics as the main beneficiary, it is not astonishing that the NACA developed a keen interest in these results. The former Paris scientific attaché, William Frederick Durand, had served as chair of the NACA before he went to Paris. After his return from Paris to Washington, he presented the Executive Committee of the NACA with first-hand impressions about aeronautics in Europe. Another NACA member with European experiences was Joseph S. Ames, a physics professor at Johns Hopkins University in Baltimore. In 1917, Ames had led another NRC mission to Europe. In April 1919, he proposed that a representative of the NACA should be stationed in Paris "with the duty of collecting information from the French, British, and Italian Governments, and such data from Germany as he could, and transmitting same to this committee." The proposal to send a NACA representative to Paris in order to gather scientific intelligence in Europe was originally made by William Knight, a First Lieutenant of the Army's Air Service. He had served under Benjamin C. Foulois, the commander of the Army's Air Service of the American Expeditionary Forces in France during the war. With this experience, Knight was the obvious candidate for this task. To supervise this activity, the Committee on Publication and Intelligence was established under Ames's chairmanship, together with the Office of Aeronautical Intelligence and a foreign office in Paris. The committee's task was defined quite comprehensively as "the collection, classification, and diffusion of useful knowledge on the subject of aeronautics, including the results of research and experimental work in aeronautics done in all parts of the world." The major part of this mission was supposed to be performed by the "Paris Office," as the NACA's foreign office in Europe was called for the sake of brevity [175].

Within few weeks the plan was realized. In the summer of 1919, Knight established an office in Paris. He proposed that he be assigned to the U.S. embassies in Paris, London, and Rome as Aeronautical Attaché, but the State Department refused to attribute such an official diplomatic status to the "technical assistant of the NACA," as the NACA's representative in Paris was officially designated. Unofficially, however, the American embassies were advised to provide Knight with "all appropriate courtesies and assistance in the discharge of his duties" [176]. Knight further insisted that his mission be more officially acknowledged, but he was put off to some time in the future "after your office has gotten fairly established and the character and necessity of your work clearly demonstrated" [177]. One of the reasons to deny the NACA's representative in Europe diplomatic status was the rivalry with the military attachés who regarded aeronautical intelligence as part of their own mission – and the NACA as an organization was subordinate to the U.S. military: "The collection of information for the Army and Navy will always be a function of the Army and Navy Departments," Knight was told by a colonel, "through their Military and Naval Attachés, and such function cannot be taken up by a civilian advisory organization. Did you ever hear of a child supporting his parents?" [178]. A general demanded that the Paris Office be shut down again, but Knight and Ames found the support of General Pershing for the NACA's foreign mission, and so the military finally put up with the unwelcome rival provided "that Mr. Knight was not soliciting nor giving military information" and "that he was confining his activities to purely technical and scientific matters." One military attaché even conceded that the Paris Office could provide "information which would be invaluable to the Military Intelligence Section of the General Staff at Washington, which otherwise might not reach them" [179].

The rivalry with the military attachés was not the only subject of dispute. Before the Versailles Peace Treaty, it was particularly problematic to establish relations with the former enemy, Germany. "I asked General Pershing if he thought that by going to Germany in the near future I could hope to obtain information on aeronautical work there," Knight reported to Washington in August 1919, but Pershing advised him "not to go for a few months yet at least" [180]. Although Knight postponed his first trip to Germany, he began to establish contact with German aeronautical scientists by correspondence. In November 1919, he introduced himself to Prandtl as the "Technical Assistant in Europe to the U.S. National Advisory Committee for Aeronautics and explained that he had been "accredited, so far, to the British, French and Italian Governments, for the exchange of information between those Governments and the Government of the United States, through the U.S. N.A.C.A. In the near future, I hope to be able to do the same in Germany." In order to make his intent of information exchange explicit, he sent Prandtl the NACA's Annual Report of 1917 and offered to provide "any particular information of a technical or scientific nature about our aerodynamical work in the States." In response he hoped that Prandtl would assist him in procuring the Technische Berichte. He was also interested in obtaining information about the new Göttingen wind tunnel of which he had learned via a French aerodynamicist [181].

Prandtl responded in a four-page letter and in German "because I do not master sufficiently the English" and "in order to make sure that I express myself unmistakably." He welcomed the suggested exchange of information and provided a short description of the new wind tunnel; furthermore, he promised to help Knight obtain the Technische Berichte. He also expressed as a personal wish that Knight transmit his best regards to the aged Gustave Eiffel with whom he had come to "a very cordial relationship" in 1913, although he admitted: "The ugly war has destroyed so many sympathies so that I do not know whether Mr. Eiffel is willing to renew a relationship with a German. Therefore, I do not want to address him myself" [182]. Eiffel's reaction was as anticipated: "Je considère qu'il n'est pas douteux que, suivant l'opinion unanime des savants Français et notamment de l'Académie des Sciences il est impossible de songer, avant longtemps, à entretenir des relations personelles avec les savants Allemands. Aussi quelle que soit la très haute estime dans laquelle je tiens tous les travaux d'un homme aussi éminent que Monsieur Prandtl, je chercherai à me procurer ce qu'il publiera par voie d'achat et à en profiter de mon mieux." Knight transmitted this passage to Prandtl literally and "without any comment." His own status as "guest of the French nation" prevented him from undertaking further steps, as he wrote to Prandtl [183], who to some extent had already expected such a response. Nevertheless he was disappointed. He had hoped, Prandtl confided to Knight, that Eiffel "in the philosophic mind which comes with age lift himself above the opinion of his compatriots who let the hatred win over reason. Now I see that this is not the case" [184].

This correspondence mirrors on the private level how the international scientific community at large was split into hostile camps after the war. The infamous "War of the minds" ("Krieg der Geister"), as the propaganda of intellectuals for the cause of the combatants on both sides of the war has been called, became transformed into a "Cold War in Science" for several years after 1918. To some extent, the Entente's discriminatory international postwar-politics via the International Research Council (IRC) were rooted in America, where George Ellerly Hale had first drafted the plan for the IRC as an inter-allied war-time organization of national research councils modeled on the American NRC [185]. Knight's contact with Prandtl, therefore, together with his offers for an exchange of information, was a delicate matter with respect to the inter-allied science politics after the war. Nevertheless, Knight regarded the exchange of information with the German professor as just the beginning. He hoped that he could soon "come to Germany with official credentials to the German Government, when I hope our exchange of technical data and information will be accomplished in a more complete and efficient manner than can be done at present in a more or less unofficial way," as he wrote to Prandtl in January 1920 [183].

Counteracting the official boycott was not merely the result of a personal internationalist attitude, to which Knight repeatedly confessed. During its formative years, the NACA had more to take than to give from German scientists in the field of aeronautics. Knight and Ames were particularly interested in learning more about the new Göttingen wind tunnel and airfoil theory. In February 1920, Knight was authorized by his superiors in Washington to visit Germany [186]. Ames wrote a personal letter to Prandtl in which he expressed his joy "to note that you are willing to send us some copies of your latest works." The exchange proved useful also for domestic purposes: "Dr. Ames reported encouraging progress in the work of the Paris and Washington offices of the Office of the Aeronautical Intelligence," read the Minutes of the Executive Committee discussion at the NACA. "As evidence of the increased activity of the Paris Office, he submitted samples of German technical reports recently transmitted by the Paris Office" [187]. In April 1920, Knight paid his first visits to German aeronautical research laboratories. Besides Prandtl, he visited 14 German experts of aeronautical research, to which he became introduced by Prandtl with letters of recommendation. His subsequent report was highly appreciated by the NACA's Executive Officer and distinguished among other conflicting accounts of German aeronautics as "the first report that has contained authentic information" [188].

Soon after Knight's visit, Prandtl received a letter from Hunsaker, who was touring Europe at the same time in his capacity as representative of the NACA's subcommittee on aerodynamics. Hunsaker informed Prandtl that he was authorized to enter into a contract with him in order "to obtain your services in presenting an authoritative survey of the recent German work in Aerodynamics, both theoretical and experimental. You are considered to have made important contributions yourself, and to be the best man to give such a survey" [189]. The details of the contract were left to negotiations during Hunsaker's visit in Göttingen in July 1920. Prandtl asked Runge—who had translated Lanchester's Aerodynamics into German and combined virtuosity of the English language with a keen interest in the development of aeronautics—to be present during the negotiations, because Prandtl himself had a poor knowledge of the English language. Also present was Munk, who was eager to meet Hunsaker because he intended to emigrate to America. The result of the negotiations was that Prandtl should write a report "on the state of the art of hydrodynamics as applied to predicting the aerodynamic forces on bodies shaped like airplane wings and airship envelopes" [190]. For this report, Prandtl would receive an honorarium of US$800, which corresponded to about 8,000 marks – roughly half of the annual salary of a German professor.

Financial considerations were not unimportant, particularly during the rapidly progressing inflation in the early 1920s. The AVA was in financial straits because contracts from industry were not expected. Prandtl, therefore, suggested further contractual collaborations with the NACA. German salaries were rather low by American standards and the AVA had no shortage of well-trained personnel, Prandtl argued, and therefore, it would be rather inexpensive for the NACA to issue experimental and theoretical aerodynamic research contracts to the AVA [191]. Prandtl had already made a similar proposal to Knight, but the NACA regarded such a collaboration with the former enemy as too far-reaching. Washington informed Knight that if the NACA issued contracts to perform research outside its own facilities, it was advisable "to support those in need of encouragement in this country" [192]. When Prandtl made the same proposal to Hunsaker, he first received no answer;

later Hunsaker apologized: "I am sure I do not know what to reply as to your question whether the National Advisory Committee for Aeronautics has any more investigations which might be handled at Göttingen. I am inclined to think that with the demand for economy on the part of the politicians our Committee will be somewhat restricted for funds in the coming year" [193].

Knight and Hunsaker were not the only emissaries who visited Prandtl in the summer of 1920. Another traveling NACA representative in Europe was Edward P. Warner, chief physicist of the new Langley Memorial Aeronautical Laboratory (LMAL), established by the NACA near Hampton in Virginia. Warner was particularly interested in studying the experimental techniques developed in German aeronautical laboratories. "It is appropriate," he reported from his travels, "that any discussion of aerodynamical work in Germany should begin with Göttingen and with Prof. Prandtl, where the first serious work of the kind was undertaken, before the war, and where the most extensive and interesting results have been obtained both in respect of wind tunnel testing and of purely mathematical investigations." Half of his 12-page report dealt with the Göttingen wind tunnel and the balances used to measure aerodynamic forces. Although the focus was on the experimental research in the AVA, Warner also mentioned the "Prandtl theory of wing action, together with the work along the same lines by Munk and Betz," because they provided "a practical tool for the engineer" [194]. By this time, Wladimir Margoulis, a Russian aerodynamicist who had emigrated to France and now assisted Knight in the Paris Office, had already procured a summary report on Prandtl's airfoil theory based on Prandtl's publications in the proceedings of the Göttingen Academy of Science. The NACA published Margoulis's summaries in subsequent issues of its Technical Notes in July and August 1920 [195]. Another NACA report was published in November 1920 on the Göttingen wind tunnel testing [196]. From these reports, most aeronautical engineers outside Germany learned for the first time about the experimental and theoretical aerodynamic research achieved in Prandtl's laboratory during the war.

Knight paid another visit to Prandtl in the fall of 1920. Half a year later, he was relieved from his duties as the NACA's foreign intelligence officer "at the earnest request of the Army Air Service." Apparently he was a source of friction with the military attachés. His successor was John Jay Ide, a former naval officer who knew better how to avoid clashes with his military rivals than Knight. "This change of personnel," a military intelligence officer wrote to the NACA, "will be heartily welcomed by all Military and Air Attachés in Europe." But when Ide, like his predecessor, proposed that he be accredited in a more official manner, the unsolved issue about the status of the NACA's representative abroad caused new trouble. The State Department kept denying official diplomatic status. "This is a rather delicate matter to handle," explained the NACA's Executive Officer as he turned down Ide's proposal. The rivalry with the military attachés also remained a bone of contention. Almost two years after Ide had started his mission in Paris, he informed the NACA officers in Washington about a correspondence among military attachés that had come to his attention: they hoped "that the wave of economy may reach even the National Advisory Committee for Aeronautics and, by cutting off their appropriations, eliminate their representative in Paris who is merely the fifth wheel to a wagon." But Ide had contacts in the aeronautical circles in Europe "not possessed by any other American," as was acknowledged in Washington, and the NACA would not give up such a "tremendous asset" [197], [170, p. 75].

For Prandtl, the change of personnel in the Paris Office did not matter. Knight had rendered him a last service shortly before he was relieved by organizing a "journal circle" in order to provide the German aeronautical community with the most recent literature in the field – literature that otherwise would have been difficult to obtain during the early postwar era [198]. Knight kept in close contact with Prandtl by correspondence for many more years, and his successor continued the friendly relations between the Paris Office and the German scientists by frequent visits and a regular exchange of technical publications. "At Göttingen I visited the Laboratory and spent the day with Prof. Prandtl and Dr. Wieselsberger," Ide reported about his first visit. In a personal letter to the Executive Officer of the NACA, George Lewis, he described the cordial atmosphere of his visit in more detail. He invited his German guests to his hotel for dinner, and they spent many hours in a relaxed mood with Moselle wine. "I am pleased to note that you were successful in meeting Professor Prandtl," Lewis responded. The NACA was eager to nurture the relationship with Prandtl, although Lewis advised Ide that he should pursue it without any official commitments in order to avoid political trouble [199].

The friendly relations between Prandtl and the NACA prevailed throughout the 1920s and 1930s. Besides the regular exchange of material with the Paris Office, Prandtl corresponded with Knight, Ide, Hunsaker, Ames, and Warner on a host of private, political, technical, and scientific matters. Disregarding the official boycott, a regular, but informal, exchange of aerodynamic know-how between the USA and Germany via the Paris Office came into full swing within only a few years after the war. Via the travels of the NACA's "Technical Assistant" the results of the Göttingen AVA became known not only to the NACA, but also to the English, French, Italian, and other European aeronautical research facilities along Ide's itineraries. In addition to this informal transfer of knowledge, the NACA's Technical Reports and Memoranda became a most widely appreciated means to internationalize aeronautical knowledge all over the world. In retrospect, it seems ironic that this process was initiated by an office dedicated to intelligence in the interest of the USA in order to catch up with Europe in aeronautics. Today, intelligence is regarded mainly as secret information gathering, but back then, it was considered in a broader sense and comprised the collection of information and its distribution. It is even more paradoxical that this internationalization reached its climax when German science was officially isolated from the international scientific community, and aviation, in particular, was affected by the Versailles Peace Treaty.

The internationalization effected by the USA after the First World War was not restricted to aeronautical sciences. In more basic sciences, too, the USA was eager to catch up with Europe. In the early 1920s the Rockefeller Foundation started to fund European research centers and to accelerate the spread of new scientific results by an international fellowship program. Wickliffe Rose, the founder of Rockefeller's International Education Board (IEB), reported in November 1923: "Higher mathematics in the United States has had its development in the main since 1890. America is behind Germany, France and Italy at the present time." The Rockefeller Foundation was "primarily interested in American education and in the 'national interests' of the United States," a recent study on Rockefeller's initiative to internationalize mathematics concluded. Scientific internationalism was "a strategy to pursue genuinely American goals." The Rockefeller Foundation, too, maintained an office in Paris in order to establish relations with European scientific centers. Its mission resembled that of the NACA office, and as far as Germany was concerned, Göttingen was also a major target for Rockefeller's goals. When Rockefeller emissaries pondered the funding of Göttingen mathematics, they regarded the proximity of Prandtl's institute as an advantage. Although Prandtl is "more of an engineer than a physicist," a report remarked, he "is also professor in University—mathematics department." The Rockefeller representative came to the conclusion that the "presence of Prandtl's lab[oratory] in Göttingen is a factor of strength in the general scientific situation there." As a result, Rockefeller financed the construction of a new mathematics institute in Göttingen and encouraged traveling research fellows to sojourn there. Among the Rockefeller fellows of the 1920s who were particularly attracted by Prandtl was, for example, the Cambridge student Sydney Goldstein, who would later rise to prominence in theoretical fluid dynamics in England [200, pp. 18, 37, and Appendix 6].

## 4.2
### Standardization

The more aerodynamicists learned about work in other laboratories, the more they became aware of the need for standardization of their experimental and theoretical procedures. Beginning in 1920, British aeronautical engineers conceived a project for comparing and subsequently standardizing wind tunnel data [201]. Earlier wind tunnel tests at the British National Physical Laboratory (NPL), Eiffel's Laboratory in Paris, and the Massachusetts Institute of Technology in the USA had revealed to what extent the measurements differed from facility to facility. Under the headline "Standardization and Aerodynamics," the American aviation magazine *Aerial Age* started a series of articles in June 1920, which revealed the opinions of the directors of the leading aerodynamic research laboratories in Europe that there was a need for international coordination of their work. The spokesman for internationalizing and standardizing aerodynamics was Knight. He proposed summoning the leading international aerodynamicists to a conference on standardization. During the two years of his term as the NACA's representative in the Paris Office he became aware of how the tangle of methods rendered the measurements of different laboratories incompatible with one another. For example, wind tunnel measurements of the same wing profile in Eiffel's Laboratory and at the NPL resulted in different polar diagrams. Knight therefore expressed the concern "that the present state of things is fraught with danger to the Science of Aerodynamics. As a matter of fact, when these divergencies are brought before the public, and especially before airplane manufacturers, as they must inevitably be, confidence in the work of the laboratories will be utterly shaken." Only a conference dedicated to the standardization of aerodynamical research could avoid further damage. In his plea for such a congress, Knight argued that political obstacles should be disregarded and representatives of all leading aerodynamic laboratories should participate – including German scientists. He criticized the tendency "to snub and to pretend to ignore the wonderful progress made by the Germans in aerodynamics during the war" [202, pp. 3 and 7].

Knight had started already in the summer of 1919 to plan a "Congress of representatives of Aerodynamic Laboratories for the standardization of the work performed in such laboratories," preferably in Paris and under the guidance of the NACA, but he met with resistance among his own colleagues [203]. "I do not believe that it would be wise for us to take the lead of calling such a congress at the present time, as the National Advisory Committee for Aeronautics really have nothing to show in the way of research work up to the present time," Warner responded to Knight's plan. "In any case it seems to me that we should not attempt to initiate and control a conference which would sit in Paris" [204].

But Knight was not so easily discouraged and asked his European contacts to support his initiative. In October 1921 he published Prandtl's comment on the issue of standardization in *Aerial Age* in an attempt to elicit the comments of other leading European aerodynamicists in order to demonstrate "that all that is needed to bring about results is to take the lead in bringing them together and letting them decide something which will be agreeable to everybody and especially to manufacturers and designers of aircraft who are the only ones for whom the research work is done in the laboratories and books and for whom technical reviews are published." Prandtl supported Knight's plea with technical details. For example, he pointed to the various methods of creating the stream of air in wind tunnels: "In order to accomplish any comparative results in wind tunnel tests, it is of prime importance to have the air currents comparable," Prandtl argued. It was critical, for example, to keep the wind eddies at a minimum. In Göttingen, they had paid much attention to this problem and found that with a special grid formed like a honeycomb and attached in front of the conical narrowing passage before the test chamber, the current of air was most homogeneous. In order to achieve high Reynolds numbers as in real flight, large model sizes had to be used, with the consequence that air flow was modified due to the vicinity of the tunnel walls, but such disturbances could be accounted for computationally: "The modern wind theory allows a calculation of the influences exerted by the walls of the test tunnel or by the limitations of the free air stream." Another problem that called for international coordination concerned the mounting of models. Prandtl argued that part of the discrepancies between the measurements at the NPL and Eiffel's Laboratory was due to different mountings, which caused disturbing effects in the wake of the models: "As known, any disturbing element on the suction side entails quite a drag which increases with the angle of incidence while any disturbing influence on the pressure side brings a decidedly lower and with increasing angle of incidence a decreasing drag" [202, pp. 10–13].

In a subsequent issue, Knight introduced Prandtl's protégé Theodore von Kármán, "a most brilliant scientist who has been prominent in the development of aeronautics in Austria during the war and who is now at the head of the Aachen aerodynamic laboratory." Kármán condensed the most important items requiring standardization into three categories: coefficients, methods of measurement, and definitions and symbols. None of these were of mere academic importance. Kármán pleaded for the widespread use of dimensionless coefficients because they enabled the application of the laws of mechanical similarity. He mentioned the study of heat transfer in gases and fluids as an example in which the use of such coefficients led to a better understanding. With respect to measurement techniques, he emphasized like Prandtl, how important it is for the achievement of comparable test results in wind tunnels that the current of air is as free as possible from turbulent eddies. He suggested developing instruments for the measurement of turbulence. Furthermore, he agreed with the proposal that in order to compare results from different wind tunnels, all laboratories should first measure simple test bodies, such as spheres and disks of equal size; in fact, "it would be best to have the same model make a round trip to every laboratory adapted for such work." As a first step, Kármán proposed convening a "preliminary conference between a few of the most prominent scientists and technical men" who should prepare the ground "for creating an international scientific aeronautical association which is the best for bringing about a much desirable cooperation among aerodynamical research workers." As a site for such a meeting he proposed a village like Bozen or Meran in Italian Tyrol [202, pp. 18–20].

Although Knight was no longer the NACA's European representative at this time, he pursued the plan of an international conference on standardization with unabated zeal. He used Prandtl's and Kármán's expert opinions as evidence that the most distinguished aerodynamicists regarded the standardization in aerodynamics as expedient and that petty resentment of German scientists should not doom this plan to failure. However, in view of the hostile French attitude towards Germany, Prandtl was pessimistic whether Knight's plan could materialized [205]. The official U.S. attitude was "against mixing in anything that is international, or that can in any way be tied to the Peace Treaty." The USA had not ratified the Versailles Treaty but concluded a separate Peace Treaty with Germany and Austria in July 1921. Official international commitments were considered delicate political affairs, although Lewis admitted in a letter to Knight's successor that the NACA was "extremely interested in standardization" [199]. However, Knight no longer felt obliged to comply with the official position of the U.S. government after his dismissal and was more determined than ever before to pursue his plan even "against the narrow minded nationalistic tendencies of some European nations and the selfishness of the United States" [206]. In a sequel to his previous articles, he published in the February 1922 issue of *Aerial Age* the expert opinion of Giulio Costanzi, an Italian aerodynamicist "well known among aeronautical scientists on account of the important research work done by him at the Royal Aircraft Establishment in Rome." In March, the series was continued by an article by the Russian aerodynamicist and former research collaborator at Eiffel's Laboratory, Wladimir Margoulis. Through to September 1922, four more pleas followed, authored by the directors of aerodynamic laboratories in Rome, Vienna, Amsterdam, and Washington. In December, Knight concluded the series with another passionate call for action: "At present such a divergency exists between experimental results obtained in various wind tunnels, when no such divergency should exist, that the confidence of aircraft manufacturers and designers in the usefulness of wind tunnel research work is badly shaken" [202, pp. 29–65].

Knight's initiative, however, was doomed to failure. The NACA finally declared that it is "not willing to take the initiative in the forming of an International Congress for the Standardization of aerodynamics." The British

Aeronautical Research Committee also withdrew its support because it did not consider "that the time is yet come for the proposed congress of representatives of aeronautical research laboratories." A preliminary international meeting held in Paris in November 1921, the First International Congress of Aerial Navigation, failed to meet its goal because "American and British laboratories were conspicuously absent" and "German Laboratories were not allowed to join," as Knight criticized. A proposal by the director of an aerodynamic laboratory in Madrid to grant the Eiffel Laboratory the authority to publish guidelines for the international performance of wind tunnel tests was commented by Knight with the following remark: "Frankly, we fail to see that the matter is so simple as Mr. Herrera seems to think." Under the headline "The Moral of a Sad Story," Knight presented a sober account on the state of affairs: 1) all leading experts in the area of aerodynamics regarded the lack of international collaboration as a serious problem; 2) the chaotic tangle of symbols, definitions and modes of representation hinders the use of research results; 3) there is no shortage of readiness for international collaboration on the part of the scientists and engineers themselves; 4) because most laboratories are state-controlled, an initiative to surmount the present obstacles has to start from one or more governmental aeronautical institutions; 5) any effort in this regard has to take into account the clauses of the Versailles Peace Treaty; 6) under these conditions no "truly international agreement" may be achieved; 7) comparative model tests in different wind tunnels performed so far had been undertaken without participation of German aerodynamic laboratories and are therefore only of limited use. If there was an organization that could lead aerodynamics out of this impasse, it was the NACA. This organization possessed "the assurance of the most effective cooperation of scientists of all nations (former allied and former enemy nations) who, in spite of the official taboo which separated and still separates in most European countries scientists in two groups: friendly and enemy, would have welcomed any attempt on our part to bridge the gap, in so far at least as aeronautics are concerned." As an example of setting political concerns aside for technical reasons Knight mentioned the NACA's decision to use Göttingen's mode of representing airfoil data "simply because they were the most logical coefficients to adopt." France would follow in due time. With regard to laboratories in Great Britain, however, he regarded it "very doubtful indeed if they will ever adopt symbols, coefficients and graphical methods of representations other than their own." At the very end, he once more appealed to the NACA to use its prestige as "the finest aeronautical scientific organization in the world" for an initiative towards internationalization and standardization: "Why should short-sighted and short-lived political considerations deprive this nation and our National Advisory Committee for Aeronautics of the great privilege of being able to make the first move?" [202, pp. 70–97].

## 4.3
## International Conferences

The NACA published Knight's plea together with the preceding series of articles in *Aerial Age* as a Technical Note in its series of reports – but it did not act as Knight had proposed. No international conference on standardization was held in Paris under the umbrella of the NACA or elsewhere as the initiative of another organization. In the subsequent years, the NACA issued several test series in its Langley Laboratory for the purpose of standardizing wind tunnel measurements [207–210], but the need for international agreement became more urgent by these measurements. "The actual process of standardization still lies in the future," a report concluded in 1925 after comparing the test results from Langley's wind tunnel No. 1 on the drag of disks, spheres, and certain airfoil profiles with those from other laboratories. "Even when such tests as the series above have been completed by all the laboratories concerned, an enormous amount of work will have to be done before any real state of standardization is attained" [208].

Despite widespread unanimity among the aerodynamicists about the need for standardization, the community of aeronautical scientists was still far remote from the ideal of the "international brotherhood" Knight had envisioned [202, p. 97]. In February 1923, two months after the final article on standardization had appeared in *Aerial Age*, Prandtl wrote to Knight: "After the Frenchmen have invaded our German coal region I declare that I do not wish to collaborate with a Frenchman and that I also do not wish to see a Frenchman in my Göttingen establishment, except if he declares in writing that he disapproves of the politics of his government." Although he shared Knight's view that the NACA should take the initiative for the organization of a conference on standardization, he was pessimistic whether the plan could materialize; first of all, the NACA should convince England to join this cause, Prandtl suggested. In the present political climate, however, he did not regard it likely that an official and truly international conference could be convened. More promising were privately organized meetings like a conference held in Innsbruck which was "unclouded by any political troubles." He expressed his belief " that only such a private association of scientifically interested people offers the perspective for really useful international work. All political viewpoints of official governments are only obstructive for science" [211].

The Innsbruck conference to which Prandtl alluded resulted from the private initiative of Kármán. After the war Kármán was returned to his Aachen chair with the intent of resuscitating the science of hydro- and aero-mechanics, which had progressed to new frontiers since 1914, but, as he wrote in April 1922 to Italian mathematician Tullio Levi-Civita (1873–1941), "unfortunately the personal intercourse among those who work in this area is rather sparse." Kármán cherished an international atmosphere privately as much as in his profession. He invited students from all over the world to his home where his mother and sister promoted an atmosphere of cooperation and open-mindedness. Kármán thoroughly disliked the nationalist resentment that sprang up during the "cold war" among many of his colleagues after the First World War. In his own discipline, he noticed that there were also gaps between those who dealt with the mathematical, physical, and technical aspects of fluid mechanics. Therefore, he suggested convening "a very unofficial meeting" of "about 30 to 40 gentlemen." He proposed Innsbruck as the place for this conference "because Austria is rather neutral soil, and a sojourn there would be affordable for all participants." If "such a casual meeting" would be successful, "the official circles" would soon follow this example. He asked Levi-Civita to invite the colleagues from the Romance-language- and English-speaking countries, while he would care for those in Germany, the Netherlands, Austria, Switzerland, Russia, Czechoslovakia, and Scandinavia. He emphasized that he would extend the invitation to England and France, "but I cannot judge whether in these countries the friendly attitude is far enough advanced so that an invitation would not be rejected" [212]. Although the plan materialized, and the conference was held in Innsbruck in September 1922, it was not regarded as a breakthrough to overcome the hostility between the major former wartime enemies. The conference proceedings list nine participants from Germany, three from the Netherlands, six from Italy, two from Sweden, and one from each of Norway and Poland [213]. Knight did not regard the Innsbruck conference as a model for the proposed international conference on standardization: "American, British and French scientists did not answer the call of their German and Italian brethren, not because they did not want to, but because they could not on account of the unfortunate preponderance of political considerations over other considerations of higher nature" [202, p. 96].

For those who participated in the Innsbruck conference, however, it was regarded successful enough that Kármán and his colleague from the Netherlands, Johannes Martinus Burgers (1895–1981), felt encouraged to plan a sequel "mechanics congress" in Delft [214]. Burgers was director of the Laboratory for Aero- and Hydrodynamics at the Technical University Delft and exerted a role in the Netherlands similar to Prandtl's in Germany. He is regarded as "the 'father of fluid dynamics' in the Netherlands" [215]. Burgers and Prandtl had corresponded with another since 1919. In October 1923, Burgers asked Prandtl to support his and Kármán's plan to convene the conference in Delft by participating in a preparatory committee and by signing a common conference call. In order to achieve greater international weight, Burgers proposed that the call also be signed by American, English, and French scientists [216]. Prandtl agreed in principle, but "because of political feelings" made his own signature dependent on a clause:

> "I have always defended the idea that the common cause of scientific interests should bring together again the scholars of formerly

hostile nations for the best of their peoples. But I had to revise this view, as far as Frenchmen and Belgians are concerned, because of their illegal invasion into the Ruhr area, and the never-ending oppression of the German people since then by Frenchmen and Belgians; therefore I am forced to declare that I do not wish to participate in a committee side by side with representatives of the French and Belgian nations as long as the present politics of oppression holds on. I also do not wish to be in touch with a Frenchman or Belgian personally, except if he condemns the present politics of his government in a clear manner. I ask you, therefore, to include me as a member of the preparatory committee only if for some reason the Frenchmen will not participate" [217].

Prandtl was not alone with such resentment. Richard von Mises (1883–1953), professor of applied mathematics at Berlin University, had the same misgivings. "Now von Mises and I, independently, have arrived at the same conclusion," Prandtl informed Kármán, "that at the present time we cannot agree under any circumstances that our names appear side by side with compatriots of the French Republic as members of the committee."

Kármán tried to mediate. Scientists from the neutral countries, England, America and Italy had agreed to participate without any conditions, he responded, but the French had declined, "some of them with an expression of regret," he added. With the absence of French members, Prandtl and Mises's stipulation became obsolete as far as the preparatory congress committee was concerned. For the conference itself, however, Prandtl insisted that he did not wish to have contact with French or Belgian scientists. Kármán argued that this would be "practically unfeasible" and impossible from a moral point of view. But if the congress would in fact take place it would be an important step "for the official recognition of German science in the whole world." He regarded it "completely wrong" to abstain from participating under these circumstances [218]. Burgers, too, was disappointed by Prandtl's attitude, but he hoped that the major part of the problem was resolved after the French had declined or ignored the invitation to participate [219].

Due to the "self-elimination of the Frenchmen" Prandtl considered "the difficulties which had existed for us Germans" as overcome and declared his consent for his name to be added to the congress committee. "Of course I only meant that I do not wish to collaborate with a Frenchman in the committee or elsewhere on a personal basis," Prandtl softened his tone. "If one or another Frenchman will be present at the conference, this will not bother me" [220]. Mises, too, declared his consent to participate in the congress committee. The other members came from England (L. Bairstow, W.G. Coker, A.A. Griffith, R.V. Southwell, G.I. Taylor), Norway (V. Bjerknes), Austria (P. Forchheimer), Italy (T. Levi-Civita), Czechoslovakia (T. Pöschl), Russia (A.A. Friedmann), the

USA (J. Hunsaker), and Switzerland (E. Meissner). The congress was held in September 1924 in Delft. One hundred and five Dutchmen, 54 Germans, 14 Englishmen, seven Russians, four Poles, three Belgians, three Italians, three Norwegians, three Scots, three Czechoslovakians, three Americans, two Austrians, two Swedes, one Australian, one Bulgarian, one Canadian, one Egyptian, one Romanian, one Turk, and even one Frenchman attended. Altogether there were 214 participants from 21 countries – a remarkable demonstration of international scientific cooperation only six years after the war [221].

The Delft congress was the first in a series of International Congresses of Applied Mechanics. The second was held in 1926 in Zurich. Afterwards, these congresses were convened in four-year intervals: 1930 in Stockholm, 1934 in Cambridge, England, and 1938 in Cambridge, USA. After the Second World War the tradition of these conferences was resumed under the umbrella of an official organization, the International Union of Theoretical and Applied Mechanics (IUTAM) [222]. However, the quest for internationalism among scientists was not free of its share of challenges. The correspondence among the organizers during the 1920s and 1930s provides evidence for the ensuing struggle involving national issues, which more than once interfered with the scientific internationalism presented in the public declarations. During the preparation of the third mechanics congress, for example, a controversy arose because Belgian scientists intended to use this occasion to celebrate the hundredth birthday of Belgian independence and therefore voted to convene the congress in Liège rather than in Stockholm. The international congress committee regarded this move as an effort to exploit scientific internationalism for national political goals. "In total contrast with the Belgians, who seem to regard a political event as a good reason to call for a scientific meeting, we think that no occasion would be less appropriate for this purpose," the Dutch members of the committee wrote to their Swedish colleague Carl Wilhelm Oseen (1879–1944). They suggested that he should insist on the proposal to convene the congress in Stockholm [222, p. 12]. Burgers regarded the Belgian plan "as extremely dangerous for the international collaboration achieved so far" and asked Prandtl for support in order to strike it down [223]. "We German members of the international congress committee," Prandtl assured Burgers, "have corresponded about this issue and are unanimous in so far that our position must side with yours" [224].

**4.4**

**Applied Mathematics and Mechanics: A New International Discipline Between Science and Technology**

Apart from their debatable demonstration of scientific internationalism, the "mechanics congresses," as they were called for short, signaled the emergence of a new disciplinary identity among researchers in fluid mechanics. On a national scale the first instance of this new attitude could be observed in 1920 at a conference of natural scientists in Bad Nauheim. It was discussed, Prandtl informed Kármán, "that at the next meeting, a division for applied mathematics or something similar should be convened in close contact with the divisions of physics and mathematics" [225]. The opportunity to establish the new specialty encompassing features of mathematics, physics, and technology came in September 1921 in Jena at the joint meeting of the Deutsche Mathematiker-Vereinigung, the Deutsche Physikalische Gesellschaft, and the recently founded Deutsche Gesellschaft für Technische Physik. As was proclaimed at this meeting and published in the first issue of a new journal, the Zeitschrift für angewandte Mathematik und Mechanik (ZAMM), the specialty of applied mathematics and mechanics should be given greater weight and presented in a more coherent manner. The initiators of this move were the same as those who would later participate in the committee of the international mechanics congresses: Prandtl, Kármán, and Richard von Mises. Since 1920, they (and a few others like Hans Reissner, Mises's fellow professor of mechanics at the Technical University in Berlin) kept an intensive correspondence on the professional and institutional ramifications involved with the establishment of applied mathematics and mechanics as a new discipline between physics, mathematics, and engineering. They expressed their conviction that this field was distinct from traditional engineering sciences or technical physics, and they would not feel appropriately represented by the Verein Deutscher Ingenieure (VDI) or the Deutsche Gesellschaft für Technische Physik. They regarded applied mathematics and mechanics as a specialty for which there did not yet exist an appropriate organizational umbrella. The result of their collaborative effort was, besides the foundation of the ZAMM, the establishment of a new professional society, the Gesellschaft für angewandte Mathematik und Mechanik (GAMM) [226, 227]. Year after year, Prandtl acted as the chair of the GAMM.

With the new journal ZAMM (1921), the GAMM (1922) as a new organizational umbrella, and the new tradition of international mechanics congresses in Innsbruck (1922—later called the zeroth International Mechanics Congress) and Delft (1924), applied mathematics and mechanics rapidly developed its own momentum. Its most important subfield was hydro- and aerodynamics, the theme of the Innsbruck congress, but other mechanical specialties such as plasticity, elasticity, and strength of materials were also encompassed. In addition to besides hydro- and aerodynamics, the Delft congress featured two other divisions of "rational mechanics" and "elasticity theory." Burgers's declared intent was to broaden the scope of the Innsbruck conference so that "the whole mechanics" was addressed – not arbitrarily, but with the focus on "technically important problems," as he wrote to Prandtl [216]. By the mid 1920s, a new international community of mathematicians, engineers, and physicists surfaced: its representatives were the members of the congress committee for the international mechanics conferences; its mother disciplines were still called physics, mathematics, or engineering, but the profile of the new applied mechanics and mathematics community presented them with a distinct identity.

Prandtl's local setting in Göttingen also changed by the mid 1920s. In 1925, he became the director of a larger institute of fluid dynamics. Since 1911 Prandtl had entertained the hope of including in his research more fundamental scientific problems of fluid mechanics in addition to technologically motivated aerodynamics (see Chapter 2). When he received a call to the Technical University of Munich in 1920, the Kaiser Wilhelm Society averted Prandtl's departure by fulfilling his wish for a "Kaiser-Wilhelm-Institut für Aero- und Hydrodynamik," as it was titled in the original proposal. However, the new institute did not materialize immediately. Headlines such as "Prandtl turns down the Munich call," "Prandtl accepts the renewed call to Munich," and "Prandtl's call back to Göttingen," followed the year-long tug of war between Göttingen and Munich. In the economic situation after the First World War, "all efforts failed to obtain the required money from the Reich's Ministry of Finance," the director of the Kaiser Wilhelm Society recalled in his memoirs. Finally, the Göttingen University offered a honorary degree to a private sponsor "if he enables the development of an important specialty in Göttingen and to keep a highly esteemed colleague here by erecting a hydrodynamical institute." The deal worked out. With the additional private funds of the sponsor, the new Kaiser-Wilhelm-Institut was erected in 1924 beside the buildings of the AVA. In July 1925, the new "Kaiser-Wilhelm-Institut für Strömungsforschung, verbunden mit der Aerodynamischen Versuchsanstalt in Göttingen" was formally inaugurated [67, pp. 240–243], [228, pp. 80–88].

Prandtl had always stressed the importance of combining applications with fundamental research. He regarded his specialty – applied mechanics—from the perspective of both a scientist and an engineer. When he was perceived by others as a representative of aerodynamics, he had mixed feelings. For example, when *Aerial Age* nominated him to be elected to their editorial staff, Prandtl first hesitated to accept because he regarded himself more as a fluid dynamicist than as a representative of aeronautical interest; nevertheless, he finally accepted his election [229]. The right balance of scientific and technological interests was also the subject of discussions at the International Mechanics Congresses. A proposal at the Zurich Congress in 1926 to replace "Applied Mechanics" by "Technical Mechanics" in the title of the conference was rejected; such a proposal had already been discussed for the Delft Congress two years earlier. The Dutch representatives in the international congress committee argued, "Applied Mechanics" should be kept because "it is very important that the connection with the contiguous fields of mathematics and physics be maintained." After the Second World War this tendency was made more explicit when the new umbrella organization called itself the "International Union of Theoretical and Applied Mechanics" [222, p. 13].

To have established applied mechanics as an new international discipline, with fluid dynamics as its most important specialty, does not mean that the various national traditions were easily brought into accord. Politics was not the only hurdle on the way towards an international practice in science; the scientists themselves were often unable to overcome their mutual traditions developed within their national environments. The much cherished "international brotherhood" among scientists in general, and the experts on fluid mechanics in particular, was a myth. Prandtl's unforgiving attitude with regard to French participation at the first mechanics congresses illustrated the politically motivated hostility; Max Munk's case presents us with an example of incompatible national traditions on the level of practical aerodynamic work.

## 4.5
### Internationality in Practice: Max Munk at the NACA

Max Munk emigrated to the USA in 1921 and became employed by the NACA. In view of the restrictions to aviation in postwar Germany, engineers who had chosen aeronautics as the subject of their professional work saw an uncertain future. "The situation seems to be terrible there and all Germany seems to be willing to move to the U.S.A.," an aerodynamicist at the NACA on commented one of Knight's reports from Germany [230]. Munk was in such a situation. But in contrast to others, the NACA had a keen interest in his employment. "Dr. Munk is at present Aerodynamical Expert of the Zeppelin Company in Germany," Ames informed the Executive Committee of the NACA in November 1920. "His employment would probably be the cheapest and most effective way of obtaining a vast amount of information developed in Germany during the war and not published" [231]. Despite the NACA's interest, however, it was not easy to obtain the official permit for Munk's employment. "My American prospects are not so bad," Munk reported to Prandtl after Hunsaker had sent him a comforting letter, "the main problem is my German nationality" [232]. The process ultimately required two signatures from the president of the USA for Munk's transfer to the NACA: one for the permit to enter the country and the other to employ him in a government agency [233, p. 75].

At the NACA, Munk was received with high expectations. Munk would be "a very useful man in the Committee's office," Ames promised his colleagues, particularly "to draw general conclusions from the work of other people" [231]. Frederick Norton, a physicist at the NACA's Langley Laboratory, suggested as "subjects for Dr. Munk" that he should first of all "write a clear and practical paper on the Prandtl theory with examples showing its application to a number of cases taken from American wind tunnel tests" [234]. When Munk began his work in March 1921 in the NACA's Washington office he was eager to meet the expectations of his new employer. He did not, however, regard himself as a general problem solver but wished to direct the NACA's aerodynamic research more actively. Munk was employed barely a month when he proposed that the NACA should build a new compressed-air wind tunnel. The Reynolds number is proportional to the product of air pressure and model size, and for small models, large Reynolds numbers like those of airplanes in free flight could be obtained if the pressure were increased [237]. The Executive Officer of the NACA forwarded Munk's proposal to the Langley Laboratory and added "that the idea of constructing a wind tunnel of this type has gained in favor among those interested in the project in Washington" [235].

However, when Munk paid a first visit to the laboratory in order to familiarize himself with the work of his colleagues and to discuss the design of the compressed air wind tunnel, Norton complained: "While I have a great respect for his ability in aerodynamics, I have the feeling that he is not giving us all of the information that he is able to." He also expressed doubts whether Munk's proposal was as advantageous as his Washington superiors believed, and suggested "another method of obtaining large values of Reynolds's number on tests of wings which I believe eliminates all of the disadvantages enumerated for the compressed air wind tunnel." Norton's alternative method consisted of testing wing models and other test objects in free flight by suspending them below an airplane on long wires. "By knowing the speed of the airplane, the pull on the wires and the angle at which the model is deflected backwards, the lift and drag coefficients can be easily computed," he argued. "This method not only gives the proper Reynolds number but it also gives a velocity, a size, a degree of turbulence and an unrestricted body of air which is identical with that in which the full sized airplanes fly, so that there can not be any doubt as to the applicability of the results to the full sized machine." Apart from the disagreement on technical details Norton feared trouble if Munk permanently moved to the laboratory: "If Dr. Munk stayed at the Field as I believe he desires to do it would be extremely difficult not only to fit him into the organization but to prevent friction between him and the officers around the Field" [236].

But at the Washington NACA headquarters Munk's expert opinion was rated higher than Norton's criticism. The "Variable Density Tunnel (VDT)" was built and started operation at Langley in October 1922. It was widely acknowledged as evidence for the NACA's rise to international prominence for wind tunnel testing. By the mid 1920s, the VDT was used for a systematic series of profile measurements, the so-called M-sections, after Munk. A few years later, the NACA assumed the leading role in the world for its catalogs of wing profile measurements [233]. Munk also published a new method for calculating the lift, load distribution and pitching moment of thin wings. His theory focused on the mean line of a wing section (i.e., the locus of points halfway between the upper and lower surfaces) and was in good agreement with experiments for not-too-thick wings of almost arbitrary shape ("thin-airfoil-theory") [238–240].

The NACA engineers' resentment about Munk's performance was noticeable from the very beginning of Munk's employment, but as long as he was based at the NACA's Washington offices and his contacts with the Langley Laboratory happened as an occasional exchange of letters and memos, each party went its own way. However, a growing tension could be felt even behind the sober official correspondence: "Dr. Munk will take absolutely no suggestions of any kind," Norton once complained when he wished to change a minor detail in the design of the VDT [241]. Munk was perceived by the Langley engineers more and more as an arrogant German, and Munk's correspondence with his former professor in Göttingen shows that this reproach was not altogether wrong: "As a scholar one has here fewer competitors, but also a much smaller audience," Munk wrote to Prandtl in August 1921. "The civilization here is further than in Europe, but the mental culture (Geisteskultur) is somewhat retarded." Furthermore, he was jealously concerned about the priority of his ideas. When Prandtl was astonished that Munk did not mention Margoulis, who had propagated the idea of the variable density tunnel earlier than Munk, he received this response: "I invented the high-pressure tunnel entirely by myself and therefore did not mention Mr. Margoulis. Whether he can say the same about himself I do not know. In the capacity of Mr. Knight's secretary he had a chance to take a look at one of my letters which concerned this subject matter" [242]. Even an admirer of Munk (Robert T. Jones) admitted that "he was perhaps a little difficult to get along with" [243].

When Munk moved to the Langley Laboratory in 1926 as director of the Aerodynamic Division, the engineers' animosity toward the new boss turned into an open revolt. As a consequence, Munk resigned. The details of the revolt against Munk are not known. The official historical account of the Langley Laboratory remarks on this episode as follows: "Munk was unusual in the Langley setting. The first thing that any group of Americans would have noticed about him, once hearing him speak, was that he was a foreigner. No doubt his thick accent and unfamiliar inflections made him seem more eccentric than he really was. What was worse in the early 1920s – a time of rampant nativism – Munk was a German, a 'hated Hun,' only recently the enemy of the United States and its allies." Munk's failure is interpreted as a "clear instance of nonadaptation between different national cultures of science and engineering, or as a case in point showing how 'culture shock' may affect technology transfer" [233, p. 91].

# 5
# A "Working Program" for Research on Turbulence

Despite different national cultures and mutual resentment, scientific internationalism was more than a mere ideology. Prandtl's desire for collaboration
with colleagues in other countries was not completely abolished by political
events like the Ruhr crisis. It sprang from a sincere scientific interest to exchange and communicate new research results after years of forced isolation
from international scientific life.

Furthermore, with the focus on aviation-related wind tunnel testing, fundamental research in fluid dynamics had been pushed into the background
during the war years. Even before the war, the rise of aeronautics involved
a stronger emphasis on applied aerodynamics than on basic problems in hydrodynamics. It was not by chance that Prandtl demanded as early as 1911
that his institute be expanded to an institute of hydro- *and* aerodynamics.
When his wishes became fulfilled in 1925 with the new Kaiser Wilhelm Institute (KWI), Prandtl wrote to the president of the Kaiser Wilhelm Society
that this institute should be named "Institute for Fluid Dynamics" (Institut
für Strömungsforschung), because this designation "encompasses everything
which will be done in the institute" [67, p. 248]. Although the new institute
nominally comprised the former AVA and Prandtl served as director of the
entire establishment, his focus was on the fluid dynamics institute; the AVA
was directed by Betz, Prandtl's deputy.

The emphasis on basic fluid dynamics, however, should not be interpreted
as a turning away from practical applications. This becomes evident if we examine the research on boundary layers and turbulence more closely, because
these are the two research fields that dominated Prandtl's scientific interests
more than anything else.

*The Dawn of Fluid Dynamics: A Discipline between Science and Technology.* Michael Eckert
Copyright © 2006 WILEY-VCH Verlag GmbH & Co. KGaA, Weinheim
ISBN: 3-527-40513-5

## 5.1
## Turbulent Pipe Flow

Prandtl often approached basic problems from a practical angle. Data on fluid resistance for flat plates or in pipes, as measured for decades in hydraulic laboratories or institutes for ship-building, offered plenty of problems to test the boundary concept. In 1908, Heinrich Blasius had shown for the first time in his doctoral thesis that this theory in fact offered the prospect of tackling age-old practical problems (see Chapter 2). If this theory could be extended so that it did not only apply to laminar flow, new insight could be gained for turbulent friction and turbulent boundary layers. After finishing his dissertation, Blasius worked as an engineer at the Berlin Testing Establishment for Hydraulics and Ship-Building (Versuchsanstalt für Wasserbau und Schiffbau), where data on fluid resistance from many practical areas were at his disposal. In 1911 Blasius wrote to Prandtl that he was processing these data in such a way that they were comparable to another by the law of mechanical similarity. For this purpose he had to display the coefficient of fluid resistance as a function of the Reynolds number. If the law of mechanical similarity was valid, Blasius argued, he could investigate "to what extent it is disturbed by the roughness." He referred to "Froude's experiments" as evidence that mechanical similarity applied to surface friction on plates. Now he aimed for a broader analysis of "hydraulic friction" by extending the investigation to the vast amount of pipe flow data [244].

Blasius processed the data of fluid resistance in pipes and found that the so-called head-loss[1], i.e., the loss of pressure over a certain distance, varied as $R^{-1}$ (where $R$ is the Reynolds number) for laminar flow, and as $R^{-1/4}$ for turbulent flow [245].

These empirical laws became the basis for the further development of boundary layer theory. Prandtl and Kármán derived from these laws in the 1920s general laws for turbulent surface friction. They concluded that the mean velocity in the turbulent boundary layer increases as $y^{1/7}$, where $y$ is the distance from the wall of a flat plate. It was likely already discussed during the First World War how Blasius's power laws for turbulent pipe flow could be extended to the case of turbulent surface friction along flat plates, because in 1917, Betz and Wieselsberger were involved with investigations

---

1) The head-loss may be expressed as $h = \lambda \frac{l}{d} \frac{v^2}{2g}$, where $l$ is the distance over which the pressure loss is measured, $d$ is the diameter of the tube, $v$ is the flow velocity, and $g$ is the gravitational acceleration. $\lambda$ is the dimensionless coefficient of friction whose dependence on the Reynolds number ($R$) Blasius attempted to determine. For the flat plate, the corresponding frictional coefficient was found proportional to $R^{-1/2}$ for laminar flow, in agreement with the theoretical result of Prandtl and Blasius' calculations for the laminar boundary layer; for higher Reynolds numbers, Blasius could not arrive at a conclusive law on the basis of the available data.

of the resistance of fabric-covered plates. The goal of these measurements was to distinguish the resistance of an obstacle due to the roughness of its surface ("skin friction") from that caused by its shape ("form drag" or "pressure drag"). In wind tunnel tests only the total drag was determined; by additional pressure measurements in front of and in the wake of an object, the form drag could be measured separately, so that by subtracting this drag form the total resistance, the skin friction could be obtained. Wieselsberger published the results after the war in the first volume of the *Ergebnisse der Aerodynamischen Versuchsanstalt zu Göttingen* and concluded on the basis of "a recently obtained formula by L. Prandtl" that even for very smooth surfaces (fabric that was varnished several times) the skin friction was caused by a turbulent boundary layer. A summary of these results was published in 1922 as a NACA report [246, pp. 124, 136], [247].

Skin friction due to turbulent boundary layers became a research theme in other laboratories as well. Thomas Stanton, a former collaborator of Reynolds, performed many experiments for determining turbulent wall resistance at the National Physical Laboratory in England [248]. In Delft, B.G. van der Hegge Zijnen, a doctoral student under Burgers, developed a novel technique for measuring the air velocities in the boundary layer along a flat plate in a wind tunnel. He used a hot-wire instrument with which he was able to measure wind velocities as close as 0.05 mm from the wall. This method was based on the principle that the electric resistance depends on the temperature; if a thin metal wire heated by an electric current is exposed to a stream of air, the extent to which it cools off depends on the velocity of the air stream; the velocity of the air stream, therefore, can be electronically monitored by accounting for the ensuing changes in electrical resistance. This method had already been described before the First World War, but only with the use of extremely thin wires (with a diameter of approximately 0.015 mm) and sophisticated electronic circuits did it become feasible for measurements of velocity fluctuations in boundary layers [249]. At the first international mechanics congress in Delft, Burgers presented to his colleagues Zijnen's result that the 1/7-law was found to agree with the measured data to an accuracy of 2% [250].

Despite this experimental confirmation, Prandtl and Kármán's 1/7 law for the turbulent boundary layer along a flat plate was entirely based on an extrapolation of Blasius's data from 1913, which in turn were based on older hydraulic pipe flow measurements. Both for the transition from laminar to turbulent flow and for fully developed turbulence, new data appeared desirable. Wind tunnel experiments were considered less appropriate for this purpose because of the problems of discerning skin friction from form drag; furthermore, even with guiding vanes and honeycombs the circulating air of a wind tunnel was itself turbulent to some unknown degree, and so the boundary layer of an object became affected by the wind tunnel's own turbulence.

For these reasons Prandtl regarded pipe flow experiments better suited than wind tunnel tests for the investigation of turbulent boundary layers. In 1919 Prandtl invited Ludwig Schiller, a physicist from the University of Leipzig, to perform systematic experiments on pipe flow in his Göttingen institute, particularly in order to cover the range of Reynolds numbers where the transition from laminar to turbulent flow takes place [251–253].

In addition, Prandtl charged a doctoral student, Johann Nikuradse, with systematic pipe flow measurements. Nikuradse investigated the turbulent flow of water through pipes of circular, rectangular, and triangular cross section with smooth inner walls and found that "Prandtl's 1/7 power law" is in "good agreement" with the measured data [254]. He also attempted to present a visual elucidation of turbulent flow by modifying Ahlborn's method: water was pumped through a 6-m-long open water channel at a speed of roughly 9 cm/s and was made turbulent by forcing it through a narrowing passage into the channel, a camera could be moved on rails over the water surface and take pictures, and the fluid motion was made visible by sprinkling aluminum powder on the water. At different velocities of the camera, different vortical motions were made visible (see Fig. 5.1) [254, pp. 42–43].

Another doctoral student (Fritz Dönch) experimented with an "air channel" (no closed-circuit wind tunnel but a straight rectangular box) with adjustable walls in order to analyze accelerating or retarding flows [255]. With this work, Prandtl intended to find support for his theoretical results about turbulence, but pipe flow remained the most important phenomenon for the experimental investigations of turbulence. In 1928 Nikuradse presented new results on skin friction in smooth pipes that he had performed in a larger device in the new Kaiser Wilhelm Institute as an extension of his doctoral work.

The scope of these experiments included very high Reynolds numbers. It became obvious that the "1/7 law" was not valid over the entire range of Reynolds numbers from the onset of turbulence to the highest measured Reynolds numbers; the exponent decreased "from 1/7 in the range of Blasius with its upper limit at about $R = 100,000$ to roughly 1/10 at $R = 3 \times 10^6$" [256]. In another update of his pipe flow experiments in 1932, Nikuradse remarked that Prandtl no longer believed in a power law [257, p. 15]. In the same year, Prandtl presented a report on turbulent friction in pipes and along plates in the fourth volume of the *Ergebnisse der Aerodynamischen Versuchsanstalt zu Göttingen*. Instead of the "1/7 law," the data on turbulent skin friction along smooth walls were now fitted by a logarithmic law [258, pp. 18–29].

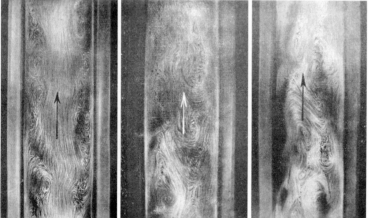

**Fig. 5.1** Water channel used in Prandtl's institute in 1925 for visualizing turbulence. At different camera velocities (*from left to right*: 7.45 cm/s, 8.33 cm/s, 8.5 cm/s; flow velocity = 9 cm/s) different vortical motion becomes visible.

## 5.2
## Prandtl's Research Program on Turbulence

The quest for a "universal wall law," as the formula for turbulent skin friction was called, cannot be overestimated as the guiding motive during the early years of turbulence research. Although the challenge to come to grips with the phenomenon of turbulence was primarily theoretical, Prandtl did not regard research on turbulence as a mere academic exercise. Both its underpinning with experimental data and its prospects for technological applications rendered turbulence as much a matter of practical engineering as an academic science. Only from the dual perspective of academic and practical goals does the full scope of Prandtl's research program on turbulence become apparent. Experiments on pipe flow, visualization of vortical motion in water channels, theoretical derivations of "universal laws" for the turbulent boundary layer, and a host of other research in Prandtl's university institute, the AVA, and the new Kaiser Wilhelm Institute during the 1920s and early 1930s appear at first sight unrelated to one another, but in fact, they are evidence of a coherent and persistently pursued research program on turbulence. The years from 1920 to 1934 were recalled in Prandtl's school as "the creative years of turbulence research." Throughout Prandtl's career after the First World War, turbulence was the predominant research topic, as 35 publications of Prandtl and 22 doctoral dissertations under his supervision on various aspects of turbulence clearly demonstrate [259].

Prandtl's curiosity about the phenomenon of turbulence arose early on with his attempt to understand the process of vortex-shedding along a flat plate in terms of the boundary layer concept. In his Heidelberg paper in 1904 he speculated how an irrotational fluid becomes vortical through a spiraling surface of discontinuity, which emerges at the edge of an obstacle [64, Figs. 3 and 4]. A few years later, in a manuscript titled "Turbulence I: vortices in laminar motion," he developed this concept further – long before he published on turbulence for the first time [260]. The headline "Turbulence I" suggests that there were probably subsequent early attempts to understand turbulence which were not preserved. The next manuscript on related problems dates from 1914 and addresses the "frictional flow around a cylinder" [261]. After the war, he resumed the challenge of turbulence more vigorously, as numerous pages of a manuscript from 1919 and subsequent years demonstrate [262]. As with the beginnings of airfoil theory during the decade before its publication in 1919, little of these early studies on turbulence became known outside his own circle of pupils before Prandtl began to publish on it in the 1920s.

According to a draft from the year 1916 for a "working program on the theory of turbulence" (see Fig. 5.2) Prandtl discriminated between the "onset of turbulence," i.e., the transition from laminar to turbulent flow, and "accomplished turbulence," i.e., fully developed turbulence [260]. Prandtl's early efforts dealt with the former. It was also addressed as the "stability problem" because the traditional approach to analyze the transition from laminar to turbulent flow followed a well-established method in theoretical physics in which a small oscillation is superimposed on a stable process in an attempt to find out whether the resulting oscillation exponentially decays to a stable level or grows to make it unstable. Since Reynolds' experimental and theoretical investigations in the 1880s, mathematicians and physicists attempted to determine by this "method of small oscillations" when a stable laminar flow became unstable; the resulting critical Reynolds number would then be iden-

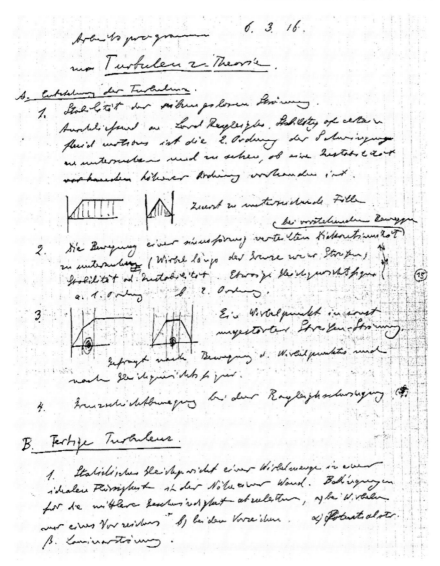

**Fig. 5.2** Outline of Prandtl's working program on turbulence.

tified as the transition from laminar to turbulent flow. However, when this procedure was applied to the Navier–Stokes equations, insurmountable difficulties emerged. As Lord Rayleigh had shown in 1887, a small disturbance between adjacent fluid layers with linear velocity profiles can give rise to instability even in the absence of friction. If friction was taken into account, further problems emerged. The simplest case of a so-called Couette flow, where a linear velocity profile is created in a fluid between a fixed and a moving wall, seemed intractable. In 1908 William M. Fadden Orr and Arnold Sommerfeld independently derived a perturbation equation (the Orr–Sommerfeld equation) for this case from which the transition could be derived – in principle. But when the intricate calculations were carried out, no critical Reynolds number was found. In other words, laminar flow could not be made turbulent in theory – in striking contrast with everyday practical experience. In 1920 a review article summarized the futile effort from the past decades to determine a critical Reynolds number and designated it as the "turbulence problem" [263].

Prandtl publicly presented the results of his early studies on turbulence for the first time in 1921 at a physics conference in Jena: "We, i.e., Mr. O. Tietjens who under my supervision carried out the calculations, investigated the stability and lability of laminar flows, as they arise along a flat wall in case of small viscosity, and we followed the method by Lord Rayleigh neglecting friction" [264]. For Prandtl's student Oskar Tietjens (1893–1971) the calculation of the "boundary layer motion with Rayleigh oscillation" was the subject matter of his doctoral dissertation [265]. Tietjens found that small oscillations render the laminar flow instable, "contrary to the dogma," as Prandtl commented on this result at the Jena conference. "We did not quite believe this result at first and checked it three times by independent and different methods. Each time it resulted in the same sign, which means instability" [264].

Prandtl's presentation met with harsh criticism and gave rise to further discussions. "I hope that you no longer feel a resentment from the debate last autumn in Jena," Prandtl wrote to Sommerfeld's pupil Fritz Noether, the author of the review on the turbulence problem. He admitted that his response to Noether's criticism "must have sounded somewhat brusque" [266]. Noether replied that he had "only wished to express my expert opinion on the reported subject matter," but he insisted that his criticism "has not changed with the printed version of your talk" [267]. In his report on the turbulence problem Noether did not regard Rayleigh's method as a valid approach. In a letter to Prandtl, he argued that he did not consider Rayleigh's results pertinent "because they do not approach the right limit with vanishing friction" [268]. Because Prandtl's analysis was based entirely on Rayleigh's procedure, Noether's criticism meant a fundamental denial of Prandtl's approach. This debate confirmed once more what Prandtl had expressed in a letter to Kármán in 1921 in this manner: "Turbulence seems to be haunted by a particularly wicked devil so that all mathematical efforts are doomed to failure" [269]. So far, "the turbulence problem" was that all theories failed to yield a transition to turbulent flow up to the highest Reynolds numbers—now the opposite case resulted from Tietjens's calculations: There should not even exist a laminar boundary layer because the slightest disturbance would make it unstable even at the lowest Reynolds numbers – in obvious contrast to experimental observations "that below a critical Reynolds number of about 1,000

no turbulence occurs." Tietjens believed that his theory failed "in this essen-
tial regard" because the calculations were based on assumptions that did not
correspond with real physical processes. "Particularly the assumption of the
buckled velocity profile of the main flow is supposed to be responsible for
the unsatisfactory result," Tietjens concluded [265, p. 214]. Ludwig Hopf, another Sommerfeld pupil with a record of pertinent publications on turbulence,
had expressed the same suspicion earlier at the Jena conference, when he had
criticized that the investigation was based on unrealistic velocity profiles "of
straight lines, which have a kink at certain arbitrary sites" [264, p. 694].

In May 1922, Hopf prepared Prandtl for "perhaps a sensation" in the theory
of turbulence because Sommerfeld had informed him "that one of his infant
prodigies (Heisenberg, 5th semester) had calculated the absolute dimensions
of the Kármán vortex streets and allegedly also mastered the problem of turbulence" [270]. Werner Heisenberg investigated the stability of the so-called
"Poiseuille flow," a flow with a parabolic velocity profile between two parallel plates; it became the subject matter of his doctoral dissertation [271]. Other
than the linear profile of the Couette flow, the parabolic profile resulted in a
limit of stability, but Heisenberg's calculations involved methods of approximation that were difficult to justify. However, Hopf regarded Heisenberg's
theory as another attack against the "dogma," and whetted Prandtl's appetite
because it seemed to "confirm completely all your views which you presented
in Jena" [272].

Prandtl was pleased, although he expressed "some qualms" about Heisenberg's approximations [273]. Not at all pleased was Noether, who had found
no critical limit of stability in his own earlier analysis of Poiseuille flow. He
did not trust Heisenberg's mathematical methods. With regard to Prandtl and
Tietjens' work, he argued that the problem is due to the difference between
"ideally smooth" and "practically smooth walls." He published a new account of the "turbulence problem" and argued "that in an ideally smooth tube,
no turbulent flow is possible." Prandtl immediately objected to this view, although he admitted that his mathematical abilities were "far from sufficient
to understand your calculations completely," as he wrote to Noether in a personal letter [274]. Their exchange of opposing views was carried out publicly – but it ended without a solution [275].

Heisenberg's dissertation and the dispute with Noether motivated Prandtl
to begin a new effort on the stability problem. Rather than for a velocity profile of buckling lines, he now posed as the theme for another dissertation "the
oscillations of a profile composed from a parabola and a straight line," but
"the work got stuck because the doctoral student failed," as Prandtl wrote to
Hopf in July 1926 [276]. A new doctoral student, Walter Tollmien (1900–1968),
was entrusted with this problem. In 1929 Tollmien was able to report success:
the calculations resulted in a diagram similar to phase diagrams (see Fig. 5.3),

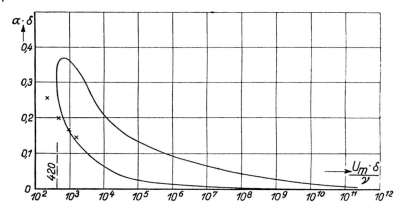

**Fig. 5.3** Tollmien's diagram on the stability problem.

in which the state of the system was dependent in an intricate manner not exclusively on the Reynolds number but also on the magnitude of the disturbance. Each state was represented by a pair of variables – the "wavelength" of the disturbance and the Reynolds number. The unstable states reached like a tongue into a sea of stable states. Although "the confrontation of our results with experiments" was still not satisfactory, as Tollmien admitted, the calculations meant a breakthrough in understanding the riddle of the onset of turbulence [277].

## 5.3
### The Mixing Length Concept for the Fully Developed Turbulence

Prandtl's "working program" from the year 1916 also contained a germ for future investigations on fully developed turbulence. Although he had no mathematical clue how to proceed, Prandtl's visualization of the creation of vortices starting from the boundary layer and their "diffusion" into the adjacent flow dictated his goal: "Calculate statistically the coefficient of diffusion," he formulated the forthcoming challenge on a sheet of paper dated 18 February 1916 [261]. Since Blasius's 1913 paper on turbulent friction based on empirical data on hydraulic resistance, Prandtl regarded pipe flow as a useful starting point for the study of turbulence. He derived further empirical clues from film sequences showing complex vortical motions emerging from the walls of an open water channel (see Fig. 5.1). But the high expectations of deriving a statistical turbulence theory from these observations were unfulfilled for many years. Neither pipe flow data nor vortex images suggested how to come to grips with fully developed turbulence in terms of a theory.

Ten years after formulating the goal in his working program, Prandtl admitted at the Second International Mechanics Congress in Zurich in 1926: "What I would like to address is the 'big problem of fully developed turbulence': an inner understanding and a quantitative calculation of the processes by which vortices—despite their frictional attenuation – give rise to ever new ones, and a determination of the mixing strength that results from the competition between the attenuation and creation of vortices, will therefore not be solved so soon." Although he could not solve the "big problem," Prandtl extrapolated some new insights about these processes from the empirical data on turbulent friction in pipes and open channels. Nikuradse's doctoral dissertation, for example, provided among other data diagrams on the distribution of flow velocities in non-circular tubes from which Prandtl concluded that there should exist "secondary flows" that transport momentum into the edges. Such secondary flows could only result from forces caused by a "turbulent mixing motion," Prandtl argued, but he could not relate this motion to the velocity profile. He hoped, however, that "in an experimentally controlled 'phenomenological' manner it is always possible to analyze theoretically the average motion in a given turbulent flow." Reynolds had derived equations for the averaged turbulent motion (see Chapter 1) in which certain terms could be interpreted as additional forces – the so-called Reynolds stresses. These forces were independent of the viscosity of the fluid and resulted in a specific type of turbulent friction. However, its similarity with viscous friction suggested that the Reynolds stress should be regarded as an "apparent friction" caused by an exchange of momentum between adjacent fluid layers. Prandtl defined a "mixing length" along which a "fluid package" ("Flüssigkeitsballen") loses its momentum in the transverse direction of the main flow [278].

By introducing the concept of a "mixing length," Prandtl was able to reformulate the Reynolds stress term so that it could be used as a starting point for calculating specific cases of turbulent motion. "Recently I occupied myself a great deal with the task to arrive at a differential equation for the average motion of turbulent flow derived from quite plausible assumptions and appropriate for various cases," Prandtl wrote in October 1924 to Kármán. The clue was "a length adjusted to the boundary conditions," he revealed to his former pupil [279]. A jet of air, for example, which is emitted through a nozzle into still ambient air, broadens as a consequence of turbulent mixing of air in the transverse direction of the jet. In this case, the mixing length was a characteristic distance over which turbulent air from the jet imparted momentum in the transverse direction of the jet into the still air. When he first published his concept, Prandtl compared the mixing length to a "braking distance" [280]. Further assumptions for the mixing length were necessary in order to use the concept for calculating specific cases, so that at first it seemed as if one unknown was simply replaced by another one. But before the concept of mixing length was introduced, problems like the broadening jet of air were completely inaccessible for quantitative evaluation. With the further assumption that the mixing length was proportional to the distance from the nozzle, the turbulent broadening of a jet of air could be calculated and compared with experimental data [281].

The solution of problems by means of the mixing length concept involved the introduction of new parameters that had to be determined from experiments, like the constant of proportionality for the distance from the nozzle in the broadening jet problem. This implicated a new facet for the relation of theory and experiment because beforehand unsuspected features of turbulent phenomena called for closer investigation. A doctoral student of Betz, for example, analyzed the turbulence caused by deflecting a stream of air along curved walls. He determined experimentally the "apparent kinematic viscosity," i.e., the friction caused by turbulence. In order to compare it with the theoretical result according to the mixing length approach he had to introduce a new quantity, which subsequently became useful as a measure for the degree for turbulence due to the bends of Göttingen-type wind tunnels [282]. Similar investigations were performed in water channels with inclined walls [283]. The mixing length approach could also be applied to calculate the turbulent wake behind obstacles. The problem was analogous to Tollmien's investigation of the broadening of a jet. Prandtl made it the theme of a doctoral work for Hermann Schlichting (1907–1982), who thereby became initiated into the Göttingen turbulence program [284]. The "big problem" of fully developed turbulence was not solved by the mixing length approach, but, as Schlichting noticed almost half a century later, "none of the many 'theories of turbulence,' conceived by other authors, especially the statistical theory of turbulence, succeeded in replacing it by something substantially superior with regard to the calculation of turbulent flows" [285, p. 304].

## 5.4
## A Kind of Olympic Games

Schlichting's evaluation hints at the competitive attitude with which fluid dynamicists pursued the riddle of turbulence. A fierce rivalry developed during the first decade after the First World War between Prandtl and his protégé Kármán, who also made turbulence research a part of his working program in Aachen. Prandtl's success with the mixing length concept presented at the International Congress of Applied Mechanics in Zurich in 1926 made him feel "that Prandtl had won a round," as he recalled in his autobiography, "and I came to realize that ever since I had come to Aachen my old professor and I were in a kind of world competition. The competition was gentlemanly, of course. But it was first-class rivalry nonetheless, a kind of Olympic Games, between Prandtl and me, and beyond that between Göttingen and Aachen. The 'playing field' was the Congress of Applied Mechanics. Our 'ball' was the search for a universal law of turbulence" [85, p. 135].

The competition had started in February 1921 when Kármán presented his "Dear master, colleague, and former boss" with a derivation of the 1/7 power law for the turbulent surface friction in a long letter . He duly recalled that Prandtl had occasionally mentioned such a derivation to him earlier, but claimed that he had not understood Prandtl's arguments then and therefore "reconstructed it and built on this basis a kind of 'turbulent boundary layer theory.' " On page 5 of his letter he arrived at "the practical question: Having been silent for quite a while, I would like to publish something now, after so much of my efforts has failed (...) But I would like to ask you whether you have already published your 1/7 law so that I can refer to it, or whether you intend to publish it soon. Furthermore, is my derivation the same as yours or not? (...) My calculation is entirely in the same vein as my earlier methods, and thus. I regard it as grown upon my own manure heap; however, after you have fertilized the humus with the 1/7 law" [286].

Prandtl claimed his priority by responding that "for a pretty long time, say since 1913," he had known a formula to derive the wall friction law for the plane flow along a plate if it is known for pipe flow. He admitted that Kármán was "peremptorily further advanced" than he was with regard to the turbulent boundary layer; he had similar things planned for the future. However, he had "already at earlier times attempted to calculate boundary layers in which I had assumed a viscosity enhanced by turbulence, which I chose for simplicity as proportional to the distance from the wall and proportional with the velocity in the free flow." A "big question" is now "how we proceed with the publication. I do not want to be guilty that your work appears belatedly." Prandtl had other urgent obligations, and he could not present his own paper before April. "If you do not want to wait so long it would probably be the most appropriate to publish your derivation yourself; beyond the 'fertilization,' it is indeed your own intellectual property. I will see if I can gain recognition by my own right with my different derivation. Ultimately, I can get over it if the precedence of publication has gone over to friendly territory" [287].

Kármán published his derivation without further delay in a paper titled "On laminar and turbulent friction" [288]. He also presented it in the following year at the Innsbruck conference [289]. Prandtl's derivation appeared only in 1927 in the third volume of the *Ergebnisse der Aerodynamischen Versuchsanstalt* [290]. By the published record, therefore, Kármán won the first round of this contest. However, he acknowledged that the suggestion was first presented "by Mr. Prandtl in an oral communication in autumn 1920; the publication appears with his consent, whereas my derivation is somewhat different from his" [288, p. 238]. Prandtl often presented new ideas first to his students in seminars or lectures before he worked them out for publication – as we have seen with the origin of the airfoil theory. The 1/7 law for the turbulent friction of a flat plate presumably originated in a similar way: Prandtl explained to his students how Blasius's formula for turbulent friction in pipe flow may be used as a starting point for deriving the equivalent formula for a flat plate. It is plausible that Prandtl arrived at this idea soon after Blasius's paper had appeared in 1913—and then presented it to Nikuradse and others whom he involved in his working program on turbulence. Nikuradse acknowledges in a footnote of his dissertation that "the derivation was presented by Prof. Prandtl in a discussion in Göttingen on November 5th, during the winter semester of 1920" [254, p. 15]. Despite his later publication, therefore, Prandtl's precedence in introducing the 1/7 law seems undisputed.

In Kármán's paper, "On laminar and turbulent friction," however, the derivation of this law was only a side-effect of a larger attempt "to present the basic idea of Prandtl's boundary layer theory as simply as possible both from a mathematical and a physical point of view." Kármán's declared goal was to render the boundary layer theory amenable for solving "complicated problems with simple mathematical means" [288, p. 233]. By integrating over the thickness of the boundary layer, he derived an equation of the incoming and outgoing momentum – a procedure Prandtl's student Karl Pohlhausen had just applied in his doctoral dissertation [291] in order to calculate the laminar boundary layer. Kármán's extension of the Pohlhausen procedure was also applicable for the turbulent boundary layer. The derivation of the 1/7 law was only a first example by which the new Pohlhausen–Kármán procedure, as it became known, proved its utility. Kármán also calculated the velocity profiles in laminar and turbulent boundary layers along flat plates from which he derived skin friction coefficients and compared them with Wieselsberger's data published in the first volume of the *Ergebnisse der Aerodynamischen Versuchsanstalt*.

The rivalry between Kármán and Prandtl, therefore, soon extended to experiments as well. While Nikuradse the turbulent flow along smooth walls investigated in Prandtl's laboratory, Kármán initiated experiments with rough walls at his Aachen institute. His Aachen colleague Ludwig Hopf, who had begun his own career with experimental turbulence investigations in Sommerfeld's institute [292], conceived a program in 1922 for systematic experiments on "hydraulic roughness" [293], and Kármán trusted a doctoral student with a first series of such measurements [294]. "I learned from Pohlhausen that the roughness experiments proceed well," Prandtl wrote to Hopf in May 1922, "if it is not asking too much, I would be very grateful for an occasional preliminary communication of the resulting data" [295]. Hopf promised to send the requested data as soon as they were available [270]. Neither Prandtl nor Kármán, however, succeeded to derive new theoretical insight from these data, but they were only a prelude to further investigations about the roughness problem: "The two dissertations mentioned in the recent issue of the ZAMM are again pretty good works," Prandtl congratulated his rival in Aachen in July 1928 [296]. The measurements of velocity profiles along rough walls reported in one of these dissertations lent support to Prandtl's mixing length concept [256] such that systematic investigations for the flow in tubes with rough inner walls then became a major part of the experimental turbulence research program in Göttingen. Prandtl derived a formula from "a simple similarity consideration" with which the then unpublished results could be fitted: "The entire tube problem has found a very general solution by combining few empirical data with theoretical conclusions," he reported in May 1932 [297].

In the meantime, Kármán had worked out an alternative theory for the fully developed turbulence, which arrived at practically the same results as Prandtl's mixing length concept. Like his Göttingen rival, Kármán assumed that turbulent friction is caused by a transfer of momentum transverse to the direction of the main flow, but he did not make the assumption of a mixing length. Instead he postulated that the transfer of momentum at fully developed turbulence happens everywhere according to the laws of mechanical similarity. A comparison with Prandtl's theory resulted in a formula for the mixing length which contained only one dimensionless factor and left no room for additional assumptions. Kármán's constant, as it was called, depended only on the average magnitude of the turbulent velocity fluctuations. For a flow between parallel plane walls, Kármán's theory resulted in a logarithmic wall law. The 1/7 law, which was believed to represent the universal wall law for turbulent friction, was now regarded only as an approximation. Kármán checked his wall law with Nikuradse's most recent experimental data on turbulent skin friction from Göttingen. When the data confirmed his formula, he was "in a state of exultation. I felt sure I would win this round from my old teacher." Four years after Prandtl had presented his mixing length concept to the international community of fluid dynamicists in Zurich, Kármán was thus able to present a result of comparable importance at the International Congress of Applied Mechanics in Stockholm. But "since Prandtl had been cooperative in letting me have his unpublished experimental data, I felt that I should put my cards on the table before him rather then play them at Stockholm before the eyes of the scientific world." When he told Prandtl that he had news to report, Prandtl invited him to give a lecture before the Göttingen Academy of Science. "Prandtl was crestfallen," Kármán recalled the response when Prandtl became aware of what he had achieved [85, p. 137]. When the theory was published in the proceedings of the Göttingen Academy, there was little evidence of the fierce rivalry. Kármán's result appeared as just another confirmation of "Prandtl's heuristic mixing length approach" and as an "excellent agreement of the resistance law with empirical data" [298]. However, when Prandtl abstained from delivering a paper at the Stockholm conference a few months later and left the stage to Kármán, it was obvious who had won this round of their competition.

A new outburst of rival sentiments occurred two years later, when Kármán learned from a preprint of a conference at the Hamburgische Schiffbauversuchsanstalt (Hamburg Ship Building Establishment) that one of Prandtl's research collaborators had portrayed the recent progress concerning turbulent skin friction largely as a Göttingen accomplishment. He was "mildly amazed" and hoped for an apology in the final conference proceedings: "How I feel hurt!" he complained to the organizer of the Hamburg conference. "For the next fifty years your publication will be the standard text for practical engineers; they will only know about Göttingen and Prandtl" [299]. "Presumably it escaped you too," Kármán wrote to Prandtl, who had been present at this conference, "that this presentation offered nothing which was not already in my own presentation in Stockholm in 1930." He regarded Prandtl "a model of a just man," but he did "not quite trust your lieutenants who understandably do not know other gods besides you. They wish to claim everything for Göttingen." He was not comforted by Prandtl's remark that their institutes often produce the same results at the same time. "This was the case in 1921 when we calculated the resistance of a smooth plate simultaneously on the basis of Blasius's law for pipe flow resistance. And this has happened now again, 10 years later," Prandtl explained this parallel as a consequence of their friendly competition, "for when we see one another we always discuss unsolved problems and receive the spur to focus on just these things from these discussions" [300, p. 87]. But Kármán regarded Prandtl's "nice words" as a weak excuse and insisted that Prandtl admit the precedence of his work. He would "agree if the plate formula becomes known under our both names," he responded, "but to be excluded from the banquet would be too hard." Kármán was particularly concerned that his contribution could not be appropriately acknowledged in the forthcoming fourth volume of the *Ergebnisse der Aerodynamischen Versuchsanstalt zu Göttingen*. If his merits were not mentioned in this "standard treatise for the practitioners" he feared that "my role in this issue would remain buried forever" [301].

Prandtl did his best at a reconciliation. Together with his "lieutenant," he added an appendix to the final conference proceedings in which Kármán's merits were mentioned. And "there is of course reference made to your work" in the *Ergebnisse*, Prandtl assured Kármán [302], who felt now a little embarrassed: "I hope that no nasty taste remains from the debate in the mouths of those involved," he responded [303]. Prandtl was relieved that his prodigy student and colleague now was "to some extent mollified" and apologized once more that he had contributed to Kármán's hurt feelings. But under the surface the rivalry kept smoldering. "In order to make you understand me completely," Prandtl broached the subject again, "I will briefly expose to you my attitude concerning this issue." Thus, he introduced a three-page account of how he approached the problem of turbulent skin friction from the perspective of his mixing length concept since 1927, and how he had engaged Nikuradse to verify his formulae experimentally. Kármán's theory, too, relied on Nikuradse's experiments "so that our calculations became more and more similar." In Kármán's version, however, a coefficient remained undetermined. For this reason "practitioners who are not willing to immerse themselves into the theory" could make nothing of it, with the result that "our ready-made formula for practical use became more popular than your theory" [304].

The rivalry between Aachen and Göttingen should not obscure the fact that Prandtl and Kármán shared much common ground. After the First World War, Kármán had returned to Aachen with ambitious plans for which Prandtl's Göttingen institute provided the model. With a new series of communications, the *Abhandlungen aus dem Aerodynamischen Institut an der Technischen Hochschule Aachen*, Kármán spread the results of his institute as his Göttingen rivals did with their *Ergebnisse*. Kármán's assistant Karl Pohlhausen was a doctoral student under Prandtl before he arrived in Aachen. Even when the rivalry reached a climax, there was a vivid exchange of research results between Aachen and Göttingen. In 1930 the common ground between both centers became further consolidated when Wieselsberger was called to Aachen to fill newly established chair for Applied Mechanics and Fluid Dynamics (Angewandte Mechanik und Strömungslehre) [305]. Even in America, where Kármán became director of the Guggenheim Aeronautical Laboratory of the California Institute of Technology (GALCIT), contact was maintained by occasional exchanges of advanced students (see Chapter 9). Tollmien, for example, spent a three-year sojourn at the GALCIT; he used this opportunity for a comparison of Kármán's similarity hypothesis with Prandtl's mixing length approach [306].

Kármán and Prandtl also shared a common understanding about the peculiar character of their discipline. When in 1920 Prandtl received a call to Munich, they arranged that Kármán should succeed Prandtl in Göttingen if the move materialized. As an "engineer *and* mathematical physicist" Kármán had the same double qualification as he himself, Prandtl wrote to Aachen [307]. He did not identify himself and Kármán as theoretical physicists, but as representatives of technical mechanics—a discipline with the prospect of fruitful applications in engineering. Although this discipline involved sophisticated theoretical concepts, these were never far from practical uses, unlike "the wire-drawn applications of Bohr's rule to all possible atomistic construction," as Prandtl remarked on the recent advances of theoretical physics [308].

After more than a decade of ambitious and rivaling research on the most advanced topics of fluid mechanics in Göttingen and Aachen, Kármán, Prandtl, and their doctoral students embodied a species of practical theorists which became typical for modern fluid dynamics. Kármán and Prandtl were eager to see the results of their research on turbulence acknowledged by practical men. Otherwise, if their rivalry had been a competition about precedence alone, Kármán could have been satisfied that he had presented the new logarithmic formula for turbulent skin friction before the Göttingen Academy of Science in 1930 and at the International Mechanics Congress in Stockholm. But, as he wrote to Prandtl, "Who reads the proceedings from the Göttingen Academy and the Stockholm Congress?" [301]. He was mollified only when Prandtl assured him that his merits were also mentioned in the Göttingen *Ergebnisse*, where aeronautical engineers would look for the new research results in aerodynamics, and in the proceedings of the Hamburg conference, which were addressed to engineers concerned with ship-building. Their zeal to reach not only theorists also became apparent when Prandtl and his collaborators presented their recent results in the *Handbuch der Experimentalphysik*. Theoretical papers like Tollmien's "turbulent flows" and "boundary layer theory" were presented side by side with Tietjens's "visualization of flow forms" and Betz's "micromanometer," although only the latter really met with the expectations of experimental physicists. At the same time this volume illustrates Prandtl's central role in the emergence of fluid dynamics as a modern twentieth-century discipline: nine out of eleven articles originated from members of Prandtl's Göttingen circle [309].

## 5.5
## Wind Tunnel Turbulence

A good deal of turbulence research after the First World War was related to the flow in pipes and open channels. Such flows were of natural interest in hydraulics and in ship-building. Applications in aeronautical engineering also suggested themselves, as Wieselsberger's skin friction experiments on smooth and rough surfaces have shown. For Prandtl, Kármán, and their pupils, a close relationship between hydro- and aerodynamics was all the more natural because they had experimented with water canals and wind tunnels since Prandtl had made the boundary layer concept the subject of experimental investigation (see Chapter 2). Among these investigations, Prandtl's trip-wire experiment, performed on a sphere in a flow of air, received renewed relevance in 1919 when Hugh L. Dryden (1898–1965), a doctoral student of Ames, extended Prandtl's analysis from spheres to cylinders. In 1921, now in his capacity as director of a new "Aerodynamical Physics Section" at the National Bureau of Standards in Washington Dryden informed Prandtl about recent research in various laboratories and concluded from the available data "that there is ample evidence to show that the usual expression for resistance is not valid for experiments in wind tunnels on cylinders and spheres." Dryden regarded the more or less turbulent flow of air in wind tunnels as responsible for this failure. "I would like to know your ideas as to a proper method of numerically defining the turbulence of tunnels and your idea as to the physical conception of the turbulence," he asked Prandtl. "It seems to us here at the Bureau of Standards that the most important wind tunnel problem is a study of turbulence and its effects, and we would be glad to hear of any experiments or theoretical discussion bearing on the subject" [310].

Dryden's argument was a natural extrapolation from the trip-wire experiment, which had shown how a small bump on the surface of a sphere can give rise to turbulence and result in a drastic change of resistance. If this happens with a bump on the surface, then it could also happen as a consequence of a bumpy air. Thus, the challenge arose: how does one measure and control the bumpiness – i.e., the turbulence – of the flow of air in a wind tunnel? "The influences of the vorticality must be extensively studied, this is my opinion too," Prandtl agreed with Dryden. It was for this very reason that he had stressed the importance of building the new Göttingen wind tunnel so that the air flowed as vortex-free as possible. Prandtl attributed "little scientific value" to older Göttingen measurements on the resistance of cylinders, published in the *Technische Berichte*, because they were obtained during the war "with quite a crude device." He sent Dryden the results of new experiments on the resistance of cylinders, which were made "very carefully," but he could not provide an experimental method for measuring the wind tunnel turbulence. In his opinion, it would be difficult to design such instruments, because they must detect very fast variations of flow velocities [311].

Research on wind tunnel turbulence became a major issue in other aerodynamic laboratories as well. Researchers from the Massachusetts Institute of Technology, for example, modified the air flow in the test chamber of a wind tunnel by introducing wire nets of different mesh size behind the honeycomb. They measured the lift and drag of various test bodies whose data were known from measurements in other laboratories and inferred from the deviations the degree of turbulence in their modified test chamber. Rather than making the air flow more turbulent, however, "the introduction of a wire screen into the air stream of a tunnel tends to produce a less turbulent flow," they found. "It is also shown that the turbulence tends to die out more rapidly downstream as the screen becomes finer." However, the degree of turbulence could only be accounted for in a qualitative manner: "The presence of vorticity in the air flow of a wind tunnel is unquestionable, and can be visualized by the introduction of a series of narrow silk streamers into the air stream" [312].

At the NACA wind tunnel turbulence was an issue that had caused some concern as early as 1921: "How reliable will the results be from the compressed air tunnel?" Norton asked when the first plans of Munk's VDT were discussed (see Chapter 4). In this wind tunnel the air was deflected in narrow bends, with few precautions taken to calm the flow, so that "the question of turbulence, which undoubtedly has a very great influence" deserved attention [236]. In 1924 NACA engineers compared the results of drag measurements on spheres obtained in various wind tunnels with data from free fall experiments. None of the wind tunnels built so far "can even approximate the nonturbulent condition prevailing in the atmosphere," they concluded and suggested "that an extensive study be made of the effects of scale and quality, or 'intensity,' of turbulence" [207, p. 485]. In a subsequent NACA report, data of various test bodies (spheres, disks, cylinders, and certain wing profiles) as measured in the No. 1 wind tunnel of the Langley Laboratory were published. This was a conventional wind tunnel of the "Eiffel type" in which the air was much less turbulent than in the VDT. The measurements were performed with the utmost precision – and yet they could not claim universal validity: "The data collected here must be considered, primarily, as data concerning the tunnel, and not the models tested there" [208, p. 219].

The most vigorous attempts to measure the degree of turbulence in wind tunnels were made at the National Bureau of Standardsby Dryden and his collaborators. In 1926, they reported on the results of experiments performed in cooperation with the NACA over the past two years. With unprecedented precision, they measured the resistance of cylinders in an airflow which was made turbulent by different wire meshes. In addition, they recorded the temporal variation of pressure in the air stream with such precision that they could discern the rotation of the propeller blades as one cause of pressure variations. Beyond a growing awareness of the difficulties encountered with high-precision wind tunnel measurements, however, these investigations did not immediately result in a quantitative measure for the degree of turbulence [313]. A decisive breakthrough was only made by using hot wire anemometers. After two more years of tedious experimenting, Dryden and his colleagues reported success. The apparatus was "very bulky, far from portable, and in many respects inconvenient to use," but the results aroused great expectations. Velocity fluctuations in the air flow could be measured with a spatial and temporal resolution that was beyond the reach of other methods [314]. The method also suggested a definition for the degree of turbulence in a wind tunnel: "The turbulence at a given point is taken to be the ratio of the square root of the mean square of the deviations of the speed from its mean value to the mean value. The turbulence is a mean fluctuation taken in a definite manner and expressed as a percentage of the mean speed" [315, p. 152]. By this time, there were three wind tunnels at the National Bureau of Standards, and Dryden's measurements resulted in a different degree of turbulence for each tunnel. Furthermore, the result depended on the position within the test chamber of the tunnel. The prospects of standardizing wind tunnels according to these measurements were sobering: "It will now be appreciated that wind tunnels cannot be standardized in the sense originally intended," Dryden and his collaborator argued. "It is not possible to determine one or more correction factors by means of which results on a new model may be corrected to be comparable with the results of some standard tunnel" [315, p. 166].

In particular, for non-streamlined bodies like spheres and disks, where the transition from laminar to turbulent flow resulted in drastic changes in drag, the data obtained from different wind tunnels could differ considerably from one another. For streamlined bodies, wind tunnel turbulence also affected the test results if skin friction contributed noticeably to the total drag, i.e., for rough or appropriately formed surfaces. The effect of wind tunnel turbulence on wing profiles was less conspicuous as long as the models were so small that the transition to turbulence was out of reach. But that changed as soon as wind tunnel tests became feasible at higher Reynolds numbers, such as with the NACA's VDT. "You will be interested to know that recent observations in the variable density wind tunnel at Langley Field show pronounced effects of turbulence on the aerodynamic characteristics of airfoils," Dryden wrote to Prandtl in March 1931 [316]. The effect was most pronounced with thick profiles [317]. Prandtl agreed with Dryden: "The question of turbulence in wind tunnels is certainly one of the most important problems in wind tunnel practice." He was particularly impressed with how Dryden managed to record the fast turbulent velocity fluctuations ("up to 4000 cycles per second") of the air flow in the wind tunnel. "If it is allowed, it would be important for me to hear what kind of physical phenomena you will use for this measurements" [318]. Dryden revealed that they had improved the hot wire technique in his laboratory and now hoped to reach a precision "where any desired characteristic of the turbulent flow can be measured, perhaps only with great difficulty and complicated apparatus, but surely and with reasonable accuracy. However, a theory of the effect of turbulence on the transition from laminar to eddying flow is needed to guide the measurements" [319].

The latter remark shows how wind tunnel turbulence, a vital issue of aeronautical practice, provided a new impetus for fundamental theoretical research on turbulence. Dryden hoped that Prandtl's boundary layer theory would offer the clues for a better understanding of the turbulence problem. "The discussion of the experiments described in this paper will be phrased in the language of the boundary layer theory of Prandtl," he introduced his NACA report on wind tunnel turbulence, "and as there is no one article to which the reader may be referred for the necessary information, it is desirable to state briefly the elements of this theory" [315, p. 151]. Prandtl's boundary layer theory was thus spread among aeronautical practitioners—the readers of NACA reports – who would otherwise have paid little attention to theoretical concepts. Once more, theory met practice because it offered prospects to solve problems of crucial practical importance; however, by the early 1930s, the problems related to turbulence proved more intricate than Dryden could have anticipated (see Chapter 10).

# 6
# Aerodynamics Comes of Age

Basic science and applied technology shared a common interest in turbulence research. Theoretical physicists like Heisenberg, practical engineers from the Hamburg ship-building establishment, and aeronautical engineers engaged in wind tunnel testing regarded turbulence as a challenge – although from quite different perspectives. Most other fluid mechanics topics, however, became almost exclusively regarded as subject matter of engineering science. Novel theories like Prandtl's airfoil theory fell outside the realm of physics: they belonged to the domain of engineering and applied mathematics. Aerodynamics did not enter the spectrum of twentieth century science as a physical subdiscipline but as the "basic engineering science of airplane technology" [112].

   How came this structure about? The emergence of engineering science is a secular process, beginning in the nineteenth century in the wake of the Industrial Revolution and extending far into the twentieth century. Various local and national traditions resulted in different forms of institutionalization. This process, furthermore, was not the same for different branches of engineering. The history of technology, like the history of science, provides a host of literature that cautions us to beware of sweeping generalizations [320]. Even if we restrict our analysis to those branches of engineering in which fluid dynamics plays a central role, there is no coherent pattern. Branches like mechanical engineering and ship-building became institutionalized earlier than aeronautics. In view of such diversity, it seems wise to focus on the case of aerodynamics in Germany in order to analyze the process of institutionalization in some detail. Prandtl's towering position as director of the AVA and as the intellectual center of a growing school suggests that he played a major role in this process.

## 6.1
### How Aerodynamics Became Institutionalized at Technical Universities

In 1921 Prandtl recalled that when he was a student, aerodynamics was "something entirely unknown in the curriculum of a technical university"

*The Dawn of Fluid Dynamics: A Discipline between Science and Technology.* Michael Eckert
Copyright © 2006 WILEY-VCH Verlag GmbH & Co. KGaA, Weinheim
ISBN: 3-527-40513-5

[321]. He started lecturing on aerodynamics in 1909 after he was asked by Göttingen University to teach aeronautics both in lectures and in seminars. This move marked the introduction of aeronautics to the curriculum of technical universities. Other German technical universities soon followed. In 1910 the Aachen Technical University offered a "Training in theoretical and technical aerodynamics" [322]. Similar programs were already initiated before the First World War at technical universities in Darmstadt, Brunswick, Munich, and Berlin. As in Aachen, the initiative came from professors and industry sponsors, who also founded an association of early flight enthusiasts in 1912, the "Wissenschaftliche Gesellschaft für Flugtechnik, WGL" (Scientific Society for Aviation). The journal of this society, the *Zeitschrift für Flugtechnik und Motorluftschiffahrt, ZFM*, also offered to publish aerodynamic research results related to aeronautics [120, pp. 83–88].

However, these pre-war initiatives did not instantaneously establish aeronautics as a new engineering science. The foundation of the aerodynamic laboratory at the Aachen Technical University illustrates these haphazard beginnings. The initiative was a consequence of the joint venture founded by Reissner and Junkers (see Chapter 3), but the enthusiasm of the two professors was met with suspicion from the university. Beginning in 1906, Junkers was involved in quarrels with the president of his university about his industrial activities because they allegedly distracted him from his obligations as a university professor. Tired, after years of quarrels, he resigned from his university position in January 1912. Although by this time, aerodynamics was included as a new specialty in the curriculum, the university was reluctant to establish a new chair for it; the university accounted for the discipline merely by adding aerodynamics to Reissner's designation. In 1913 Reissner followed a call to the Technical University in Berlin, and the new aerodynamic laboratory at the Aachen Technical University became deprived of its founders immediately after its establishment. When Theodore von Kármán came to Aachen as Reissner's successor, he noticed that there were not enough students to justify a separate chair for aerodynamics. Mining and other branches of mechanical engineering offered more secure job opportunities than aeronautics [85, 323].

The establishment of aeronautics as a new branch of engineering at technical universities spurred new interest with the start of the First World War. The president of the Technical University Hanover expressed his conviction that the technical universities "as nurseries of science" do their best to foster the development of aviation "first in a mere national patriotic sense and then also for the benefit of all mankind" [324]. But a survey performed at the request of the Scientific Society for Aviation after the war painted a bleak picture: By 1921, chairs for aeronautics ("Flugtechnik bzw. Flugzeugbau") existed only at the technical universities in Stuttgart and Darmstadt. The need was recognized in other places too, but this awareness only resulted in the establishment of teaching contracts at the technical universities in Aachen, Berlin, Brunswick, and Hanover [325]. Although there was a broad spectrum of aeronautical lectures, as observers from abroad occasionally noticed with a trace of envy [326], no official curriculum was available for aeronautical engineering. In Prussia, a first study plan for aeronautics at technical universities was initiated in 1924, but it left ample leeway for implementation. Aeronautics did not yet figure as a self-contained engineering specialty, but engineering students could choose it as an elective for their final diploma examination [327]. At the Technical University in Berlin-Charlottenburg, for example, the ministry's plan gave rise to a new curriculum for engineering students of ship-building, which belonged to the faculty of machine building. Beginning in 1925, these students could choose aircraft construction as the main subject for their diploma examination "provided that they have acquired sufficient knowledge of the basics of ship-building, electrical technology and general fundamental concepts of machine building" [328].

The major journal of aviation in Germany, the *Zeitschrift für Flugtechnik und Motorluftschiffahrt*, published regular surveys on what lectures students at technical universities and colleges could choose if they were interested in aeronautical engineering. By the end of the 1920s, a student who wished to become an "academic airplane engineer" would be offered the appropriate lectures at the technical universities in Aachen, Berlin, Brunswick, Danzig, Darmstadt, Hanover, Karlsruhe, Munich, and Stuttgart. During the first semesters, the program was the same as for machine building, electrical engineering, and ship-building; after the pre-diploma examination ('Vordiplom') special lectures and seminars on aeronautics and related subjects had to be studied; the final degree required a diploma thesis, practical exercises, and an oral examination. The details differed from university to university. The examination regulations at the Technical University in Berlin, for example, prescribed an exercise in practical design such as designing an entire aircraft (Entwurf eines Luftfahrzeugs) or designing an airplane engine (Luftfahrzeugkraftanlage), a theoretical aeronautical study (theoretische Untersuchung aus der Luftfahrttechnik), a workshop- or economic/technical study ("werkstatts- oder wirtschaftstechnische Arbeit"), or some other detailed study of an optional subject (e.g., mechanics, statics, physics, meteorology, etc.). The oral examination would cover aircraft construction, machine building, general manufacturing, fluid mechanics, and an optional subject. The practical exercises were given highest priority: "It is suggested that students spend 18 to 20 months for the entire practical training," the study order recommended, among which 26 weeks should be spent in a general machine building workshop before the pre-diploma examination, and 52 weeks in a factory of motor production, automobile construction, in an airplane factory or at an airport [329].

A comparison of aeronautical lectures and seminars at the technical universities in the early and late 1920s shows how rapidly the sciences related to aviation expanded during the decade after the First World War. In the early 1920s, the survey of aeronautical lectures printed in the ZFM found a place on one page; at the end of the decade it extended over three printed pages. In the years after 1931, the journal abstained from printing the detailed lecture titles and merely presented summaries, subdivided into feature articles on aeronautical specialties such as aircraft construction, airplane engines, radio, aviation medicine, aviation law, aviation traffic, meteorology, navigation, and fluid mechanics. Each specialty was presented together with a list of technical universities where a student would find a pertinent lecture or seminar. Lectures on aircraft construction, for example, were presented in the winter semester of 1931/32 at the following technical universities: Aachen, Berlin, Brunswick, Danzig, Darmstadt, Hanover, Karlsruhe, Munich, and Stuttgart; in addition, this specialty was taught at technical colleges in Bingen, Frankenhausen, Hamburg, Ilmenau, Köthen, Konstanz, Mittweida, Oldenburg, Strelitz, Weimar, and Wismar [330]. Besides the quantitative growth of aeronautical teaching, the surveys also indicated that aeronautics as an engineering specialty, comprised a set of diverse specialties. Aerodynamics, which was often referred to under the more general headline of "fluid mechanics," was merely one of these. But under the spell of aeronautical engineering, fluid mechanics was rapidly expanding as a new subject of teaching.

Textbooks on flight-related fluid mechanics are another indicator of the formation of this new engineering specialty. As early as in 1919, a professor of the Technical University Hanover published a textbook titled *Flugtechnik* [331], a "first comprehensive account of what the science of flight may credit as benefits of the war," as a reviewer commented, "*the* textbook on flight technology!" The book "will help to spread and deepen the science of flight" [332]. In 332 pages, the basic knowledge of fluid dynamics were dealt with as far as they addressed the aerodynamic lift and drag. Other chapters examined problems of stability and the strength of materials. Although the textbook did not include details of Prandtl's airfoil theory (which was published at the same time in the proceedings of the Göttingen Academy of Science) it offered as much as a contemporary practical airplane designer would find useful for a basic understanding.

A few years later, *Aerodynamik* appeared, authored by two leading representatives of the former Flugzeugmeisterei (FLZ—see Chapter 3) as part of a handbook series on airplane technology. The authors aimed to achieve a comprehensive account of the theoretical foundations of flight "because their practical importance is beyond doubt after the successes of Prandtl's airfoil theory, and a closer association of these theoretical and physical considerations with basic ideas of airplane design is to be expected in the future," [333, p. V]. This textbook also exploited the harvest of aeronautical research and development from the past war. It made extensive use of the material presented earlier in the *Technische Berichte*. A reviewer praised it as "a legacy" of the FLZ which a reader would study "as a laughing heir, however filled with sincere mourning about the vanished greatness" [334].

A number of other textbooks on aerodynamics appeared in the 1920s – and addressed ever-increasing and diverse interest. Richard von Mises's *Aerodynamik des Fluges*, first published in 1919, appeared in a third edition in 1926 [335]. In the same year Hermann Glauert's *The Elements of Aerofoil and Airscrew Theory* appeared, the first English textbook largely dedicated to Prandtl's airfoil theory [336]. In 1929 it appeared in a German translation. In 1927 and 1929 *Grundlagen der Fluglehre* addressed the interests of readers with predominantly practical goals and little mathematical underpinning [337]. Students who wished to learn the mathematics would find the required knowledge in Mises's book or in the *Aerodynamik des Fluges. Eine Einführung in die mathematische Tragflächentheorie* [338]. Another more mathematically oriented treatise, *Einführung in die theoretische Aerodynamik*, was authored by a professor for aeronautics from the Technical University Darmstadt in 1927 [339]. Students who intended to deepen their understanding by including more fundamental knowledge from fluid mechanics would also find appropriate literature. In 1931 Prandtl published *Abriss der Strömungslehre* [340]. In the same year the renowned Handbook on Experimental Physics edited four volumes on *Hydro- and Aerodynamik* [309].

Judged by lectures and textbooks, therefore, fluid dynamics may be regarded as well-established by the beginning 1930s, particularly as a subsidiary specialty for aeronautical engineering. Nevertheless, students of this new specialty would not easily find an appropriate employment because the German aviation industry only slowly recovered from the decline after the First World War [341]. During the mid 1920s the decline appeared to be over and a restructured aircraft industry with fewer but more powerful firms arose, but the boom was soon replaced by the economic crisis at the end of the 1920s. An advisory organization for academic professions (Deutsche Zentralstelle für Berufsberatung der Akademiker) issued in 1929 a leaflet, "The airplane engineer," with a warning not to choose this profession: "In view of the present situation of the airplane-, motor- and subsidiary parts industry, particularly in Germany, as well as the financial situation of the Reich and the states, we must generally warn against embracing the profession of an airplane engineer without particular zeal and talent, because despite the development during the past years, the offer of manpower now already exceeds the demand" [342]. The rate of graduation in this specialty at the technical universities decreased; in 1932, a low of only six new aeronautical engineers per semester from all technical universities together was reached [120, p. 229].

## 6.2
## Glider Flight

The rise of aeronautical engineering also had a strong emotional component. Many aeronautical engineers were motivated to study this specialty because it was closely related to aviation as a sport. When the restrictions of the Versailles Treaty forbade the production of motorized airplanes during the early years after the First World War, glider flight was revived as an activity where aeronautical enthusiasm could be expressed beyond political limitations. The initiative originally emerged among flight enthusiasts of the journal *Flugsport*, but soon attracted wider circles, among them former flight aces, industrialists, professors and students from technical universities, and representatives from the ministry of transport. Beginning in 1920, they organized annual contests on the "Wasserkuppe," a hill in the Rhön. The mood in which this activity was pursued is amply expressed in the words of August von Parseval, the pioneer of airships and professor at the Technical University in Berlin: "The goal that the ancients had before their eyes, which the old master Lilienthal did not attain but put within tangible reach, to exert the powerless gliding flight of birds, to float like them effortless in the ether, this goal is not yet completely reached; but one has come so close to it that a total mastery of this activity is only a question of time." Parseval, however, had not only ancient dreams of mankind in his mind when he praised the sport of glider flight. It provides also "an excellent training of pilots for motorized flight and will help us to keep the interest for aviation in Germany vivid and to bridge the dead time which we have to endure under the pressure of the Entente" [343].

Professors at technical universities like Parseval also regarded glider flight as a challenge for the engineering profession. According to Parseval's colleague August Pröll from the Technical University Hanover the "enforced rest for [aircraft] construction" should be used to prepare for the time when the demand for engineers would rise. "In aerodynamics, where theory and reality contradict one another so often, it is necessary to demonstrate to the students vividly by self-performed measurements the limits of speculative research." In order to bring theory and practice together, he suggested establishing aerodynamical laboratories with wind tunnels at the universities and providing the students with practical experience in the form of glider flight [325, p. 165]. Glider flight offered an opportunity to become acquainted with a variety of aeronautical subfields, from airplane design to meteorology. Many technical universities implemented Pröll's suggestions. At many technical universities students formed "academic flight groups" (Akademische Fliegergruppen, or Aka-Flieg) that designed their own gliders. The Rhön contest became a kind of fair. Prizes were awarded for the best glider flights, i.e., for the maximal flight duration and the longest distance between the sites of start and landing [344].

Despite its sportive character, glider flight also mirrors the professionalization of aeronautical engineering during the 1920s. The first glider designs still evoke memories of the early days of flight, when the shape of birds was looked to as a model for designing airplanes. Gustav Lilienthal, for example, Otto Lilienthal's brother, propagated the concept of an extremely curved wing because he believed that a bird is lifted and carried forward on a constantly forward rolling vortex, which forms at the leading edge of the wing (see Fig. 6.1) [345]. The prestige of his famous brother "misled some enthusiasts of glider flight to take his assertions for proven facts and they designed wings which were doomed to failure," one glider pilot recalled [344, pp. 123–124]. Gustav Lilienthal also attempted to exert his influence in academic teaching. In 1920 he asked Prandtl for a recommendation to be employed as a lecturer at the Technical University in Berlin. In 1916 Prandtl had still regarded Lilienthal's studies as "worth pursuing further," but in 1920, he considered them outdated: "I believe that the present state of your studies does not let us expect an advancement of contemporary flight technology," he responded to Lilienthal's request. "I therefore ask you not to expect my support for your application for the lectureship at the Technical University" [346].

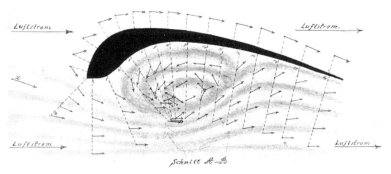

**Fig. 6.1** Gustav Lilienthal's concept of lifting vortex rolls underneath a curved wing.

Glider flight also let other aerodynamical misconceptions come to the fore. A heated debate addressed the causes of dynamic glider flight. The apparent effortless soaring of a seagull or an albatross gave rise to speculations of how wings interacted with the air so that no muscle power was required to lift and propel the bird. In 1921 Friedrich Ahlborn surveyed a number of attempts to explain this phenomenon. According to Ahlborn's own explanation the invisible power for constant soaring was provided by "pulsative wind forces." He believed that "turbulent forces of the wind" exert a similar effect as the active flapping of wings [347]. Albert Betz had proposed a similar explanation and proposed a test of this explanation by measuring the fluctuations of air speed in the vertical direction of the wing's motion [348]. Prandtl also contributed to the debate. It is "not impossible that some birds exploit the rapid wind fluctuations according to the fish-tail principle," he argued, and speculated about a possible application of this principle in glider plane design, "perhaps by an appropriate construction of elastic wing profiles." However, he cautioned that before designing elastic glider wings one should study the phenomenon experimentally: "It is very likely that the wind fluctuations responsible for this effect do not have the right order of magnitude from which an appreciable advantage is to be expected." Prandtl regarded it more likely that soaring birds "mainly benefit from rising currents of air" [349]. Ahlborn, however, regarded the fish-tail principle as the most likely explanation. He had experimented as early as 1896 with fish-tail propellers in the form of elastic plates attached to the hull of a boat. Ahlborn christened this boat "autonaut" because it was supposed to propel itself through the water by exploiting the up and down motion of the waves. He believed that the excellent flight performance of some early airplanes, such as the "Tauben," was also due to the elastic behavior of the wings. "I cannot see why it is necessary to perform more experiments with unmanned models after such experiences with elastic propulsion," Ahlborn criticized Prandtl's caution [350].

The belief in a pulsative effect was shared by many glider flight enthusiasts, although they disagreed when it came to the details of explaining dynamic glider flight. Kármán attempted to explain it in terms of a "fluctuation theory." He did not regard Prandtl's assertion of rising currents of air as a sufficient explanation. Kármán illustrated dynamic glider flight with a mechanical analogy: if a sphere is allowed to move freely along a rail that has arbitrary up and down slopes, then it is possible to force the sphere upwards by appropriate horizontal back and forth motions of the rail. In another model he used a spiraling rail mounted off center on a disk: by rotating the disk at an appropriate speed, the sphere on the rail was forced to move upwards [351]. Such models were meant not only as an analogy for the riddle of dynamic gliding. They were also a didactic means to entice students to consider the aerodynamic problems of glider flight – an issue that became a major part of Kármán's teaching at the Technical University in Aachen. "Professor von Kármán allowed us to use the entire workshop of the Aerodynamical Institute," his assistant Wolfgang Klemperer recalled the beginnings of the "Flugwissenschaftliche Vereinigung Aachen," one of the first glider flight groups at a German technical university [352]. Klemperer won the first Rhön contest in 1920 with a self-designed glider, the "black devil" (Schwarzer Teufel), in which he achieved a flight duration of almost two-and-a-half minutes and a flight distance of 1,830 m.

Klemperer's success did not result from "dynamic gliding," and Kármán's "fluctuation theory" was soon forgotten again. The practical experience of the Rhön flights showed that the prevailing mode of glider flight was "static" rather than "dynamic," as defined in the contemporary sense [353]. In 1922 Prandtl did not even mention dynamic gliding when he drew the lessons of the recent Rhön contest [354]. In 1926 Klemperer concluded that there was still no certain proof of success for dynamic gliding, which he now defined more generally as the exploitation of gusts of wind *and* the exploitation of steady wind layers [352]. (Under the latter definition, dynamic glider flight remains an issue of research today). Rather than exploiting gusts the glider pilots over the hills of the Rhön learned to exploit the rising currents of air caused by thermal differences. In 1922 the Great Rhön Prize was awarded to the Hanover student group with their "Vampyr" for achieving a flight duration of 3 hours and 10 minutes; they also won the prize for the largest flight distance, more than 10 km [355]. The best planes were not designed such that Lilienthal's "rolling vortices" would carry them forward or Ahlborn's "pulsative forces" would cause the wings to flap, but according to recent insights from airfoil theory and wind tunnel tests. The designers were advanced engineering students, some of whom would later rise to prominence [356].

Gliders like the Hanover Vampyr were designed according to the most recent aerodynamic knowledge. The Vampyr's wing profile was tested in the Göttingen wind tunnel and optimized for a minimal drag-to-lift value. Because the total drag depended both on the shape of the profile (profile drag) and the planform of the wing (induced drag), students who wished to design optimal gliders for the Rhön contests had to be aware of Prandtl's airfoil theory. The higher a "quality number $c_a^3/c_w^2$" (with $c_a$ and $c_w$ being coefficients of lift and total drag) the lower the sinking velocity of a glider. This figure, for example, one criterion for optimal glider design, was derived by Prandtl in 1921 from airfoil theory [357]. Another consequence of airfoil theory was the choice of a high ratio of span ($B$) to depth ($T$) of the wing in order to minimize the induced drag. The Vampyr's span-to-depth ratio was 10, compared to a value of 2.6 of Otto Lilienthal's gliders. The technical director of Junkers's airplane factory, Otto Mader, used this comparison in 1924, together with examples of birds with good and bad gliding abilities (see Fig. 6.2) in order to illustrate how recent aerodynamic knowledge contributed to the progress of airplane design [358].

Gliders were of no economic interest for the Junkers factory, but Mader's presentation was not on gliders but more generally on airplane construction. The audience was the Association of German Engineers (VDI), which hints at another aspect of the coming of age of aeronautical engineering: the relationship of the aviation industry to academic aerodynamics.

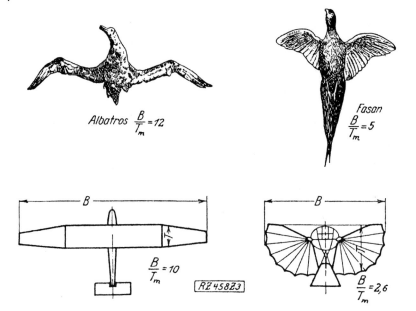

Albatros $\frac{B}{T_m} = 12$

Fasan $\frac{B}{T_m} = 5$

$\frac{B}{T_m} = 10$

RZ 45823

$\frac{B}{T_m} = 2,6$

**Fig. 6.2** Based on their span-to-depth ratios, the "badly flying pheasant" was compared to the "king of the sea-gliders, the albatross." The Vampyr, developed by the Hanover glider group and winner of the Rhön contest, was associated with the albatross , while the pheasant was associated with Lilienthal's "still imperfect glider".

## 6.3
### Kármán and Junkers: The Beginnings of Industrial Consulting in Aeronautics

The rise of civil aviation and the clandestine development of military aviation abroad offered splendid prospects for airplane manufacturers who had survived the dire postwar years. The restrictions in the wake of the Versailles Treaty did not prevent them from focusing on the construction of big motorized airplanes. As long as political obstacles existed, airplane construction was displaced to other countries [341]. Junkers, for example, conceived in December 1920 a "program for finding ways and means to create work and sales possibilities," of which one section was titled "Bypassing the restrictions." Junkers envisioned two possibilities: "fabrication abroad" and "fabrication in Dessau for Entente (...) or neutral countries" [359, p. 198].

But as long as the restrictions were in effect, glider flight was the only legitimate way to foster aeronautical interest in Germany. Because of the involvement of leading academics like Prandtl and Kármán, and the integration of glider design in aeronautical teaching and research at many technical universities, this sport also served as a proving ground for new aeronautical concepts that might result in new applications for future airplane design. It is not accidental, therefore, that airplane industrialists paid attention to the glider sport although it was rather marginal as an industrial activity by itself. Junkers became one of the earliest and most generous sponsors of the Rhön glider contests. Another industrialist, the Aachen manufacturer of car bodies and railway wagons, Georg Talbot, became infected by the glider obsession through Kármán. Talbot donated money to Kármán's glider group and provided workshop space for them in his factory. In return Talbot was awarded an honorary doctoral degree by the Technical University Aachen. "Doctor, to walk with you is a great honor – and profitable," Kármán thanked Talbot quoting a verse from Goethe's Faust. With Talbot's money Kármán even founded a firm, the "Aachener Flugzeug G.m.b.H.," and sold a few gliders. But the business turned out to be a flop, "and I soon abandoned my first attempt to become an industrialist," Kármán recalled later [85, p. 101].

Kármán's excursion into the world of business and industry, however, was not entirely futile. Junkers, who still possessed a private research laboratory in Aachen (see Chapter 3), bought Kármán's moribund company in 1923—and with it came Kármán and Klemperer as advisors. With Junkers as the only shareholder, the firm was renamed the "Aachener Segelflugzeugbau G.m.b.H. (Sef)." Its business goal was defined as "not oriented towards profits but merely the research in the area of glider flight technology" [360]. But from the very beginning "other work in the area of aviation" was envisioned too [361]. Kármán proposed to extending the firm's range of work from gliders to motorized small airplanes, for example. He regarded such planes as "the motorbike in the air"; he also regarded the development of helicopters as a promising business for the future. Although Junkers did not share these views and urged Kármán to pay more attention to economic considerations, he granted his advisor the "total freedom of decision" for the direction of the Sef [362]. In practice, the Sef was run by Klemperer, because "v. Kármán himself is not interested in directing mere constructive works," as was remarked in a protocol after Klemperer left the firm by the end of 1924. Under Klemperer's supervision, for example, the Sef designed a small two-seater training airplane. Organizationally, the Aachen Sef was run as a subsidiary department of Junkers's Dessau research division. After Klemperer's departure, Junkers's interest in the "wooden airplane construction" of the Sef declined noticeably. He decided that "the Aachen workshop operation should be cut back as soon as possible," without, however, dissolving the Sef completely. Kármán suggested "that Junkers agrees that the Technical University [Aachen] assumes responsibility for the workshop under the explicit provision that he can use it for his own purposes at any time" [363]. Hardware and part of its personnel were transferred to Dessau. "Mr. v. Kármán will stay on a contractual basis as a collaborator with us, but without the need to keep up the present big operation of Sef," the main office of the Dessau Junkers firm described the result of this reorganization in January 1925 [364].

Afterwards, there was hardly any mentioning of the Sef. "For Junkers there was nothing left except paying back the debts of the workshop." So ended the chapter on Junkers and glider flight in the history of the Junkers factory [359, p. 164]. But what appears as a failed strategy was in fact a clever move in the Dessau firm's policy. Junkers spared no costs to invite Kármán to Dessau in order to request his advice. The Aachen professor then was hosted as Junkers's personal guest in his villa and had to "do nothing more than talk over aerodynamics problems," Kármán later recalled. "For two weeks of this 'consultation,' he offered to pay me a sum which was half of what the government paid me for teaching a full year" [85, p. 110].

To what extent Junkers benefited by Kármán's consulting in terms of money is difficult to estimate, but it is obvious that Kármán's advice resulted in tangible applications of the design of Junkersairplanes, such as slotted wings. Kármán's student, Gustav Lachmann, and an English engineer, Frederick Handley-Page, had discovered independently that the stalling of flow at high angles of attack is prevented by slots in the wing. In 1922 Kármán presented Junkers with the design of a "canard with slotted wings" as one of several proposals for which he expected they "appear appropriate to find your interest." At that time the Aachen aerodynamicists were not yet contractually linked with the Dessau firm, but Junkers experimented with a modification of Lachmann's principle – two narrowly adjacent wings; so Junkers was indeed interested in Kármán's proposal. Perhaps it was this proposal rather than the general glider obsession which prompted Junkers to take over the Sef. Kármán suggested that the "lift increase by Lachmann's slot" should be tested "in practice and under the simplest and most clearly defined real-flight conditions" [365]. Preceding this proposal were wind-tunnel experiments performed by Klemperer in Kármán's institute at the Technical University Aachen [366].

The slotted-wing principle was also tested in the Göttingen wind tunnel (see Fig. 6.3) [162, Heft 2, pp. 55–65]. Although there was no quantitative theory which accounted for the effect, Lachmann provided an explanation in terms of the boundary layer concept: the air blown through the slot across the upper wing surface makes the boundary layer there turbulent, which prevents the flow from separating from the wing. Using Kármán's formulae for the thickness of the turbulent boundary layer, he was able to estimate the power which was transferred into lift generation by the slots [367, pp. 184]. By the end of 1924 Junkers authorized the Sef to construct a test plane with a "double wing" as a lift-increasing device [363]. Although Lachmann's slotted wing and Junkers's double wing were not the same, both made use of the same principle – and the double wing became a characteristic feature of Junkers airplanes long before the principle was routinely used in other airplanes in the form of start- and landing flaps.

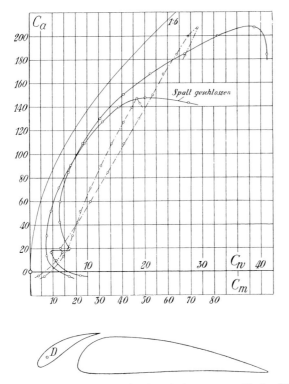

**Fig. 6.3** Polar diagram of a slotted wing measured in the Göttingen wind tunnel.

Kármán also provided his expert opinion for Junkers in patent fights and served as a consultant for political decisions within the firm. In June 1928, for example, he wrote a report on the situation of aviation in Japan "with regard to a possible engagement of the Junkers works." Kármán based this report on his recent experiences as a consultant for Kawanishi, a Japanese airplane manufacturer who had invited Kármán for an extended sojourn in 1927 in order to get his advice on the construction of a wind tunnel. During his conversations with the Japanese industrialists he became aware that "they were primarily interested in making Japan a power in world aviation" [85, p. 131]. He was able to provide Junkers with a feel for the Japanese airplane industry: one has to discern the difference between the "big corporate groups directed by politicians" and the "self-contained airplane factories," he introduced his report. "The three corporate groups, which deal with all kinds of industrial activities related to military projects of the government, i.e., ship-building, weapons, submarines, airplanes etc., are Mitsubishi, Mitsui, and Kawasaki." He then described how each one was related by license contracts with European partners and where he saw their virtues and weaknesses, respectively.

He found it regrettable that Junkers's representatives in Japan favored Mitsubishi and proposed negotiating instead with Kawanishi as Junkers's Japanese partner [368]. Although Junkers did not follow Kármán's advice in this case, because the negotiations with Mitsubishi had "already resulted in an accomplished agreement," he took note of his recommendations "with a very particular interest" in order to decide how the relations with Japan should be pursued in the future [369].

Besides Junkers and Kawanishi, Kármán was also an advisor for the Zeppelin Works at Friedrichshafen and other firms with some interest in aerodynamics. Other colleagues of Kármán also became more and more engaged in industrial consulting. Albert Betz, for example, once became involved as consultant of Heinkel in a patent dispute with Kármán about the slotted wing patent [85, p. 113]. If industrial consulting by academic scientists is regarded as another criterion for an engineering science, aerodynamics clearly met this requirement by the end of the 1920s.

## 6.4
### Profile Measurements

Beyond such external characteristics, however, the coming of age of aerodynamics is most clearly revealed by its use. What was the relationship between supply and demand in aerodynamics, if we identify the Göttingen AVA as the former and the aircraft industry as the latter? The war-time tripartite relationship between science, military, and industry, in which contracts between airplane manufacturers and the Göttingen aerodynamicists were encouraged and mediated through the Berlin FLZ (see Chapter 3), no longer existed in the 1920s. The military also disappeared as a source of funds for the expansion of research facilities. The immediate postwar years were a dire period for Prandtl and his colleagues. Negotiations among the Association of German Airplane Manufacturers, the Kaiser Wilhelm Society (KWG) and other state institutions resulted in securing the further survival of the AVA, but the galloping inflation made this survival difficult. "The AVA had to see how to exist day by day" ("Die AVA mußte also weiterhin von der Hand in den Mund leben"), these years are described in the history of this institution [67, p. 220]. Junkers offered Prandtl his individual help if such was required, but he regarded it an obligation of the state and the entire airplane industry to save the AVA [370].

The AVA survived these dire years mainly because of industrial contracts. Even when its survival was secured, the AVA had to rely on such contracts for its further development. However, when the restriction of the Versailles Treaty for civil aviation ended in 1926, the German aircraft industry boomed and provided the AVA with ample contractual work. Additional support came in the form of test contracts from airplane manufacturers who received government subsidies from the Ministry of Traffic (Reichsverkehrsministerium, or RVM). The foundation of the new Kaiser-Wilhelm-Institut für Strömungsforschung in 1925 also helped the AVA. "The total revenue of the combined research institute, which consisted mainly of the budget of the KWG, the charges of the contractual tests for private firms, and the work for the RVM, displays a considerable growth," a historical account of the AVA commented on the financial situation in the second half of the 1920s; "no trace was left from the concerns how to persist" [67, p. 260].

It is remarkable how leading airplane manufacturers evaluated the use of the AVA during these dire years when its very existence depended on the readiness to offer help. Junkers considered the survival of the AVA as "very important in the national interest" because it was "the only purely scientific aerodynamic research establishment in Germany." This was an internal evaluation within the Junkers factory preceding an official judgment requested by the RVM from the airplane industry on whether the AVA deserved to be kept alive. The director of Junkers's research division even suggested that they should donate money for the AVA's survival themselves if necessary [371]. This evaluation is all the more noteworthy because the Göttingen research up to this time was regarded as of little importance for airplane design. The AVA "had never been leading from a technological perspective," the Junkers memorandum remarked on the contribution of the Göttingen aerodynamicists to the war effort. "A work like the systematic investigation of thick profiles would never have originated from the initiative of the Göttingen facility because of their technological advantages. Its collaborators did not regard it as their duty to look for new aviation technology." Although they diligently fulfilled their contractual obligations, the pursuit of these contracts lacked methodical planning. "This must now be created," the memo demanded, all the more because the AVA was "scientifically unmatched" and foreign countries could not compete with a comparable institution. As a particular scientific virtue, the memo praised the Göttingen airfoil theory as a usable tool for "practical technology" after it was shown that the theory agreed well with experimental results and that "promising works are in progress" [372].

Junkers's view that the Göttingen aerodynamicists did not pursue profile measurements according to methodical planning contrasts with Munk's remark in the Technische Berichte, which claims that in addition to the contract work for the airplane factories tests were "oriented towards the accomplishment of certain results in a systematic manner" [135]. For example, it is not entirely true that they did not investigate thick profiles, as the communication of test results of profile No. 198 illustrates [136]. However, it is true that such profiles were only occasionally investigated. For Junkers, thick profiles, like the one seen in Fig. 6.4, were mainly interesting because of their structural virtues and because they were a central feature of the Dessau airplane design.

From the Göttingen perspective, thick profiles were not given as much attention. Prandtl himself admitted later that Göttingen's profile measurements during the First World War were insufficient: "As far as the 'thick wing' is concerned," he explained this neglect to a Junkers biographer, "we had a very low wind speed in the first wind tunnel at which the thick wings came off badly" [373].

**Fig. 6.4** A section of the thick wing of Junkers's "Eiseneindecker (Iron-monoplane) J-2" built in 1916 (Source: Deutsches Museum, Munich).

Increasingly, Prandtl and his collaborators were eager to demonstrate after the war that their institution was able to perform methodical and systematic investigations that were useful for airplane design. First of all this required a suitable way of making the results of these investigations available. Before the war, a series of communications in the *Zeitschrift für Flugtechnik und Motorluftschiffahrt* served this purpose. During the war the Technische Berichte of the FLZ took over this task. From 1919 to 1922 the Göttingen aerodynamicists resumed the tradition of publishing special communications in the *Zeitschrift für Flugtechnik und Motorluftschiffahrt*, but they soon reached the decision that this was no longer an appropriate means of communicating their results. Beginning in 1920, they edited a new series of communications, the *Ergebnisse der Aerodynamischen Versuchsanstalt zu Göttingen*. It aimed at making available their research results available "to the expert community in a more comfortable manner than was the case in the past," as Prandtl introduced the series [162, I, preface].

The major part of the 140-page volume of the first issue was dedicated to the investigation of wing profiles, performed in the large wind tunnel with its 2 m diameter test section (see Chapter 3) in order to update older data, which were obtained in the first wind tunnel. The new wind tunnel allowed measurements at Reynolds numbers two orders of magnitude larger than those of the first tunnel. The transferability of test results on models to real-scale conditions was the major problem for practical airplane designers. The law of similarity demanded that the same Reynolds number had to be achieved in the wind tunnel as in free flight – a requirement that was difficult to fulfill with the trend toward ever-larger airplanes (because the Reynolds number is proportional to the size). Even if the size of the tunnel were large enough so that larger models could be tested, wind tunnel turbulence (see Chapter 5) presented another problem. In both regards, the new Göttingen tunnel surpassed other wind tunnels, but the influence of the Reynolds number on the aerodynamic forces upon a wing had to be analyzed carefully in order to provide airplane designers with a feel for the reliability of the profile data. Airfoil theory was used to independently evaluate the quality of the data. For example, the theory predicted an ideal polar for a wing of span $b$ and planform surface area $F$ in the form of a parabola, $c_w = c_a^2 \frac{F}{\pi b^2}$, where $c_w$ and $c_a$ are the induced drag and lift coefficients, respectively. This parabola was drawn in the polar diagrams together with the measured polars of a profile in order to indicate how well the actual data approached the theoretical limits. A comparison of wings with different aspect ratios made clear that the difference of the total drag (as measured in the wind tunnel) and the induced drag only depends on the shape of the profile; this difference was called "profile drag" for short. As a general tendency, it was noted that the profile drag decreased with increasing Reynolds number, and the polars approached the limiting parabola of the induced drag [162, I, pp. 37 and 71].

Another enquiry addressed the problem of the wing's planform. Airfoil theory predicted a minimal induced drag if the lift distribution was elliptical over the span (see Chapter 3), which implied an elliptical planform. However, as Betz had shown in his doctoral dissertation, the dependence of the aerodynamic forces on the exact shape of the planform was not very sensitive as long as the ratio of planform area to span was the same. For the regular profile measurements in the new Göttingen wind tunnel, a "normal wing" with a rectangular planform (with a span of 100 cm and a depth of 20 cm) was used; with this choice, it became expedient to verify Betz's theoretical conclusion and experimentally test how the profile data depended on the chosen planform. A series of measurements on wings of five different planforms with the same profile confirmed the theory: "a systematic variation due to the contour cannot be detected," Prandtl and his collaborators concluded [162, I, p. 65].

With such reassurance, the profile measurements at the AVA became a regular activity. They were performed "almost exclusively on behalf of airplane firms, flight associations and private persons," as a review in 1926 reported [162, III, p. 33]. The review was communicated publicly without naming the contractors; the anonymous communication of the profile data was similar to the wartime reports in the Technische Berichte; the profiles were numbered consecutively, but there was no systematic variation of parameters from profile to profile. Nevertheless, the frequent measurements of very thick profiles (perhaps as a consequence of Junkers contracts [374]) and a special series of measurements on so-called Joukowsky profiles reveal more systematic planning than existed during the war. The measurements of Joukowsky profiles (see Fig. 6.5) resulted from the AVA's own initiative. They were designed mathematically by conformal mapping and were "relatively comfortably accessible to theoretical investigation" because Kutta–Joukowsky theory allowed one to calculate the lift of such profiles for infinite span [162, III, p. 13]. Together with the Göttingen airfoil theory, these results could be used to calculate lift and induced drag for a finite span. The only difference between experimentally determined polars of Joukowsky profiles and the theoretical data was due to the difference of the total drag (as measured in the wind tunnel) and the induced drag (the only drag accounted for by airfoil theory); in other words, for Joukowsky profiles, the profile drag could be directly determined. Another virtue of these profiles was that thickness and curvature could be varied systematically, so that it became clear how lift and drag depended on these parameters [375].

Between 1921 and 1932, the Göttingen aerodynamicists published four issues of the *Ergebnisse der Aerodynamischen Versuchsanstalt* with polars of 723 profiles. This catalog symbolized the coming of age of aerodynamics as a new engineering science in the sober language of data sheets and diagrams – but most pervasively, nonetheless. At a time when most airplane firms did not yet possess their own wind tunnels, these data provided crucial information for airplane design. Even those few industrialists like Junkers, who owned a specialized aerodynamic laboratory with a wind tunnel, could not have afforded to perform such extensive tests. The Göttingen profile catalog incorporated theoretical and experimental knowledge about the forces exerted on wings, which airplane designers would not have obtained at the time from other sources. For example, in order to determine the power needed to propel an airplane of a given weight at a desired speed in horizontal flight, one first has to choose the coefficient of lift from the equation "lift = weight." The polar diagram of a chosen wing profile relates this coefficient of lift to a drag coefficient, which in turn allows one to calculate the resistance, i.e., the force

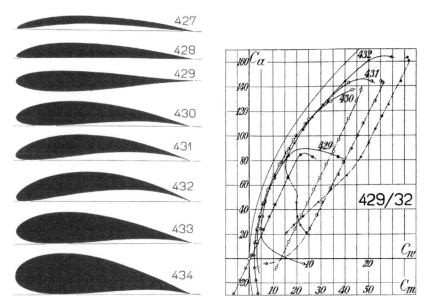

**Fig. 6.5** Joukowsky profiles and their associated polars [162, I, pp. 80 and 99].

that the airplane's engine has to provide in order to surmount the drag. Based on this knowledge the power can be calculated from the formula "power = force × velocity." The relation between drag and lift which is expressed in the polar diagram for each profile, therefore, is decisive for determining not only the shape of the wing but also for the choice of the engine of an airplane.

Beyond this basic requirement for the design of new airplanes, the Göttingen profile catalog was important in a more general way. In 1929 the authors of an aeronautical engineering textbook concluded that among all profiles tested so far, "only relatively few turned out to be useful in practice" [337, vol. 2, p. 37]. In other words, by the end of the 1920s, the large variety of possible profiles was narrowing to a set of practical profiles. This selection process resulted from the collective experience contained in the Göttingen data sheets. Due to the anonymity of the communication in the AVA's profile catalog, competing airplane manufacturers benefited from each other's experience without giving away their own designs. The Joukowsky profiles, which did not result from contractual tests but from the AVA's own aerodynamic research, were also investigated in the collective interest of aviation.

A similar strategy may be observed at the NACA. Beginning in 1925 with its M-series, the NACA contributed to the collective knowledge for wing design by communicating systematic series of profile measurements in its Technical Reports [376], [233, pp. 65–95]. However, after a decade of profile tests at the AVA, the NACA, and other laboratories the available data further strengthened the awareness among the international community of aeronautical scientists and engineers of the comparability of their measurements. "The results obtained in the different laboratories are not directly comparable, because of the differences in the methods of testing," a report of the U.S. Navy's Bureau of Aeronautics commented in 1930 after a comparison of profile data from Göttingen with measurements in American wind tunnels (including MIT, McCook Field, and Washington Navy Yard) [377]. Three years later the NACA published a series of profile data from recent measurements in the Variable Density Tunnel [378]. By increasing the pressure to 20 atmospheres, it became possible to achieve Reynolds numbers as high as 3,000,000—almost ten times higher than in the Göttingen wind tunnel and close to those for large full-sized airplanes in free flight. But wind tunnel turbulence, which was a particular problem in the VDT, limited the comparability of these data with measurements in larger wind tunnels. "It may be stated that lift measurements in the full-scale tunnel on a series of Clark Y airfoils show larger values of maximum lift than those obtained in the variable-density tunnel, and there seems to be good reason for believing that the turbulence in this tunnel is less than in the variable-density tunnel," an engineer warned the NACA's research director about the conclusions to be drawn from these measurements [379].

## 6.5
### Airfoil Theory

The goal of a standardized technique for aerodynamic measurements that would result in internationally comparable wind tunnel data was still in the future. Nevertheless, the example of profile measurements illustrates how fast aerodynamics was ripening into an engineering science during the 1920s. Polar diagrams became a common language among aeronautical engineers. Although they merely displayed the data concerning the drag and lift of a certain wing profile in a graphical diagram, the measurements with which these data were obtained involved a multitude of experimental procedures and theoretical methods which – like a language – had to be acquired by long periods of experimentation. The notion of profile drag, for example, relied on the notion of induced drag, which in turn was based on airfoil theory, i.e., the concept that accounted for the creation of wingtip vortices. Profile drag itself was explained as a sum of two parts: form drag (due to pressure differences in front of and behind the wing, which give rise to vortex-shedding) and skin friction (due to the surface roughness of the wing, which influences the boundary layer). How these various drag contributions (and therefore the polars themselves) depended on the Reynolds number was an unsolved research problem, whose solution required the results from research on boundary layers and turbulence taking place at the time. Talking to one another in the language of polars, therefore, required not only the use of an established practice of wind tunnel measurements but also a common understanding of existing aerodynamic research problems, particularly with regard to airfoil theory.

Such common ground was not reached immediately. Until the mid-1920s the official theory of aerodynamic lift in England was based on the Kirchhoff–Rayleigh concept of discontinuity surfaces (see Chapter 1). The Kutta–Joukowsky explanation, which claims that lift is a result of circulation around the wing, was met with harsh criticism [380]. In 1922 Leonard Bairstow, a leading aerodynamicist from the Imperial College in London, was still convinced that careful wind tunnel measurements would prove that there was no trace of a circulatory motion around the wings. Prandtl's airfoil theory, however, which was based on the circulation concept, was found to agree with experimental wind tunnel tests [382]. In 1926 careful experimental tests in the wind tunnel of the National Physical Laboratory "provided an experimental verification of the law of Kutta and Joukowsky" [381]. Only after these tests was the Kirchhoff–Rayleigh explanation of lift regarded as wrong and Prandtl's theory accepted as valid in Great Britain.

Most instrumental for this conversion from the long-cherished aerodynamic orthodoxy to modern airfoil theory was Hermann Glauert, a Fellow of the Trinity College in Cambridge and collaborator of the Royal Aircraft Establishment in Farnborough. Glauert and a colleague from Farnborough visited Prandtl's institute in 1921 in order to learn how the Göttingen aerodynamicists performed wind tunnel experiments and to discuss applications of airfoil theory [383]. Subsequently, Glauert spread the gospel of the new airfoil theory among his British colleagues. Prandtl enjoyed "the attention the Göttingen airfoil theory has found in England," as he wrote to Glauert in October 1923 [384]. He had been offered the opportunity to persuade the English aerodynamicists himself in 1922 when he was invited to present a lecture before the Royal Aeronautical Society in London, but he had declined because of his poor English [385]. A few years later, however, he used the invitation to deliver the 1927 Wilbur Wright Memorial Lecture as an occasion "to thank the English experts for the great and lively interest they have taken in this theory and for the considerable efforts they have devoted to testing out the theory by means of experiments in the most diverse directions" [386]. In the meantime, Glauert had further prepared the ground with a textbook titled *The Elements of Aerofoil and Airscrew Theory*, which presented Prandtl's theory in such a lucid manner that it became a textbook classic for the budding community of aerodynamicists throughout the world [336].

In Germany, too, airfoil theory was met with criticism. Friedrich Ahlborn voiced strong objections against Prandtl's theory because he thought it was based on "arbitrary assumptions which do not agree with reality" [387].

Ahlborn did not oppose the circulation concept in principle, as the British aerodynamicists did, but he regarded Prandtl's elaboration of this concept as wrong. According to Ahlborn, the cause of the "theoretical" circulatory motion around the airfoil was caused by "a vortex sheet on the upper side of the wing" [388]. He based this view on direct evidence from his flow photographs, but he could not substantiate it in the form of quantitative results. In the course of the 1920s, he developed a profound aversion to theories which seemed unfounded by visual observations, such as Prandtl's boundary layer and airfoil theories. His arguments were framed in a language which diverged from the usual discourse among aerodynamicists of the 1920s, and he found few followers when he made his objections public. His notions "deviate from the usual terminology of mechanics," as a participant in a debate at the Scientific Society of Aviation (Wissenschaftliche Gesellschaft für Luftfahrt, or WGL) remarked on Ahlborn's objection to Prandtl's boundary layer concept [389].

Ahlborn's refusal to accept the modern concepts could be ignored as insignificant if he were not a renowned pioneer of experimental fluid mechanics. Ahlborn's views were based on his own flow photographs; they made sense when related to the flow regime in his water channel, but not for the test of airfoil theory where much higher Reynolds numbers applied. Prandtl tried hard to persuade Ahlborn. The Göttingen airfoil theory is a logical and consistent mathematical theory, he responded to Ahlborn's criticism in October 1924, "which today is acknowledged as correct by the best experts and in practical use not only in Germany but also in England, France, Italy, America, etc." [390]. In a subsequent letter, Prandtl protested against Ahlborn's characterization of himself as a theorist and pointed to the experimental tests to which he had subjected airfoil theory in the Göttingen wind tunnels. "We always proceeded hand in hand with experiments, and all objections you raise now were raised by ourselves ten years ago." Then he presented a list of investigations that had been undertaken in order to check the various theoretical predictions. In 1918, when the theory was published, it had already gone through a phase of thorough experimental tests. "In addition, the theoretical results have been checked experimentally in England with the same positive conclusion" [391]. But Ahlborn could not be converted. When he died in 1937, Prandtl assured Ahlborn's son that he had always respected Ahlborn as a pioneer of experimental research in fluid dynamics. "I stress this because your father often stood in opposition to my own theoretical work, and this has previously led to quite sharp controversies between us" [392].

The clash between Prandtl and Ahlborn also signals a turning point in the history of aerodynamics. Ahlborn's method of extrapolating from visual observations to more general views without due attention to theoretical fluid mechanics was a common approach of experimental research when theory and practice were still irreconcilably separated into different camps. Ahlborn's approach may be compared to other research from this era, like those of Lanchester or Lilienthal, for example (see Chapter 1). These pioneers achieved remarkable successes within the realm of their observational range; but their method could not account for extrapolations into the realm of aeronautical engineering as it was conquered in the 1920s. The experiments and theories of the new era had little in common with those in the age of these pioneers. Ahlborn's criticism appears anachronistic when confronted with the theoretical advances made at the same time by aerodynamicists of this new age of aeronautical engineering.

Airfoil theory became a complex of special theories. As far as the flow in a section of a wing with a given profile was concerned, various "profile theories" emerged. The oldest method was introduced before the First World War and relied on conformal mapping; although it was limited to so-called Joukowsky profiles (resulting from mappings of a circle into the shape of a profile), which were primarily of theoretical interest, this method was also amenable to practical applications in the form of so-called Kármán–Trefftz profiles [85, p. 77–78]. Another profile theory was Max Munk's thin-wing theory, which chose the mean line of a wing profile as the starting point [238]. A third procedure was the so-called method of singularities, by which sources, sinks, and potential vortices were arranged so that their superposition accounted for the flow field around a profile. Walter Birnbaum, a doctoral student of Prandtl's, developed this method in 1923 as an alternative to the calculation of thin wing profiles [393]. Glauert generalized this method in 1924; the presentation of this method in 1926 in Glauert's textbook became the blueprint for many subsequent engineering accounts on the theory of wing sections [394, pp. 64–79].

Birnbaum's method of singularities also opened a new avenue to extend Prandtl's airfoil theory. In its original form, airfoil theory was based on the concept of a single lifting vortex line; the new method was based on a lifting vortex sheet rather than a single vortex line. Prandtl entrusted another doctoral student, Hermann Blenk, with the investigation of this concept. The "lifting surface" method considerably extended the practical uses of the airfoil theory. The "lifting line" concept was limited to straight airfoils with a large span; the "lifting surface" concept could be applied to wing planforms of (almost) arbitrary shape. Blenk derived from his theory formulae "for the practitioner" through which deviations from the rectangular wing planform could be accounted for. The new theory also entered the practice of wind tunnel testing. In order to extrapolate test results of a measured rectangular "normal wing" to a wing of the same profile but another planform and aspect ratio, conversion formulae derived from airfoil theory had to be used. Blenk demonstrated with the example of the Göttingen profile No. 389 that the formulae based on the "lifting surface" theory could be applied for aspect ratios as low as 1:1, in contrast to those based on the "lifting line" concept, which were limited to ratios higher than about 3:1. Despite such progress, however, there were basic obstacles that prevented further applications of airfoil theory in practical aeronautical engineering. Blenk himself saw his theory's mathematical simplifications as a deficiency, leaving him with "essentially a linear theory" [395].

Although the "lifting surface" concept was superior to the "lifting line" concept, the latter was not rendered obsolete by the former. As long as wings were designed predominantly with a straight span, the sheet-like distribution of lifting vortices along the mid-line of a wing section could be approximately replaced by a single vortex line of appropriate strength located in the center, and the simpler lifting line concept could still be used. By the mid-1920s, its range of application was investigated in a variety of special studies [396–398]. The most pertinent problem for practical applications was the determination of the distribution of lift along the span of a wing with a given planform (often designated as the second problem of airfoil theory, in contrast to the first problem where calculations start with a given distribution of lift). This had been the subject of Betz's dissertation in 1919 (see Chapter 3), but Betz's result was hardly applicable for practical purposes. It was regarded as clumsy and became a target of opportunity for mathematical improvement [336, 399–401]. Only in 1931 was Irmgard Lotz, a doctoral student of Betz's, able to present a more viable solution [402].

Compared to the beginnings of airfoil theory at the end of the First World War it was unmistakable to what extent theory further approached the practice of aeronautical engineering. Yet after a decade of efforts by applied mathematicians, it was still a working theory for aerodynamic research laboratories rather than an instrument for practical engineering design. So far, airfoil theory was based on ideal fluid theory and could account for lift and induced drag only. Therefore, the profile drag was beyond its reach and had to be supplied from separate wind tunnel measurements. Even in its most developed form the theory required calculations that rendered it prohibitive for practical wing design. Irmgard Lotz's procedure, for example, could be applied only after determining "a specific constant for a given profile" by wind tunnel measurements and after specifying the wing's planform according to a required scheme. After such preliminary work, "a skilled computer can determine the required even and odd coefficients (perhaps up to 10) in 2 1/2 hours," its practical use was characterized, assuming a human computer who was experienced with the handling of slide rules [402, p. 191]. It is not surprising that airfoil theory remained a proving ground for applied mathematicians.

# 7
# New Applications

The rise of aerodynamics as a new applied science for the budding new branch of aeronautical engineering is so pervasive that it overshadows applications of fluid dynamics in other areas. Disciplines like meteorology, mechanical engineering and ship-building should not be overlooked as pacesetters of research in fluid dynamics. Prandtl's research program on turbulence after the First World War indicates a trend towards fundamental research (see Chapter 5), which might seem to contradict the growth of applied research. For Prandtl, however, "fundamental" and "applied" were not opposing categories. He regarded the two as complementary: fundamental riddles in fluid dynamics were closely related to practical problems, and applications often became a challenge for basic research.

For Prandtl, the explanation of fluid phenomena in terms of physical laws was the ultimate goal, but despite such interest in basic science, he did not classify research problems along a scale between "fundamental" and "applied," or with regard to distinct areas of application, but according to the realms of laws for certain fluid regimes. In the summer of 1923, for example, Prandtl defined the following "problem groups" in a preliminary research program for his new Kaiser Wilhelm Institute: "1) Research into the laws of flow valid for incompressible fluids with little friction (like water, or gases at moderate pressure gradients), 2) research into the corresponding laws for highly compressible fluids (gases under high pressure differences), 3) laws of flow close to the vapor point (Verdampfungspunkt), 4) laws of motion for very viscous fluids." Research problems of the first category belonged to the realm of "hydrodynamics in a narrower sense"—for example, the investigation of "vortex formation, turbulence, wave motion, etc." It was not important whether such flow phenomena were studied in air, water or some other medium. Nevertheless, Prandtl had specific studies in mind. He mentioned the study of "flow forms in rotating vessels and channels (both with regard to the laws of flow in turbines and with regard to the air and ocean currents on the rotating earth)." The second category concerned the realm of gas dynamics

*The Dawn of Fluid Dynamics: A Discipline between Science and Technology.* Michael Eckert
Copyright © 2006 WILEY-VCH Verlag GmbH & Co. KGaA, Weinheim
ISBN: 3-527-40513-5

with applications in mechanical engineering (steam turbines), ballistics, and aerodynamics. Research problems of the third group involved the phenomenon of cavitation, the formation of bubbles due to a lowering of the local pressure under the vapor point; this phenomenon occurred with rapidly rotating ship propellers or in water turbines and was known to cause great damage. The fourth category addressed such diverse phenomena as the "lubrication of machine parts" and the "locomotion of tiny animals (e.g., infusoria) in fluids" [404, pp. 153–154].

This list clearly demonstrates how broadly Prandtl conceived the scope of research in the new Kaiser Wilhelm Institute, as well as the diversity of applications encompassed in its course, ranging from biological fluid mechanics of microscopic animals to ballistics. It also shows that despite the diversity of the phenomena, there was a conceptual coherence among these research topics due to an orientation towards basic laws. Fundamental and applied interests went hand in hand in Prandtl's institute, even if a specific case clearly pointed to one orientation or another.

## 7.1
## Gas Dynamics

The investigation of flow problems where fundamental laws were used to account for the compressibility of the fluid started with problems of machine engineering by the end of the nineteenth century. When steam turbines were developed for a variety of industrial uses, such as the generation of electrical power, engineers became aware of a problem for which they could not find an explanation. The success of steam turbines was due to the fact that a jet of steam could transfer a very high momentum to the vanes of a turbine. A Swedish engineer, Carl Gustav Patrick de Laval, had invented a special nozzle shape through which jets at supersonic speeds could be ejected. Through the use of Laval nozzles (see Fig. 7.1), steam turbines could be rotated at supersonic circumferential speeds. Other nozzle shapes, however, seemed to choke off the jet at the speed of sound.

The experience suggested that for an incompressible flow through a narrowing passage, according to the continuity equation, the velocity increases with the square of the ratio of the diameters between the entrance and the exit of the nozzle. However, the flow of a steam jet through a narrowing passage choked off at the speed of sound, regardless of the pressure difference by which it was driven. Only if the passage was converging and diverging, such as in a Laval nozzle, could the choking be avoided and the steam ejected at supersonic speeds.

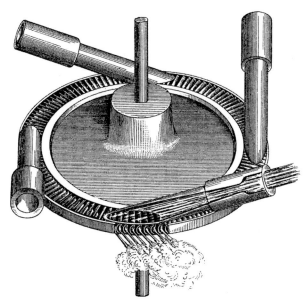

**Fig. 7.1** Laval's sketch from 1882 of a turbine, in which steam jets are emanating from converging–diverging nozzles ("Laval nozzles") and deflected by turbine blades [412, p. 2].

The cause of this phenomenon was unknown for many years after the application of steam turbines and Laval nozzles. Prandtl's first attempts to solve this riddle were motivated by a debate in the *Zeitschrift des Vereins Deutscher Ingenieure* in 1903 and resulted in an investigation of the relationship between the pressure and velocity of jets in nozzles [405,406]. A year later, he authored a review article on this topic in the renowned *Enzyklopädie der mathematischen Wissenschaften* [407]. Among other details, he observed stationary waves in the emanating jets and showed that this phenomenon is akin to the conical "Mach waves" around supersonic projectiles (see Chapter 1). Subsequently, Prandtl made the experimental and theoretical analysis of this phenomenon the theme of doctoral dissertations at his Göttingen institute [408–410]. The analogy with "Mach waves" also suggested the use of Mach's "Schlieren method" to visualize the density fluctuations in gas jets. The photographs from jets during the passage of a Laval nozzle whose inner surface was roughened in order to generate waves showed that the inner space of the nozzle was crossed by lines, which originated at the diverging flanks of the nozzle and extended beyond the nozzle's exit (see Fig. 7.2). The interpretation of the observed lines as stationary wave fronts (the images, taken with exposure times of $1/30$ to $1/10$ s, could not display transient waves) further suggested that the angles between these lines and the diverging flank of the Laval nozzle, like the "Mach angles" of the cones around supersonic projectiles, could be used to determine the speed of the jets.

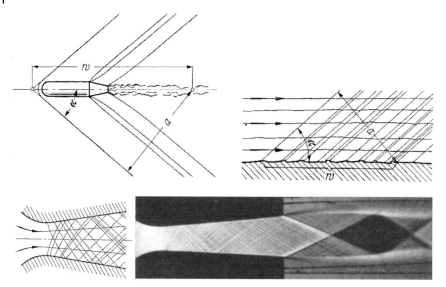

**Fig. 7.2** "Mach waves" as produced by a supersonic projectile (above, left) are also created by scratches in the wall of a Laval nozzle. The "Mach angle" $\alpha$ reveals the supersonic speed $v$ of the jet according to $\sin \alpha = v/c$ ($c =$ speed of sound) [411, pp. 947 and 952].

Together with his doctoral student Theodor Meyer, Prandtl also presented a theory for expansion waves. Such waves are created in a supersonic flow around a sharp corner; for example, when a supersonic jet leaves the exit of a nozzle into a space of lower pressure (see Fig. 7.3). Along each straight line, which originates at the corner and crosses the parallel streamlines of the jet, the infinitesimal change of pressure and velocity is the same. Such lines are called "characteristics" in the theory of differential equations. Therefore, the jet changes direction by the same angle everywhere along this line until the new value of pressure is established. Such a widening of a supersonic jet into a space of lower pressure is called "Prandtl–Meyer expansion." If the jet flows into a region of higher pressure, the edge acts as a source of compression rather than expansion waves, which add up to an unsteady shock front at an oblique angle pointing in the opposite direction. The theory of these "oblique shocks" could be derived from an earlier theory developed by Bernhard Riemann concerning one-dimensional shock waves [411].

By 1908, therefore, it had become clear that the flow through a converging nozzle was choked off because of the formation of shock waves, unless a diverging passage enabled the expansion of the jet to lower pressures at higher velocity. When the pioneer of steam turbines, Aurel Stodola (1859–1942), published a new edition of his classic textbook on steam and gas turbines in 1910, Prandtl was the obvious candidate for updating the theoreti-

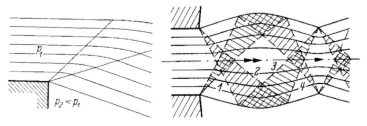

**Fig. 7.3** Expansion waves generated at a corner for a supersonic flow from a high pressure into a low pressure region [411].

cal parts on the flow of jets through Laval nozzles [412, pp. 71, 79, 90, 585]. Prandtl also reviewed the contemporary knowledge on gas motion in 1913 for the *Handwörterbuch der Naturwissenschaften* [413]. It was obvious to Prandtl and his Göttingen colleagues that research in gas dynamics offered a broad range of applications. During the last months of the First World War, Prandtl made plans for the construction of a supersonic wind tunnel at the MVA for ballistic experiments (see Chapter 3). A few years later, when the establishment of the new Kaiser Wilhelm Institute offered the prospect of the required funds, he wrote to Stodola that time had come to proceed with gas dynamics "because after the completion of the present construction work for new laboratory equipment, we will again turn towards this area" [414]. Preceding this correspondence was Stodola's request to Prandtl to "close some tangible gaps" for a new edition of his textbook on steam and gas turbines, particularly in regard to the calculation of the shape of nozzles and turbine blades. "To you, the old master of fluid dynamics we 'practitioners' have to turn," Stodola flattered Prandtl, "in order to receive the appropriate instructions about such subtleties" [415].

Stodola's request prompted Prandtl to update the theoretical knowledge obtained before the First World War with the first gas dynamic dissertations in his institute. Although these results provided a basic explanation of two-dimensional supersonic flow phenomena, they did not account for the specific problems of practical nozzle design. One of Stodola's requests, for example, dealt with the problem of how to design a nozzle so that the emanating jet is parallel. Prandtl responded that it should be possible to calculate and design the contour of such a nozzle with some effort, but he admitted that they had not proceeded yet to this point. "We have approximated it by a circular arc and scraped it according to the Schlieren image of the air jet until we considered the resulting jet satisfactory." So far he had not published more details, but he hoped to be able to calculate the problem numerically by approximating the continuous curvature of the nozzle wall by a polygonal trait of straight lines [416].

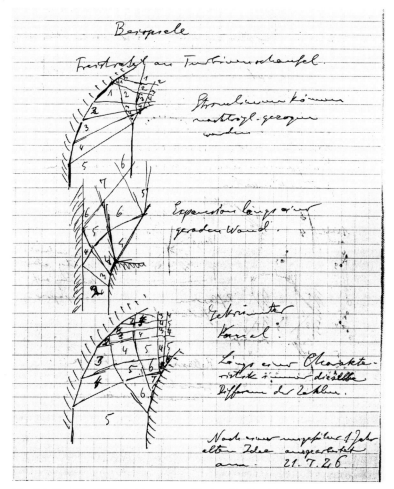

**Fig. 7.4** Prandtl's 1926 sketch of a graphical method to determine two-dimensional supersonic flows; this method became known as the method of characteristics.

Prandtl further developed this idea in his correspondence with Stodola, but he was busy at the same time with other problems (like turbulence), and he did not publish his approach for a polygonal nozzle design. He also did not update his contribution to Stodola's textbook; the 1924 edition appeared as a reprint of the fifth edition from 1910. However, the request of the practitioner of steam turbines left its mark: in 1926, Prandtl drafted a "graphical method" on a piece of paper, which may be regarded as a precursor for what became famous a few years later as the "method of characteristics" (see Fig. 7.4). He probably was already aware how important this method would become because he added to the draft the sentence that he had it developed "on 21 July

1926 according to an idea approximately one year old" [417]. As sketched in this draft, Prandtl's method accounted for a two-dimensional supersonic flow in a curved channel with polygonal boundaries by a sequence of characteristics emerging from the corners of the polygons.

Two years later, Prandtl asked his student Adolf Busemann (1901–1986) to elaborate on the details of this procedure for publication in a Festschrift at the occasion of Stodola's 70th birthday [418]. Busemann had come to Göttingen in 1925 as an advanced student with an engineering diploma from the Technical University Braunschweig. He stayed until 1931 at the new Kaiser Wilhelm Institute and focused on research in gas dynamics. Busemann also presented the method of characteristics at the annual meeting of the Gesellschaft für Angewandte Mathematik und Mechanik (GAMM) in 1928 [419] and subsequently extended it to conical flow, so that it could be used in ballistics for determining the shock waves around projectiles [420]. Prandtl also entrusted Busemann with an authoritative review article on gas dynamics for the *Handbuch der Experimentalphysik*, where he reminded the reader that gas dynamics had its roots in steam turbine engineering and recommended Stodola's treatise on this technology "directly as a textbook on gas dynamics" [421, p. 456]. In 1933, as professor at the Technical University Dresden, Busemann updated Prandtl's contribution on "Fluid and gas motion" for the second edition of the *Handwörterbuch der Naturwissenschaften* [422].

Gas dynamics was also the main research topic of Jakob Ackeret (1898–1981), who had come to Göttingen in 1921 in order to expand his knowledge of fluid dynamics. Ackeret had studied mechanical engineering in Zurich and worked as Stodola's assistant, who then recommended him to Prandtl because "he most sincerely wishes to attend your lectures in Göttingen, and, if he succeeds to win your confidence, to do research under your guidance" [423]. Ackeret became Prandtl's close collaborator during the early and mid-1920s and worked on a broad range of subjects, from glider flight to profile measurements, until he chose gas dynamics as a preferred research field. At the end of his Göttingen sojourn, Prandtl entrusted him with an authoritative review article on gas dynamics for the renowned *Handbuch der Physik* [424]. In 1927 Ackeret returned to Zurich and became chief engineer at the Escher Wyss AG, a renowned turbine manufacturer. In addition to his industrial employment, he became a lecturer at the Zurich Technical University, the Eidgenössische Technische Hochschule (ETH). In 1934 the ETH appointed Ackeret director of a new institute for aerodynamics [425].

Ackeret's and Busemann's careers, like few others, illustrate both the past and the future of gas dynamics, from steam turbine engineering to high-speed aerodynamics. Ackeret's handbook article contains an entire section on "Flow forces on moving bodies at very high velocities" in which an approximate formula is presented for the lift of a thin wing at velocities approaching the speed of sound.[1] This formula relates the result for compressible flow to the equivalent result for incompressible flow; it showed how to extend the range of the theory beyond the speed of about one third of the speed of sound – where the results of incompressible aerodynamics lose their validity – to higher subsonic velocities. Ackeret explained in a footnote that Prandtl had first presented this formula in a seminar in 1922 [424, p. 340]. When Herman Glauert saw Ackeret's presentation he wrote to Prandtl: "I have obtained this formula independently and I should be glad to learn whether your proof has been published anywhere" [426]. Prandtl responded that he had originally presented it to his students in response to a seminar by Ackeret, and Ackeret's presentation in the handbook article may now be regarded as its official publication [427]. Glauert's publication of the same result appeared in 1928 [428]. Since then this formula, which expresses a similarity law relating incompressible flow over a given profile to compressible flow (at subsonic velocities) over the same profile, is known as the Prandtl–Glauert rule.

Long before supersonic flight became feasible, Ackeret also attempted to extend airfoil theory to supersonic velocities [429]. What happened in supersonic flow around a corner in a small region of space, Ackeret argued, would have to be stretched out over the depth of the wing. According to this reasoning he regarded the surface of a wing "as an 'unwound' enlarged Prandtl corner, so to speak" so that he could apply the results of the Prandtl–Meyer expansion to the region around a wing section. He calculated the change of pressure above and underneath the wing and found that in the approximation of thin wings for supersonic velocities, the lift is almost independent of the wing's profile. For the drag he found a new contribution, a "wave drag." This contribution to the overall resistance originates from the compressibility of the flowing medium only; it is different from the induced drag (which was absent in Ackeret's theory because it was two-dimensional) and the viscous friction (because he used ideal fluid theory). As a result, he found that the lift and drag coefficients strongly depended on the velocity and that they decreased with increasing supersonic speed.

In 1928 Ackeret extended this theory so that it encompassed the transition from subsonic to supersonic velocities [430]. He obtained a kind of resonance curve for the dependence of the resistance on the velocity, with a sharp increase as the velocity approached the speed of sound, and a decrease again as the speed was increased beyond the sound velocity. This behavior agreed with observations made in ballistics with projectiles. The fastest airplanes at this time reached only about half of the velocity of sound, but the tips of the propeller blades came close to this limit, such that aviation technology "al-

---

**1)** $A_k = \frac{A_{\text{ink}}}{\sqrt{1-U^2/a^2}}$, where $U$ is the velocity of flow, $a$ is the velocity of sound; $A_k$ is the lift of the wing assuming compressible flow, and $A_{\text{ink}}$ is the lift of the wing assuming incompressible flow.

ready made some experiences which are akin to those from ballistics;" with this remark, Ackeret introduced his introductory lecture at the ETH Zurich in 1929 on the "air resistance at very high speeds" [431]. At this occasion he also introduced the ratio of the speed of flight to the speed of sound as the "Machzahl" (Mach number)—a designation that became as significant for high speed aerodynamics as the Reynolds number for hydraulics and hydrodynamics two decades ago [50, 432].

Busemann, too, became more and more engaged with gas dynamics applied to high-speed flight. In 1928 he communicated the first results on profile measurements at velocities close to the speed of sound "with regard to propellers," as he added in order to explain the practical importance of these measurements [433]. The experimental device used for these measurements (see Fig. 7.5) may be considered a prototype of supersonic wind tunnels. It consisted of an evacuated vessel into which air was sucked in at high speed during a short interval of time (about 10 seconds) and guided through a test chamber. The velocity of the inrushing air was controlled by a valve. The test chamber where the objects were exposed to the airstream had transparent walls in order to enable Schlieren images. By adding a Laval nozzle with adjustable walls, the speed of the airstream in the test chamber could be extended to supersonic velocities. By 1933, profile measurements were made up to a Mach number of 1.47 [434].

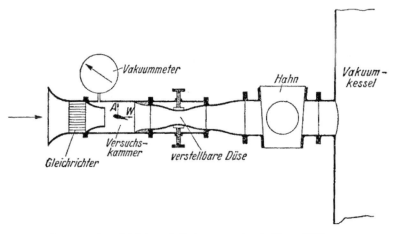

**Fig. 7.5** Cross-section of Göttingen's first supersonic wind tunnel [435, p. 282]. Air is sucked into the tube by an evacuated container (Vakuumkessel) from left to right; the air is guided through a rectifier (Gleichrichter) into the test chamber (Versuchskammer). Air speed is regulated by an adjustable nozzle (verstellbare Düse) and a valve (Hahn).

Busemann also presented the results of his profile measurements abroad at the Third International Congress for Applied Mechanics in 1930 in Stockholm [435]. Two decades after Prandtl had made gas dynamics a research topic at his institute, this area was now no longer relevant to steam turbines alone. With Busemann and Ackeret's attempts to extend airfoil theory so that it accounted for compressibility, and the new experimental techniques employed in Prandtl's KWI to obtain profiles for sub- and supersonic velocities, research in high-speed aerodynamics entered the stage as a new focus of theoretical and experimental fluid dynamics.

## 7.2
## Cavitation

The third item on Prandtl's agenda, the study of flows close to the vapor point, also became the subject matter of considerable effort. When the pressure in a flow of water locally drops close to the vacuum limit a change from the liquid to the gas phase happens, and steam bubbles are formed. At the opening of his new KWI in 1925 Prandtl explained why these bubbles were a matter of concern: "With rising pressure they collapse with much noise and cause damage on immersed solid bodies in their vicinity. This process is called *cavitation* (formation of void space)" [436].

In principle, cavitation was known as an engineering problem since the eighteenth century. Leonhard Euler's theory of pipe flow, for example, paid particular attention to sites in a conduit where the pressure becomes zero or negative. At such low pressures the fluid loses its coherence. Ackeret, who edited Euler's hydraulic papers, commented that "Euler's very explicit remarks" remained unnoticed for about 150 years, and the engineers in the early twentieth century "had to become painfully aware that cavitation is a dreadful factor for hydraulic machinery" [437, p. VII]. In eighteenth century hydraulics, cavitation was caused, for example, by the piston of a pump, which sucked in water so fast that the continuous stream of water broke in the conduit underneath the piston into a mixture of fluid and gaseous parts. The same process happens along the blades of ship propellers or the vanes of turbines when the rotation is so fast that the fluid cannot follow; i.e., it is accelerated to velocities where the pressure is lowered under the vapor pressure and bubbles form. (The process may be compared to the boiling of water: we are accustomed to bubbles forming in boiling water at a pressure of 1 atmosphere at 100 degrees°C; but when the ambient pressure is lowered, the water starts boiling at lower temperatures. Cavitation, so to speak, is a local "boiling" of water at spots where the pressure drops to zero or negative values). "It becomes an interesting question as to how fast the water can follow up the blades of a screw," a participant in a discussion at the Institution of Naval Architects commented at the end of a lecture on ship propellers in 1888 [438, p. 96]. The designation "cavitation" was coined a few years later, when tests of a new English torpedo boat showed that faster rotating propellers could not increase the speed of the boat as much as it was expected. Similar unexplained power losses were observed with water turbines.

But the more serious problem was the erosion of propeller blades and turbine vanes. It was presumed early in the twentieth century that the destructive processes were related to the cavitating bubbles, but the detailed mechanism remained a matter of speculation. In 1907, for example, it was argued in the case of an eroded water turbine that due to large pressure differences between both sides of a vane, the flow detached from the solid surface of the vane and the void space was filled with vortical water from which air emanated. "The released oxygen (which is absorbed in water to a greater amount than nitrogen) acts as an oxidizing agent against the vane, vortex threads drill out the rust and expose new spots of the material. After a relatively short period of time – in some cases already after a few months – the iron looks like a body eaten by worms" (see Fig. 7.6) [439].

**Fig. 7.6** Cavitation makes a part of a turbine look like eaten by worms [439, p. 277].

Another concern was noise. Rapidly rotating ship propellers not only fell short of expectations with regard to the speed of the ship but also produced a sound that could be unnerving for the ship's crew and passengers. For submarines, this noise was a matter of particular concern because it offered an enemy the opportunity to locate and destroy them. For this reason the German Navy charged the Göttingen MVA in the First World War with a study of the noise from submarine propellers [440]. Prandtl identified cavitation as the cause of this noise and asked his collaborator Betz to study various forms of propellers so that cavitation would be avoided or delayed to higher rotational speeds. As a consequence, Betz applied the recent knowledge of airfoil theory to ship propellers, because in both cases, the shape of the profile was crucial for the pressure in the surrounding fluid. The problem of cavitation therefore motivated a knowledge transfer from aerodynamics to ship-building far beyond the initial request to study the causes of noise from submarine propellers.

Representatives of mechanical engineering and ship-building also emphasized the pertinence of recent progress in aerodynamics. In 1924, Hermann Föttinger, a professor of ship-building at the Berlin Technical University, pointed out how important it was to know the relationship between profile and pressure if damage due to cavitation should be avoided. Like a wing, there were rapid pressure changes around the vane of a turbine, and like the former, the latter was dependent on the curvature and could be analyzed with the same theories as those developed by the Göttingen aerodynamicists for the flow of air around wings. "It would be worthwhile from an economic perspective, therefore, to deal more with these very simple and clear theories rather than to face the loss of hundreds of thousands [Reichsmark] involved with the change of large machine entities" [441]. Similar appeals were voiced by Dieter Thoma, a professor of hydraulic engineering at the Technical University Munich, who, in 1924, expressed the optimistic opinion that cavitation in water turbines had lost its former touch of mystery: "the very manifold and apparently divergent experiences have been put in order," while the field was "still veiled in deep darkness a few years ago" [442]. Abroad, too, cavitation was put on the agenda of research institutes, as the magazine *Die Wasserkraft* reported in 1925 [443]. At the same time, Prandtl invited the directors from hydraulic institutes at technical universities and industrial firms to his new Kaiser Wilhelm Institute in Göttingen for "a critical survey on the pending problems of this science." The meeting was sponsored by the *Verein Deutscher Ingenieure* and focused on practical problems [444].

More than half of the presentations at this "Göttingen hydraulics meeting," as it was called, dealt with cavitation. Seldom before was the broad range of phenomena related to cavitation exposed in such detail: "The sound of cavitation from turbine propellers is somewhat similar to the simultaneous beats of hundreds of bad piston pumps," Hermann Föttinger, for example, described the extent of the noise problem. In a steam ship propelled by rapidly rotating turbines, this noise was so unbearable "that the stewards were driven from their nearby rooms. Noise of this sort is often observed with larger water turbines, centrifugal pumps, dam valves, etc." Föttinger also reported on experiments in which cavitation eroded 15-mm-thick glass plates. This damage could not be explained, like the rusting of metal, as a chemical erosion. "A heretofore unknown and therefore unaccounted phenomenon of far greater strength must be taken into account – the mechanical effect of cavitation" [445].

During the discussion of these phenomena, it became clear that an understanding of cavitation called for new research efforts, and that Prandtl and his collaborators had already begun to jump on the bandwagon. Ackeret reported in response to Föttinger's paper on experiments in progress about the flow of water through Laval nozzles, where foam bands emerged at the narrowest part of the nozzle and extended along the nozzle walls into the diverging part. The formation of bubbles, Ackeret believed, was caused by the small unevenness of the wall; if this was the case, it should be possible to evoke this process by artificial obstacles like thin wires so that the site of bubble formation can be precisely localized. A similar line of reasoning had resulted in the discovery of the turbulent boundary layer 12 years earlier with the trip-wire experiment (Chapter 2), and scratches inside Laval nozzles were used almost 20 years earlier in order to generate Mach waves as proof of the supersonic speed of the jet in such nozzles. Based on his prior experience with gas dynamics, Ackeret must have noticed immediately that some phenomena involved with cavitation were akin to supersonic phenomena: The water pressure in the divergent part of the nozzle, for example, was similar to the steam pressure of a jet in the diverging part of a Laval nozzle. "Obviously, we have to deal with a compression shock," Ackeret concluded. "The band of foam must be regarded as a highly compressible fluid." Based on this reasoning, he argued that the collapse of the bubbles causes shock waves, which can give rise to pressures of about 1,000 atmospheres in narrowly localized regions. This confirmed Föttinger's conjecture that the damage due to cavitation had a mechanical rather than a chemical cause [446].

Research on cavitation, as pursued in Prandtl's institute, did not require entirely new methods and devices. In dire economic times Prandtl made it a virtue "to achieve as much as possible with limited funds" [436, p. 1543]. Both the experiments on gas dynamics and on cavitation made use of large tanks that were either evacuated for gas dynamic experiments or used as "windkessel" for investigations on cavitation (i.e., partly filled with water). In the latter case, the water was circulated by a pump through conduits with a test chamber. By lowering the air pressure over the water in the tank, the flow speed could be varied so that the onset of cavitation in the test chamber could be controlled.

As far as the study of cavitation damage was concerned, new experimental techniques were called for. An early attempt to describe the bubble collapse was made by Lord Rayleigh in 1917 with a theoretical investigation of the behavior of a spherical cavity in a fluid [448]. Unlike Rayleigh, Ackeret did not assume a given pressure drop at the wall of the bubble but assumed a shock wave as the cause of a rapid change of external pressure, which then results in the bubble collapse. But it was left to prove experimentally whether the damage really happened at the site of the bubble collapse. In contrast to the

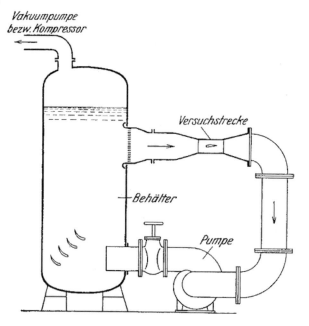

**Fig. 7.7** Cross-section of Göttingen's cavitation tunnel [447, p. 472].

stationary shock waves observed in the flow of gas through Laval nozzles, the collapse of bubbles in the test chamber of a cavitation tunnel was an extremely ephemeral phenomenon, happening within a few millionths of a second. In order to photograph the bubble collapse Ackeret experimented with a high-voltage device, which was able to deliver a series of sparks at a frequency of 100 Hz. Exposures were recorded on a film attached to the rim of a bicycle wheel, which was rotated by a motor. The developed pictures showed the collapse of bubbles like a slow-motion movie [449].

By the early 1930s, the technique was so refined that the collapse of bubbles could be pictured in unprecedented detail. The images also confirmed Ackeret's assumption (based upon his shock-wave theory) that the collapse happens within a narrow zone of compression [450]. However, an individual bubble did not collapse in a concentric manner as Rayleigh's and Ackeret's theories had assumed; the collapse happened mainly in a longitudinal direction such that "in a compression shock the rear wall bumps into the front wall," as Ackeret described it [451]. Such shocks, however, could only account for pressures of several hundred atmospheres, not thousands as in a concentric collapse. With this result, it was again an open question whether the damage due to cavitation was a mechanical effect only.

Beyond the physics of bubble collapse, which eluded the usual methods of hydraulic engineering, cavitation became the subject matter of routine testing in the 1930s. Like the wind tunnels measurements which had begun a decade earlier, by 1930, systematic measurements of profiles for ship propellers and turbine vanes were pursued in cavitation tunnels in the Göttingen (see Fig. 7.7) and in hydraulic research establishments elsewhere. The results of such measurements were presented in cavitation diagrams similar to the polars of wing profiles: lift and drag coefficients were displayed as a function of the angle of attack, with additional information of how far the zones of cavitation extended on both sides of a profile [452]. Like the wind tunnel testing of profiles, which was usually performed under contracts from airplane manufacturers, cavitation was a matter of utmost practical concern for ship-building and hydraulic engineering. Göttingen research on cavitation, for example, found immediate application in the design of ship propellers for high-speed boats and destroyers for the navy, Betz proudly reported in a review [403].

When Ackeret reviewed the state of the art in 1931 for the *Handbuch der Experimentalphysik,* he noticed that despite the obvious practical importance of cavitation, there was a lack of scientific investigation into the nature of its damaging effects. When dealt with more from a physicist's than an engineer's perspective, the treatment became highly complex because it involved a number of different processes, such as heat conduction and capillarity. "Fortunately," wrote Ackeret, from an engineer's point of view, it is only important to know how to avoid cavitation. An engineer would not need to know the details of fully developed cavitation but "simple rules" concerning its onset [447, pp. 463–464]. Thus, after only few years of cavitation research, the dual nature of this specialty as a topic for practical engineering versus physics was made explicit. Decades later, despite tremendous research activity, little has changed in this regard: engineering monographs (e.g., [454]) and physical treatises (e.g., [455]) on cavitation are divided into different camps with little overlap.

## 7.3
### Meteorological and Geophysical Fluid Dynamics

In his inaugural speech at the opening of the new Kaiser Wilhelm Institute in 1925, Prandtl mentioned meteorology as another topic "with which we are already dealing theoretically" and which he intended to explore in more detail both theoretically and experimentally. In particular, he planned to study the impact of the motion of the earth on ocean currents and wind. In order to investigate fluid motion in a rotating frame of reference on a laboratory scale, he planned the installation of a "rotating room," where all experiments on fluid motion could be performed from the vantage point of a rotating observer [278, p. 1543].

Prandtl's plan to study meteorological and geophysical fluid dynamics took shape in 1922. His interest was kindled by Vilhelm Bjerknes, who was at that time revolutionizing meteorology by the so-called "polar front theory," a catchword for new concepts according to which weather was explained in terms of cold and warm fronts produced by cyclones at the boundary of polar air masses and warmer air from more moderate latitudes (see Fig. 7.8) [62]. According to Bjerknes, the polar front was an extended and unstable surface of discontinuity between warm and cold air masses; due to its intrinsic instability, this polar front assumes a wavy contour; cyclones were interpreted as breaking waves of this front – the steeper the waves, the more variable and stormy the weather: "Along the polar front we have the most pronounced opposites of weather, the strongest winds, the sudden jumps of wind, the rapid temperature changes, along this line the formation of fog, clouds, and precipitation is happening" [457].

Prandtl was particularly interested in having a discussion with Bjerknes "about the vortical and wavy motions and about the circulation of the atmosphere" [456]. In unpublished calculations he checked Bjerknes's theory and arrived at different conclusions about the origin of cyclones. He found that the boundary between warm and cold air masses along the polar front was rather stable. If he chose the parameters so that adjacent layers of air became unstable, the instability was limited to small regions only, "giving rise to April weather but not to growing cyclones" [458, p. 38]. In other words, the system of cyclones could not be explained by Bjerknes's theory; Prandtl assumed that cyclone formation did not result from an intrinsic instability of the polar front but rather from vortical motion in the warmer air layer, which deformed the boundary when it came into contact with the polar air masses. Prandtl presented his objection in 1922 at a meeting of the German Physical Society which published only an abstract of the talk [459], but it immediately stirred the interest of meteorologists. The editor of the *Meteorologische Zeitschrift* asked Prandtl whether he would consider elaborating on the theory in greater detail [460]. The Meteorological Office in London sent Prandtl a request for a reprint. His response, however, shows that he was not certain whether his conclusions were justified. Further development of this conjecture would "possibly" confirm Bjerknes's theory, Prandtl wrote to London [461]. Under headings such as "Flows on the rotating earth," "Atmospheric circulation," or "Two fluids superposed on rotating disk," he filled numerous pages of manuscript with detailed calculations, but did not arrive at a final conclusion on the question of cyclone formation in Bjerknes's polar front theory [458].

At first sight, it seems astonishing that the short write-up on Prandtl's talk presented to an audience of physicists would arouse such interest among meteorologists. Bjerknes's polar front theory, however, signaled the advent of a

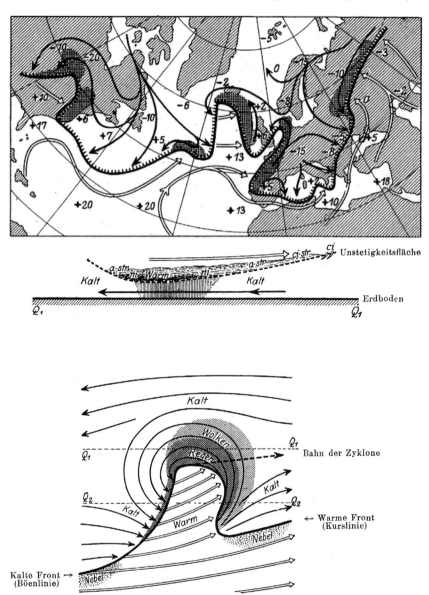

**Fig. 7.8** Bjerknes's polar front theory [457, p. 496].

new era of meteorology. While meteorology had previously been regarded more as an empirical science with little relation to fluid dynamics, Bjerknes's theory of cyclone formation made it a science based on the physics of vortices and other flow phenomena. From that point forward, fluid dynamics was regarded as a crucial element of meteorology.

Another convergence of meteorology and aerodynamics was brought about by the rise of aviation during and after the First World War. Bjerknes's own disciplinary move from hydrodynamics to meteorological fluid dynamics was to a large extent motivated by the aeronautical interest in meteorological forecasting. In the 1920s, particularly in Germany, the glider obsession further added to the mutual convergence of meteorology and aerodynamics. Even more than motorized flight, successful glider flight was dependent on meteorological knowledge. The quest for "dynamic soaring" (see Chapter 6), for example, naturally called for an understanding of atmospheric flow phenomena: "Under certain local conditions, the rising current of air becomes detached from a mountain and keeps rising," Prandtl argued at the first Rhön glider contest, "until it is brought to a halt by meteorological causes. It does not seem unlikely that, after one has attained some height over the summit, it is possible to fly over larger distances in this rising air current." Glider pilots, meteorologists, and fluid dynamicists like Prandtl joined in the challenge to achieve that goal. First of all, however, it called for better knowledge of atmospheric flow phenomena. "From the edges of mountains and rock protrusions, very unpleasant vortical eddies can emerge and extend in a tube-like shape in the direction of the wind," Prandtl wrote after the Rhön contest. "Martens reported about a maelstrom which turned him in the opposite direction to which he was steering with fully flung out rudder. These things require thorough study, and it is also necessary that the empirical material on these phenomena is carefully collected and processed; when the knowledge of the good lifting wind zones is sufficiently advanced, long-distance flights without motors in mountainous regions, from mountain to mountain, seem to become possible indeed" [354]. Harald Koschmieder, a meteorologist at the Geophysical Institute Frankfurt, was charged with collecting the empirical data of the Rhön glider contests. Koschmieder assessed and published this material in detailed reports in the *Zeitschrift für Flugtechnik und Motorluftschiffahrt* [462]. He also corresponded with Prandtl on data-gathering at the Rhön contests and various aspects of meteorological fluid dynamics [463].

Other motivations for the study of meteorological and geophysical applications of fluid dynamics were presented at the Innsbruck conference in 1922 (see Chapter 4), where a pupil under Bjerknes, Vagn Walfrid Ekman, reviewed what was known on ocean currents. In particular, Ekman dealt with the flow caused by wind at the surface of the sea: the rotation of the earth deflects the flow direction with increasing depth from the wind direction in a characteristic manner ("Ekman spiral"); at a certain depth ("Ekman depth"), it is antiparallel to the wind direction (see Fig. 7.9). A striking property of this flow is that at the surface, its direction already deviates by 45 degrees from the direction of the wind [464]. At the Innsbruck meeting, Ekman reviewed experiences from more than two decades of geophysical hydrodynamics in Norway and elsewhere. His first encounter with the problems of ocean currents dated back to Fridtjof Nansen's legendary polar expedition with the Fram from 1893–1896, when Nansen noticed that the floes always drifted by his boat at an angle to the right of the direction of the wind. "When Nansen visited us in Stockholm in 1900, we discussed the drift currents observed during his voyage," Bjerknes later recalled how he chose this theme as a subject matter for Ekman's doctoral work. "He was called in and confronted with Nansen's discussion of the problem; the same evening he had already found the distribution of the current according to the Ekman spiral" [465].

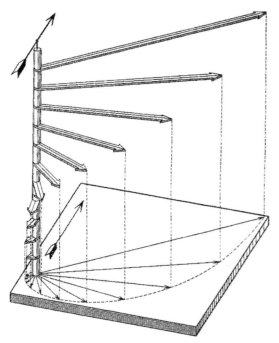

**Fig. 7.9** Ekman spiral for drift flow; the arrows indicate the flow vectors with increasing depth [464, p. 101].

In principle, Ekman's reasoning concerning the currents in the sea should also be applicable to air motion over the rotating earth, Prandtl thought after the Innsbruck conference. He wrote to Ekman that it should be possible to arrive at a theory along these lines about "the general circulation of the atmosphere" and that he intends to explore the behavior of rotating fluids to this end "both theoretically and by experiment." However, in case Ekman had also envisioned extending his theory to atmospheric circulation, he was eager to avoid sentiments of competition [466]. But Ekman reassured Prandtl that he was "extraordinarily delighted that you will tackle the problem of flows in rotating systems" [467].

Prandtl's interest in atmospheric circulation met with another plan: Ekman's theory was based on simplifying assumptions; for example, he disregarded turbulence because, as he argued in Innsbruck, little progress was achieved in this matter. In Göttingen, however, turbulence ranged quite high on the agenda of research topics (see Chapter 5), and Prandtl hoped he could advance Ekman's theory by adding new insights from turbulence research. "The recently improved knowledge concerning the flow through pipes with a rough wall make it possible, by means of certain analogies, to arrive at details about the motion of wind over the rough surface of the earth," Prandtl and his doctoral student, Walter Tollmien, argued in an article published in 1925 in *Zeitschrift für Geophysik*, the first public record of the new orientation of Göttingen fluid dynamics towards meteorological and geophysical problems [468]. Based on the 1/7 wall law derived from the mixing length concept, Prandtl and Tollmien derived formulae for how wind speed and direction varied with increasing height over the surface of the earth. Furthermore, they were able to account for the frictional force exerted by the wind at the surface, which was required, for example, for a quantitative calculation of the drift of floes. High above the surface, the analogy with pipe friction was no longer applicable. The "wall law" for turbulent friction had to be refined a few years later (Chapter 5), but that did not change the fact that Prandtl and Tollmien had achieved a breakthrough for a meteorological problem by analogy with a hydraulic problem. The lower atmospheric boundary layer with a thickness of about 100 m, where this analogy holds, became designated as "Prandtl layer" in meteorology [469].

Prandtl and his collaborators, however, did not follow up on this triumph for several years with more publications on meteorological fluid dynamics, although Prandtl's manuscripts reveal a persistent interest in this matter. Presumably he hoped to gain new insight from experiments with "fluids on the rotating disk," on which he made extensive calculations [458, pp. 58–61]. For example, Prandtl prepared an experiment with a pond filled with water which was to be kept at a low temperature in the center and heated at the perimeter. When rotated, it should allow one to study "what kind of current results, how large the exchange of heat is, etc." The same rotating pond could also be used "to imitate a cyclone," Prandtl explained in a letter to a meteorologist [471]. But the operation of the rotating laboratory caused unforeseen problems. Experimenters often suffered from a loss of balance as in seasickness [470]. What was a nuisance for the progress of experimental research, however, was a pleasure for Prandtl's daughter, who used the new facility together with friends from school as a merry-go-round [228, p. 95]. The first experiments from the "rotating laboratory – henceforth carousel – for short" were reported in 1933; they dealt with the flow in rotating channels, a problem of interest for hydraulic engineering rather than meteorology or geophysics [472].

The idea to simulate atmospheric circulation in the laboratory by experiments with rotating fluids was not new. The Viennese meteorologist Felix M. Exner had performed such experiments and described them in a meteorological textbook, as Wilhelm Schmidt, another Viennese meteorologist, informed Prandtl in 1926; Schmidt also alerted Prandtl to how precise such experiments had to be in order to justify extrapolations from the laboratory scale to real atmospheric and geophysical phenomena [474]. Perhaps Prandtl delayed his own experiments for this reason, but the correspondence with the Viennese meteorologist provided him also with valuable suggestions in another regard: Schmidt had just published a monograph on mass exchange as a basic process of atmospheric phenomena [473]; Prandtl pursued this line of reasoning "with vivid interest," as he wrote to Schmidt, because it also appeared pertinent to him "for turbulent friction and heat exchange" [471]. In his mixing length theory, he made the concept of exchange central to the description of turbulence (Chapter 5). For the time being, until the rotating laboratory became operational, Prandtl limited his meteorological studies to the exploration of adjacent fluids at different temperatures and densities without rotation [475]. In 1929, he published a paper on the stabilizing forces between adjacent layers of air at different temperatures: Turbulence may be prevented when cold air is covered over with warm air, a phenomenon that may be observed in the evening when heat radiates away from the ground and the air at some distance above the ground is warmer than the air underneath; "a smoke trail from a potato fire," Prandtl explained, "moves away horizontally without any formation of cloudlike shapes," because a laminar flow of warm air is sliding over a layer of cold air underneath without a noticeable exchange between the adjacent layers. In the opposite case, when cold air is on top of a layer of warm air, a rapid exchange takes place. A similar phenomenon happens with ocean currents, for example when freshwater covers salt water, as may be observed in the polar sea [476].

A year later, Prandtl was ready to explore this phenomenon experimentally: "Air will be blown between two 8-m-long and 1-m-wide iron plates, each with a double wall so that one can be heated with steam and the other cooled with water," he explained the planned device. "If the upper plate is heated and the lower plate is cooled, one obtains a current of air layered in a similar stable manner as the atmospheric wind, and therefore turbulence must be reduced. If the lower plate is heated and the upper one cooled, the striation will be unstable in contrast to the preceding case, and an amplification of turbulence must take place. The experimental installation is accomplished, but we cannot yet report definitive results" [477]. A similar investigation of the stability of superposed fluids at different densities was chosen as a research topic for Sydney Goldstein, who by this time spent a one-year sojourn as a Rockefeller fellow in Prandtl's institute [478]. "Remarkably soon after this,

Goldstein was to become recognized as one among the enterprising new international group, drawing its prime inspiration from the pioneering work of Prandtl," a colleague of Goldstein later remarked [479]. Prandtl also corresponded on this topic with Geoffrey I. Taylor, Great Britain's leading fluid dynamicist, and suggested that he should pay particular attention to the phenomenon of stratified flows in a salt lake when a river flows in: "Since you like to undertake extended voyages, wouldn't that perhaps be a suitable problem for you?" [480].

In 1932, Prandtl published some remarks on meteorological fluid dynamics in a Festschrift dedicated to Vilhelm Bjerknes at the occasion of Bjerknes's 70th birthday. He still had "nothing in store which is really accomplished," Prandtl admitted, but he indicated lines of research he intended to explore in the near future. He suggested the mixing length concept as an appropriate approach for studying turbulent atmospheric flow; for example, with an inversion situation, when superposed layers of air do not become mixed. He related the energy difference resulting from the different temperatures in both layers to the energy of turbulent friction between the layers and derived formulae for the velocity profile of the wind with increasing height over ground. Prandtl also addressed the problem of the earth's rotation and sketched the various mechanisms that had to be taken into account for a theory of the "general circulation in the earth's atmosphere," for which he hoped to present results "in a not too distant time" [297].

## 7.4
### The Scope of Fluid Dynamics by the Early 1930s

Although meteorological applications of fluid dynamics were not yet as numerous as traditional applications in hydraulics or flight-related aerodynamics, Prandtl's contribution to Bjerknes's Festschrift hinted at a rise of atmospheric and geophysical applications. The case of meteorology also shows that the number of publications alone is not sufficient for evaluating the importance of a research field. On the other hand, the focus on programs such as that indicated in Prandtl's correspondence and manuscripts on meteorology has to be complemented by quantitative records in order to arrive at a balanced view of the scope of fluid dynamics in Prandtl's KWI by the early 1930s. An appraisal of publications and personnel as annually reported in the *Naturwissenschaften* for each Kaiser Wilhelm Institute reveals for Prandtl's institute a total rate of about 25 publications per year from the institute's inauguration in 1925 to the year 1933; the number of "scientific employees" rose during these years from 20 to 37 in 1931, with a subsequent decline to 27 in 1933 as a consequence of the economic crisis. The total of 230 publications may be roughly sorted into these categories (with due regard to the problems such retrospective classifications bring about):

- aerodynamics (85)

- turbulence, boundary layer, vortex formation (64)

- gas dynamics (15)

- cavitation (12)

- experimental techniques (22)

- other (34)

Among the 85 publications classified under aerodynamics, 14 dealt with problems unrelated to aeronautics: "How to find the best shape of vehicles," for example, a 1922 article in the *Berliner Illustrirte Zeitung*, in which Prandtl emphasized the importance of streamlining to reduce fuel costs. The article was illustrated with images of wind tunnel tests on models of locomotives and the legendary "drop car" developed by the former airplane designer Edmund Rumpler. The addressees of such advertising efforts were car and railway manufacturers from whom the AVA hoped to acquire research contracts during a period of dire economic circumstances. Other non-aeronautical applications of aerodynamics concerned the wind pressure on buildings or the practical use of so-called Flettner rotors – rotating cylinders installed on ships instead of the masts of sails as an alternative means to exploit wind power for ship propulsion. It was long known from ballistics that a rotating cylinder experiences a force perpendicular to its axis and to the direction of flow – the so-called Magnus force. In 1923, Anton Flettner, an inventor and ship designer, had assigned a research contract to the AVA in order to analyze the practical uses of this effect. Flettner rotors were studied both with models in the wind tunnel and with an experimental ship [67, pp. 221–222, 287–290]. What makes these studies noteworthy, however, is less their importance for contemporary technological innovation (both the "drop car" and the "Flettner rotor" remained curiosities) than their wide scope – to secure contracts from public administrations, industrial firms, or other organizations like the *Studiengesellschaft für die Rheinisch-Westfälische Schnellbahn* (for a study on air resistance of trains in tunnels [482]). As a consequence, applications of aerodynamics were widely scattered and found entrance into new areas, like the studies of wind pressure on buildings which became pertinent for architecture and civil engineering [483].

In order to finance such broad research activity, more money was necessary than available from the regular budget of the Kaiser Wilhelm Institute alone. Additional funds came from ministries, industry, and other sources.

Research related to aeronautical applications was funded to a large extent by the Reich's Ministry of Transport (Reichsverkehrsministerium); the German Research Community (Deutsche Forschungsgemeinschaft, or DFG) funded a broad range of fluid dynamics projects; special research on cavitation, for example, were supported by the large Bavarian hydropower company "Bayernwerk"; research on wind pressure on buildings and other architectural structures was funded by the Prussian Ministry of Finance, the German Society of Civil Engineering (Deutsche Gesellschaft für Bauingenieurwesen), and the Association of German Engineers (Verein Deutscher Ingenieure) [484, p. 433], [485, p. 418–419]. Meteorological fluid dynamics was also financed within a new framework of the DFG's research policy with which collaborative research among several institutes was fostered and oriented toward special goals [486].

The broad spectrum of fluid dynamics is also illustrated by the organs of publication in which the Göttingen research results were communicated to the public. Aerodynamic papers related to aviation were traditionally published in the *Zeitschrift für Flugtechnik und Motorluftschiffahrt*, the journal of the German Scientific Aeronautical Society. Articles with a predominant theoretical orientation appeared in the *Zeitschrift für angewandte Mathematik und Mechanik* (ZAMM). If a theme was considered interesting for general consideration, it was published in *Die Naturwissenschaften*; engineering topics were published in the *Zeitschrift des Vereins Deutscher Ingenieure* or the *Ingenieur-Archiv*; individual special subjects were published in special journals like *Messtechnik, Metallwirtschaft, Kinotechnik,* or *Bauwelt*. Besides these technical journals the Göttingen fluid dynamicists also occasionally published in physics journals like the *Annalen der Physik, Physikalische Zeitschrift, Zeitschrift für Physik* and *Zeitschrift für technische Physik*; fellows from abroad addressed their papers to their domestic journals, like the *Proceedings of the Royal Society*. Apart from these journals, conference reports were another channel for communicating most recent research results to a variety of engineering specialties [444, 453]. The list would not be complete without mentioning reviews, textbooks and encyclopedias, like the *Müller-Pouillet's Lehrbuch der Physik* (1929), *Hütte, des Ingenieurs Taschenbuch* (1931) or the multivolume *Handbuch der Experimentalphysik*. Prandtl and his pupils contributed to this *Handbuch* from 1930 to 1932 with 15 authoritative articles; no other single publication better illustrates the leading role of Prandtl and his school in the rise of fluid dynamics and its broad scope.

# 8
# Prandtl, Fluid Dynamics and National Socialism

Hitler's rise to power had immediate and disastrous consequences for German society as a whole and for science and engineering in particular. Universities and state establishments, like the Kaiser Wilhelm Institutes, were purged of employees whose race was defined as "non-Aryan" or who were considered politically hostile to the new regime. In Göttingen the purge was most extreme: Max Born, James Franck, Hermann Weyl, and Richard Courant, for example, emigrated – but they were only the most prominent scientists driven out of the country; a list of emigrated mathematicians contains 23 names expelled from Göttingen [487, pp. 292–298]. After the purge, Göttingen's most famous mathematician David Hilbert was asked during a banquet by the Nazi minister of science: "And how is mathematics in Göttingen now that it has been freed of the Jewish influence?" Hilbert replied: "Mathematics in Göttingen? There is really none anymore" in [488, p. 36]. (A detailed account of the Nazi impact on mathematics and physics in Göttingen is presented in [489] and [490]).

Prandtl could not ignore the clear-cutting of the neighboring institutes of mathematics and theoretical physics. Although his primary institutional affiliation was the directorship of the Kaiser Wilhelm Institute for Fluid Dynamics and the associated AVA, he was also director of the university's institute for applied mechanics and in many ways was connected with university affairs. What was Prandtl's attitude regarding the purge among his colleagues, and how was fluid dynamics in Göttingen, both at the university and the Kaiser Wilhelm Institute, affected by the Nazi regime?

The question of how fluid dynamics fared under the new regime cannot be answered without closer inspection of Prandtl's own political attitude. Such inspection brings to the fore a most ambivalent relationship between science and politics. Prandtl was a well-established and widely respected member of the German scientific community as a whole and the Göttingen academic milieu in particular. When 28 mathematicians and physicists, among them such prominent names as Max Planck and Arnold Sommerfeld, drafted a let-

*The Dawn of Fluid Dynamics: A Discipline between Science and Technology.* Michael Eckert
Copyright © 2006 WILEY-VCH Verlag GmbH & Co. KGaA, Weinheim
ISBN: 3-527-40513-5

ter of protest against the dismissal of Courant, they asked Prandtl to submit this protest to the president of the Göttingen university [491, pp. 144–152]. Prandtl also tried to avert the dismissal of less renowned colleagues. Most of these initiatives were to no avail. Some of them meant that Prandtl himself became involved in political disputes. In 1934, after internal quarrels, Prandtl resigned from the directorship of the institute for applied mechanics. In another incident against his research collaborator Johann Nikuradse at the Kaiser Wilhelm Institute, who was accused to be a spy by a group of National Socialists at his institute, Prandtl first took sides with Nikuradse. The incident became an affair among rivaling National Socialist factions, because Nikuradse himself had close ties to the SS. At the request of the ministry of science, Prandtl had to fire Nikuradse's enemies; later on, he also fired Nikuradse, whom he grew to consider unreliable. As a consequence, an SS captain accused Prandtl that he had fired Nikuradse because of his National Socialist views [492, 493].

Prandtl's daughter portrayed her father as hostile to the new regime: he never became a member of the Nazi party, she recalled, and he refused to put a portrait of Hitler on the wall of his director's office. She reports an incidence when Prandtl's assistant joined the Nazi party in order to promote his career rather than because of an inner conviction – and her father found such an attitude as "nothing short of outrageous." As further evidence for his hostility towards the Nazi regime, she cites letters that Prandtl wrote in favor of Heisenberg, who became the target of fanatics who regarded modern theoretical physics as "Jewish" and prevented Heisenberg's call to Munich as Sommerfeld's successor. Prandtl also participated in a campaign against these fanatics launched by the German Physical Society in 1941 [228, pp. 123–151, 210–214]. As chair of the Society for Applied Mathematics and Mechanics (Gesellschaft für angewandte Mathematik und Mechanik, or GAMM), Prandtl is also credited as a "humanistically minded" scholar, who did not readily execute the regime's doctrines: Prandtl did not reorganize the GAMM according to the "Führerprinzip," and he expressed sympathy for colleagues who had been forced to emigrate [227].

These accounts, however, have to be balanced against other documented evidence. It is not justified to infer Prandtl's general political views from these examples and portray Prandtl's attitude toward the new regime as hostile. In November 1933, Prandtl expressed in a private letter to his brother-in-law his satisfaction that one can now "as a German carry the head higher again" [493, p. 497]. At least in one case – the dismissal of Ludwig Hopf at the Technical University Aachen – Prandtl's response to the issue of whether Hopf's merits in aerodynamics are great enough to make an exception to termination was so lukewarm that the Nazi authorities saw no reason to withdraw Hopf's dismissal [494, p. 131]. As we will see shortly, Prandtl was more than willing to place his capabilities and his institute at the regime's disposal, and the regime appreciated and honored Prandtl for this service. This is not to say that Prandtl's attitude was regarded by all Nazi authorities in the same manner. In 1937, the local bureau of the Nazi party in Göttingen characterized Prandtl as "the type of worldly innocent scientist" with little enthusiasm for "even the simplest political contexts as long as they have no impact whatsoever on his scientific activity." In order to characterize Prandtl's right-wing, but not pro-Nazi tendencies, the local Nazi representative remarked that Prandtl had pleaded in 1930 for a prolongation of Hindenburg's term as the Reich's president, and he described Prandtl as "the type of honorable, sedulous scholar of the old time, who is anxious about his integrity and reputation, whom, however, we cannot and should not spare with regard to his extraordinary valuable scientific accomplishments for the set-up of the air force" [495].

## 8.1
### Preparing for War: Increased Funding for Prandtl's Institute

In contrast to physics and mathematics, which experienced a decline as a result of Nazi politics, fluid dynamics flourished in the Third Reich. Among the first measures of the new regime was the foundation of a "Reichskommissariat" for aeronautics under Hermann Göring; in April 1933 it was officially designated as the Reich's Air Ministry (Reichsluftfahrtministerium, RLM). Prandtl hoped, as he wrote to the chairman of the Kaiser Wilhelm Society, that now the importance of research was more appreciated than before and not hindered by "wrong austerity" and that Germany would emerge from behind and catch up to other countries [496].

In 1933, Prandtl was still suffering from the shortages due to the recent world's economic crisis. At the AVA, research was financed only to a small extent by the Kaiser Wilhelm Society – despite its nominal affiliation with the Kaiser Wilhelm Institute for Fluid Dynamics. In fact, its funding came largely from contracts from industry, where there were considerable cutbacks. Even at the Kaiser Wilhelm Institute, however, which was less dependent on contractual research, research funds were frozen. Already the first signals from the nascent Air Ministry, however, indicated that change was in sight. Adolf Baeumker, who had previously administrated the Reich's Ministry of Traffic research funding, moved to the new Air Ministry and became Göring's advisor for aeronautical research planning. Baeumker was well-known and appreciated among German aeronautical scientists as a man who always had an open ear for their concerns. In the new capacity as the Air Ministry's research administrator, Baeumker immediately addressed a letter to Göttingen: "The organizational changes are in full swing," he wrote in February 1933, and announced a "boom" for future industrial aeronautics. "You cannot know how closely the bigwigs of aviation industry and traffic are collaborating. All the more I feel painfully concerned that in the area of research (...) there is only little connection with the responsible powers." Baeumker was eager to secure Prandtl's cooperation as a first step to achieving closer contact between research and politics [497].

The regime's interest in aerodynamic research soon manifested itself in terms of generous financial support: since 1928 the AVA had applied in vain for funds to build a new wind tunnel; in 1933, after a visit with Erhard Milch, Göring's undersecretary in the Air Ministry, Prandtl was informed that these funds were made available via a new "employment program" of the Nazi government. On 30 May 1933, the Göttingen aerodynamicists celebrated the AVA's 25th birthday in the presence of Milch and were elated that their establishment finally received the interest it deserved [120, p. 200]. Prandtl and his colleagues in Göttingen were not the only beneficiaries. Other aeronautical research institutes, particularly the DVL in Berlin-Adlershof, were also allotted funds to build new facilities. And this was only the beginning. Aeronautical research experienced a boom which was believed to be impossible before 1933. "Göring gave me the order in 1934," Baeumker recalled many years later, "that by 1939 Germany should catch up with the level of the big powers" [498, p. 29]. But Göring's official for aeronautical research planning did not need an order to bring his regime's goals to fruition. After the Versailles Treaty, Baeumker was "under an inner vow," he wrote in a letter in 1941, "to dedicate my future life to the German rearmament" [499, p. 14].

In March 1935 the Nazi regime publicly announced the existence of an Air Force – after years of clandestine rearmament [500]. Afterwards, the Air Ministry's measures for aeronautical research planning, too, were more vigorously advanced. Baeumker was granted the far-reaching authority of deciding on the resources needed in order to reach the goal of Germany's rising to supreme power in aeronautics. He used this authority to create new institutions. In July 1936 Baeumker established the German Academy of Aeronautical Research (Deutsche Akademie für Luftfahrtforschung), with Göring as president, Milch as vice-president, and himself as chancellor. The academy's purpose was announced to the public as "nurturing the pure intellectual relations among the leading scientists and engineers by voluntary common work in the same sense as the activity of members of the old classical academies." By that point, Baeumker had founded another organization under Göring's immediate aegis – the Lilienthal Society for Aeronautical Research (Lilienthal-Gesellschaft für Luftfahrtforschung), with himself, Prandtl, and Carl Bosch as presidents. This triumvir was meant to express the determined effort of politics, science, and industry to collaborate in making Germany a leading nation in aeronautics [501, pp. 179–182 and 237–238].

To what extent Göring's Ministry fostered aeronautical research becomes apparent from an article dedicated to Prandtl, "the old master of German fluid dynamics," by Baeumker at the occasion of Prandtl's 70th birthday. Baeumker displayed the growth of financial support for the German aeronautical research intuitions with a diagram (see Fig. 8.1) [502]:

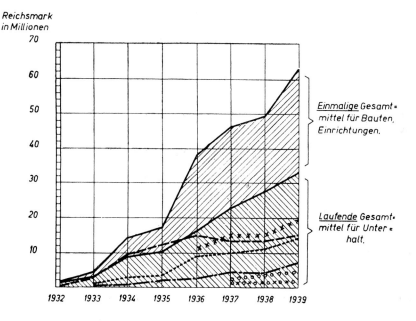

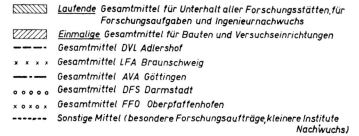

**Fig. 8.1** Expenditures for aeronautical research in Germany during the 1930s.

The published activity reports of Prandtl's Kaiser Wilhelm Institute and the affiliated AVA (which was nominally part of the KWI before 1937, but in practice was directed by Betz as a separate institution) reveal the consequences of the growing esteem of the Air Ministry: The AVA was to "an ever-increasing measure engaged with contractual work for the aircraft industry," it was reported for the period from April 1933 to March 1934; unfortunately "this favorable situation could not be fully exploited because due to the preceding long crisis the manpower and the facilities of the establishment were no longer sufficient for the strongly increased demands." The report of the following year mentioned again an increased demand for aeronautics, and this tenor prevailed in the subsequent years until 1937, when the AVA officially became a separate establishment at the request of the Air Ministry [503].

The primary role model for Germany's expansion in aeronautical research was the USA. As early as in 1925 Baeumker had returned from America "deeply moved" by the pace of technological progress. The occasion for this voyage was a "Mission of Goodwill," which Baeumker had undertaken as an official of the Ministry of Traffic together with officers from the Reich's Army. "Since then I have become a frequent visitor of the American Embassy in Berlin – until 1940, even," he recalled in his autobiography. For his research planning in Göring's Ministry, "the image of the USA with the development of the NACA" had always been before his eyes [498, pp. 10–11 and 29].

In Göttingen, too, the NACA was held in great esteem. As a result of the friendly relations with the NACA's emissaries from the Paris Office, William Knight and John Jay Ide, Prandtl and his collaborators were kept informed of international aeronautical research after the First World War (see Chapter 4). In 1929, Prandtl had used an invitation to Japan as an opportunity to voyage around the world, which he accomplished with an extended lecture trip through the USA [228, pp. 105–115]. In 1933, Prandtl was elected as a member of the Institute of the Aeronautical Sciences, an academy under the presidency of Jerome Hunsaker based in New York City, and as such, Prandtl's ties to America's aeronautical research community were even formalized [504].

The public was kept informed of foreign aeronautics through three new journals, issued by the Reich Air Ministry beginning in 1934 and addressed to specific audiences: The *Luftwelt* addressed those interested in aviation as a sport; the *Luftwehr* reported on the air forces of other countries; and *Luftwissen* informed on international progress in aeronautical research. The overall message was clear: What was allowed to the international aviation community should not be denied to Germany any longer. In regard to aeronautical research, *Luftwissen* presented to its readers as early as in the first issue a series of articles on foreign research institutions, which revealed a clear trend towards ever larger facilities. The record was held by the "Full Scale Tunnel" of the NACA's Langley Laboratory, which became operational in 1931–a giant wind tunnel powered by two 2,800 kW propellers, which created an airstream at a speed of 190 km/h in a test chamber big enough to measure the flow around entire airplanes rather than models. The second largest wind tunnel of the world was built in Chalais-Meudon in France; its test chamber had a cross-section of 8 m × 16 m and enabled tests at a wind speed of 180 km/h. Another new wind tunnel was planned at the NACA in 1934 for testing at high speeds up to 800 km/h. Such reports signaled to the readers of *Luftwissen* that in America and other countries, aerodynamics was entering the age of big science. The NACA's annual reports were translated into German, and comments praised the success of "methodical common research" in the USA and the "determination with which the N.A.C.A limits itself to the work it has been assigned, the fundamental research" [505, pp. 1, 31], [506, pp. 14–16, 41–44, 152–159] and [501, pp. 101–104]).

The trend towards big science went hand in hand with technical changes in airplane design. Systematic drag reduction by streamlining enabled the construction of larger and faster airplanes. In 1929, the NACA's Langley Laboratory received the Collier Trophy, the highest distinction of American aviation, for its drag reduction research, which resulted in streamlined cowling shapes, with the side-effect that airplane engines could be cooled more efficiently [233, Chapter 5]. Aerodynamic improvements like low drag cowling and retractable landing gear, combined with more powerful engines, accounted for flight speeds over 700 km/h, which in turn posed new challenges for aerodynamics. In 1935, an international congress in Rome was dedicated to the theme "The high velocity in aviation" [507]. The trend towards higher velocities also posed great challenges for the design of wind tunnels. The power required to propel the air through a tunnel of test-section diameter $d$ at a speed $v$ grows proportional to $d^2v^3$. The Göttingen wind tunnel from 1917, for example, with a diameter of 2.25 m and a speed of airflow of 210 km/h was driven by a motor with a power of 315 kW; to double the diameter and the speed meant a 32-fold power increase to roughly 10 MW, approximately the power supplied to a whole city [508]!

At the NACA, both the small high-pressure Variable Density Tunnel (VDT) and the Full Scale Tunnel (FST) allowed Reynolds numbers that came close to those of real flight to be reached. When plans were made for a new wind tunnel in Göttingen, a VDT concept at a medium size was given preference because it avoided the enormous power demand of a larger tunnel. Nevertheless, the dimensions of the new Göttingen "high pressure" tunnel (see Fig. 8.2), which became operational in 1936, were giant: The diameter of the fan was 7.5 m; by interchangeable elliptic nozzles the size of the cross-section in the test chamber could be varied (4.7 m × 7 m or 4 m × 5.4 m); when operated under routine conditions, the airspeed was 198 km/h and 237 km/h in the larger and smaller test section, respectively. The pressure could be increased to 4 atmospheres, but the tunnel could also be used for experiments at very low pressures less than 1 atmosphere. The required power of 1,800 kW was considerably less than that of the American and French full-scale tunnels (5,600 kW and 4,400 kW, respectively). Together with these tunnels abroad and a new wind tunnel at the DVL in Berlin with a test section of 5 m × 7 m and a power of 2,000 kW, the new Göttingen tunnel ranked among the world's largest and most powerful wind tunnels [509, 510].

Big science in aerodynamics did not only mean that the spatial dimensions and power requirements for wind tunnels were increased. It also involved specialization and diversification. By the mid 1930s, further experimental facilities at the AVA were built in order to investigate special problems. Non-stationary fluid phenomena, such as those caused by rudder movements, were studied in special water channels. A wind tunnel with an airstream directed vertically upwards was used to investigate the tailspin behavior of airplanes in free fall ("spinning tunnel"). With new tunnels in which air pressure, humidity, and temperature could be varied ("cold wind tunnels") the problem of icing was analyzed. A high-speed tunnel was built for research issued by the Air Force, Navy, and Army; it was intended mainly for ballistics, but also for the first investigations dedicated to airplane design for sonic and supersonic speed (see Chapter 10). By 1938, the AVA had grown into one of the world's largest research facilities with eight specialized institutes for wind tunnels, flight operations, low-temperature research, theoretical aerodynamics, flow machinery, non-stationary processes, high-speed problems, and instrument development. Each institute had its own director; the overall executive also was split into a technical (Albert Betz) and an administrative directorship (Walter Engelbrecht). The total number of personnel grew from fewer than 100 in 1933 to more than 700 in 1940 (see Fig. 8.3) [403].

In 1937, the AVA was formally separated from Prandtl's Kaiser Wilhelm Institute, although Prandtl remained chairman of the AVA's Executive Board. The growth of the AVA, where the bulk of aerodynamic research was concentrated, left no doubt as to what extent Göring's ministry regarded this institute as important for the preparation of war. By comparison, Prandtl's KWI without the AVA appeared small. However, such an impression is biased by the enormous expansion of the AVA. When judged by its own budget, the Prandtl's remaining institute also underwent a phase of growth during the years before the Second World War. The need for a budget increase is "in the general interest of the state," Prandtl explained to the administration of the Kaiser Wilhelm Society in 1935, "particularly in the interest of the defense of the country." Although the Kaiser Wilhelm Society raised the annual budget of Prandtl's KWI only slightly between 1933 to 1939, from about 50,000 to 77,000 Reichsmarks, its total budget tripled as a consequence of contributions from other ministries and industrial contracts from about 80,000 to 235,000 Reichsmarks [511, pp. 12 and 19].

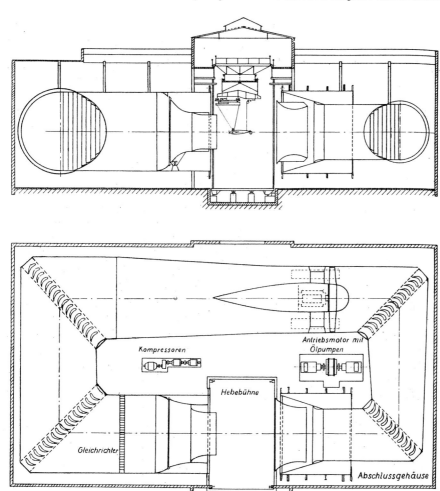

**Fig. 8.2** Vertical section and plan view of the Göttingen high-pressure tunnel; it was housed in a 50 m × 30 m building with a height of 20 m [509].

A closer inspection of the projects funded at the KWI after 1933 shows that despite Prandtl's primary interest in basic research in fluid dynamics, he was not merely a beneficiary of the regime's interest in research with some possible applications to war technologies. Research on the influence of roughness, for example, illustrates that its funding by the Navy and by the Air Ministry was

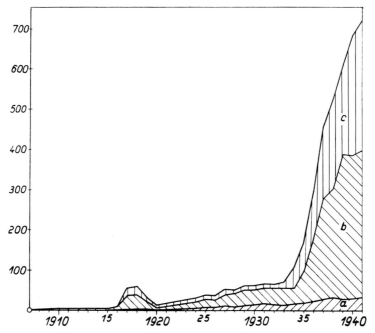

**Fig. 8.3** Growth of personnel at the AVA from its foundation in 1907 until 1941: (a) leading employees, (b) engineers and other employees, (c) skilled workers and other wage-earners [403, p. 166].

more than a generous support of fundamental research. From the perspective of the Navy, it was part of a program to reduce the drag of ship hulls and submarines. In March 1935, Schlichting negotiated the details of this program with officers at Navy headquarters in Kiel. As a result, a new water channel was built for the investigation of surface roughness as pertinent in naval uses (special rivets, superposed ship planks etc). The transfer of research results from these investigations into the practice of naval ship-building was ensured by the participation of shipyards. In 1938 the project was extended in order to account for aerodynamical roughness as well. A special wind tunnel was used in which one tunnel wall was composed of movable parts so that a pre-scribed field of pressure could be generated along the surface of an inserted test model. From a scientific perspective, the roughness program was part of Prandtl's working program on turbulence; "basic" scientific and "applied" technological research were inextricably intertwined. From an epistemologi-cal perspective, this was portrayed as an example of a "linkage of heterony-mous (military) and autonomous knowledge interests" [511, p. 24].

Despite the formal separation between the AVA and the KWI after 1937, it would be misleading to assume that both institutions were related to one another by nothing more than Prandtl's authority, and that war-related research was centered at the AVA while the KWI's main interest was basic science. When Baeumker complained after a visit in 1940 that a connecting door between both institutes was left open in violation of security regulations, Prandtl responded that "there is a strong need" for close communication because "KWI-people and AVA people collaborate in many cases in common tasks," and thus, he decided to leave this door open during daytime. He did not regard it as a violation of security because the personnel of the KWI and the AVA were checked in the same manner by the Hanover counterintelligence office—"man by man"—and because foreign researchers were no longer granted access since the outbreak of war [512]. Although the AVA was involved in military projects to a much larger extent than the KWI because of its more direct relationship with the development of new military aircraft, the Air Ministry did not foster fluid dynamics at the KWI merely as an afterthought. Science should be regarded with an eye for war-related applications, Göring declared in 1939 to the congregated scientists in an address at a meeting of the German Academy of Aeronautical Research: "Fluid dynamics has to combine with Air Force research in order to clear the basic laws for application to the Air Force and to guide the development along new paths" [513, p. 132–133].

## 8.2
## Aeronautical Science as an Instrument of Nazi Propaganda

Göring and his ministry also made use of Prandtl's reputation for propaganda purposes. His reputation both at home and abroad served to portray Nazi Germany as a country where science was still in high esteem, despite occasional ideological derailments and the purge of the Jewish scientists from the universities. There are no indications that Prandtl felt uncomfortable with this role. He also saw no contradiction between his frustrated efforts to prevent the dismissal of his Jewish colleagues and his own propagandistic mission. When Jerome Hunsaker confided to Prandtl in a letter in 1934 that he was "distressed by the news we get here of internal difficulties in Germany," Prandtl declared that the foreign press reported "very ugly falsehoods about Germany" and expressed his disappointment that Hunsaker should accept such reports as truth. "You should rather have written: 'I am pleased that now in Germany, all things are becoming better and better,'" he wrote in his response [514]. When Göring congratulated Prandtl in 1935 on his sixtieth birthday, Prandtl thanked the Air Minister with the assurance that he "will do everything for the further advancement of German aeronautics by my research and that of the institute that I direct" [515]. Shortly before, on 1 March 1935, Göring had announced publicly that Germany possesses an Air Force and declared this "day of freedom of German aviation" as a national holiday. The German Academy of Aeronautical Research chose this day for its annual meetings which were celebrated in a most grandiose manner [501, p. 181]. Prandtl must have been aware that this new "freedom" would conjure up the risk of war, because he informed Göring in a letter in December 1936 of the measures that will be taken in his institute "for the conversion of research at a beginning of war" [516].

**Fig. 8.4** Göring opens a Meeting of the German Academy of Aeronautical Research on 1 March 1938, the "day of the German Air Force." Prandtl is seated in the first row, the second from the right, next to the Minister of Science and Education, Bernhard Rust, and the Undersecretary of the Air Ministry, Erhard Milch, both in uniform [518, p. 75].

Prandtl, like many other strongly nationally minded Germans, was impressed by the Nazis' determination to free Germany from what they considered the manacle of Versailles. Aeronautics was particularly affected by the restrictions of the Versailles Treaty, and the Air Ministry could count on broad consent among aeronautical scientists and engineers. At the public conferences of the German Academy of Aeronautical Research and the Lilienthal Society, aeronautical celebrities like Prandtl were courted like never before and recruited for the national cause. Their activity before the First World War had already brought German aeronautical research to international renown, Göring praised them when he opened a meeting of the German Academy of Aeronautical Research in April 1937, but "the collapse of our empire after the end of the war has prevented this good tradition of our aeronautical research from being continued." Only "the resolution of our Führer, to restore for Germany the military authority robbed by the Versailles dictate" made it possible to follow this tradition once again. Göring urged the assembled members of the academy to focus on the challenge ahead in the spirit of "a Lilienthal and Zeppelin," a task that was "worth the highest efforts for decades and centuries" [517, pp. 139–142].

At such occasions, Prandtl often sat in the first row next to the highest authorities of the Nazi regime (see Fig. 8.4). As a member of the Executive Committee of the German Academy of Aeronautical Research and as one of the presidents (together with Baeumker and Bosch) of the Lilienthal Society, he shared responsibility for the agenda and activities of these organizations. Although it was obvious from Göring's aegis that they were directly under the purview of the Air Ministry, both the Academy and the Lilienthal Society were not immediately regarded as mere Nazi propaganda – not least because Prandtl and other renowned academics lent them an aura of apolitical scientific prominence. At Baeumker's request, Prandtl proposed a number of foreign scientists for election as "corresponding members" into the German Academy of Aeronautical Research and as "foreign members" of the Lilienthal Society. He did so fully aware of the propagandistic role of such foreign membership because in addition to the list of names, he gave the following advice: "The political attitude towards the present Germany remains to be checked in some cases" [519]. He also wrote personal letters to foreign colleagues – for example to the most senior among American aeronautical scientists, William Frederick Durand—in order to ask for their consent before they were officially elected as members. In the case that the honor was declined for political reasons, the matter would remain private and not result in a politically embarrassing issue for Göring who would officially nominate the foreign members in his capacity as the Reich's Air Minister [520].

The propagandistic role of these societies, particularly with regard to foreign countries, was similar to the role played by the 1936 Olympic games in Berlin. When Göring opened the annual general meeting of the Lilienthal Society in 1936, he first addressed "all our foreign guests," among them John Jay Ide from the NACA's Paris Office, Clark Millikan from the California Institute of Technology, and the Military and Air Attachés assigned to the Berlin embassies of the USA, China, England, France, Finland, Italy, Japan, Austria, Hungary, Czechoslovakia, and Sweden. The gathering of national and international political and scientific prominence illustrated to the readers of *Luftwissen* that Germany's aeronautical efforts were not only tolerated but even admired abroad. In 1937, the Lilienthal Society held its annual meeting in Munich, the "capital of the [Nazi] movement," again with the participation of international celebrities like Charles Lindbergh and Jerome Hunsaker from the USA. When the German Academy of Aeronautical Research met on 1 March 1939, Prandtl was awarded in the presence of numerous foreign celebrities the "Hermann Göring medal," the Nazis' highest distinction for scientific merit in aeronautics. Göring praised Prandtl as a scientist whose merits are widely acknowledged "not only in our country but also in all other countries of the world which pursue aeronautics." Few other events revealed such a close encounter of science and Nazi propaganda [501, pp. 268–276], [517, pp. 294–333], [518, pp. 389–410] and [513, pp. 133–134].

However, the façade began to crumble a few years after it was erected. In April 1938 Edward Warner addressed a letter to Baeumker in which he declared his resignation from foreign membership in the Lilienthal Society [521]. In December 1938, Durand resigned from the German Academy of Aeronautical Research, because this academy, as he explained in a letter to Prandtl, is "organized under the direct auspices of the German Government, and the present Governmental theory in Germany regarding social and political organization for the good of humanity is so remote from my own, that I do not feel I should longer remain in relation with an organization of which the titular head [i.e., Göring] is one of the highest exponents of this theory." Durand hoped nevertheless to retain Prandtl's friendship and emphasized that he would not have declared his resignation if the academy meant a sincere association "with high German scientific culture" [522].

## 8.3
## Goodwill Ambassador

Prandtl cannot have remained naive about his propaganda mission for Göring's ministry. In view of "the Durand case" Baeumker asked Prandtl to use his friendship to foreign scientists once more in order to learn in advance who would be willing to accept an invitation to participate in a forthcoming meeting: "The Academy wants to avoid being turned down under all circumstances" [523]. In a letter to the NACA's research director, George Lewis, Prandtl subsequently asked for a list of "suitable names" who would not be afraid "to participate in a meeting in which official personalities of the German government are also present, which is portrayed as horrible in the USA." Before he sent this letter he asked the Secretary of the Aeronautical Research Academy in the Air Ministry for his consent – adding that he knew "that Dr. Lewis does not like the Jews and therefore will understand my remark at the end of the letter in the proper manner" [524].

If Prandtl was hostile towards the Nazi regime in the beginning, as his daughter claimed, that attitude had changed completely after few years. In May 1937 he wrote to William Knight, the former NACA representative in Paris (see Chapter 4): "I believe that Fascism in Italy and National Socialism in Germany represent very good beginnings of new thinking and economics," and suggested that "states that do not want to fall victim to Bolshevism should start to follow similar paths, the sooner the better" [525]. Moderate sympathy, as expressed in this attitude, was better suited to perform a goodwill mission for his regime than fanatic adherence to the Nazi party doctrine. Other German scholars performed similar missions abroad as Germany's goodwill ambassadors [526, 527].

Since 1936, German scientists had to apply to the Reich's Ministry for Science and Education in order to receive permission to attend conferences abroad. Furthermore, an office was established in Berlin, the Deutsche Kongresszentrale, to deal with foreign exchange and other bureaucratic affairs. Attendance in conferences abroad, therefore, was not possible without the involvement of political authorities. In Prandtl's case, the permission to attend foreign congresses was not disputed. The International Congress for Applied Mechanics in 1938 in Cambridge, Massachusetts, became an outstanding opportunity for Prandtl to excel in the role as goodwill ambassador. He requested the permission to participate as early as November 1936 and was granted it by the Science Ministry in July 1937—with the suggestion that he contact the German Embassy in America and the local party office [528]. The Air Ministry explicitly wished that Prandtl act as the leader of the German participants [529]. Furthermore, Prandtl was authorized to present an official invitation to hold the next Congress in 1942 in Germany [530]. To this end the Kongresszentrale sent Prandtl a confidential "Leitfaden" in which detailed recommendations were formulated in order to make the proposed Congress in Germany "an instrument of German cultural propaganda" [531].

Prandtl, however, was fighting a lost cause, because the other members of the International Congress Committee, among them his exiled colleagues Richard von Mises and Theodore von Kármán, did everything to prevent Germany from being chosen as the site for the next Congress. Mises, who had emigrated in 1933 to Turkey, announced that he would bring an invitation from the Turkish government so that a German invitation could be voted down [532]. Kármán responded that "an invitation from Istanbul would be very desirable, and has a fifty–fifty chance of being accepted." He was not sure how the American members in the Committee would vote: "The people at Harvard are on our side; however, M.I.T. has a somewhat pro-Nazi leaning" [533]. The latter remark hinted at Hunsaker, who argued "that a Congress in Turkey would not be well-attended, but one in Germany would be." However, an invitation to France was also an option to which Hunsaker had no objections [534]. "After a full discussion, the Committee decided to accept the invitation of the French members," the minutes reported about the final debate. "The invitations from Prof. Prandtl and Professor von Mises were kept on file to be transmitted to the Secretary for the next Congress in the hope that these invitations would be renewed" [535].

Off the record there must have been considerable irritation about Prandtl's behavior as a goodwill ambassador for the Nazi cause – at a time when war seemed imminent because of the crisis concerning Hitler's annexation of the Sudetenland. Even Hunsaker, whom Kármán counted among those with "a somewhat pro-Nazi leaning," wrote to Prandtl after the Congress to say that he "was more disturbed than I cared to admit, by the evidence reaching us of serious war preparations" [536]. Others reacted with stronger emotions. G.I. Taylor wrote to Prandtl on 27 September 1938, three days before the Munich Agreement, in which the British Prime Minister, Neville Chamberlain, together with the French and Italian leaders, appeased

IDXP[Hitler, Adolf]Hitler by signing the annexation of the Sudetenland: "I realized that you know nothing of what the criminal lunatic, who rules your country, has been doing, and so you would not be able to understand the hatred of Germany which has been growing for some years in every nation, which has a free press" [537]. Prandtl responded a month later: "You will not really believe that, if our Führer were as portrayed by the foreign press, which is in the hands of Jews, your Prime Minister Chamberlain would have concluded the [Munich] agreement with Hitler, according to which all differences in the future should not be handled by arms but by meetings. In case the American press in its hatred sabotaged this document, I enclose a photographic reproduction of the original (exhibit No. 1)." Prandtl also enclosed, as exhibit No. 2, a speech given by Hitler, whom he described as a "man of tremendous nerve" who admittedly "made himself a million people as his bitter enemies, but on the other side eighty million as his most faithful and ardent followers." With regard to the "Jewish question" he argued: "The struggle, which Germany unfortunately had to fight against the Jews, was necessary for its self-preservation." He added to this five-page letter a number of press reports about a Hitler's visit to "the freed Sudetenland" so that Taylor could "recognize the enthusiasm of the people" [538]. When war was imminent in August 1939, Prandtl wrote to Taylor's wife: "If there will be war, the guilt to have caused it by political measures is this time unequivocally on the side of England." He defended Hitler's political goals as an attempt "to remove the final remains of the treaty of Versailles" and included more newspaper clippings [539].

Even after the attack against Poland, which provoked the outbreak of war, Prandtl attempted to persuade his foreign colleagues that Hitler's motives were sincere. On 3 November 1939, he wrote to Hunsaker: "According to England's will, unfortunately, the last action of the German government to repair the damages of the Versailles Treaty led to war. Every good German is now concerned with making known abroad the true trains of thought of the German government, which are often reproduced by the foreign press in a heavily distorted manner. I seize the opportunity to send you an English translation of a speech of the Führer in which he reports on the results of the Polish war and his plans for a new order of Europe after this campaign." He also asked Hunsaker to forward copies of Hitler's speech to other colleagues at MIT and Harvard university [540]. What Prandtl did not tell Hunsaker and other recipients of Hitler's speech was that the department of foreign affairs of the Deutsche Kongresszentrale had explicitly asked him and other goodwill ambassadors to spread Hitler's speech to their foreign colleagues, particularly to those "who possess in their country renown and influence and who are able by their own attitude and objectivity to understand the Führer's trains of thought and to spread them in their circle." Names of addressees were sent back and forth. Prandtl even requested additional copies of Hitler's speech translated to English and French. He was neither naive nor forced against his will to participate in this "propaganda mission," as he called it himself [541].

# 9
# New Centers

Despite his pro-Nazi leanings, which most of Prandtl's friends and colleagues abroad interpreted as a result of political naivety, Prandtl was regarded as the intellectual leader of the nascent international community of fluid dynamics. G.I. Taylor would not have challenged Prandtl's believe in Hitler if they had not been on such friendly terms. He introduced the passage in which he called Hitler a "criminal lunatic" by expressing his hope "that whatever happens between our countries the friendship and admiration which I, in common with aerodynamical people in other countries, feel for you will be unchanged" [537]. A few years earlier he had confided to Prandtl how much he wished that he succeeded in nominating him for the Nobel prize. However, he doubted whether "the existing method of nomination by past Nobelprizemen will ever produce a 'non atomic' prizeman" and felt "very strongly that if the Nobel Prize is open to non-atomic physicists it is definitely insulting to us that our chief – and I think that in England and USA at any rate means you – should not have been rewarded in this way" [542].

Prandtl's esteem as "our chief" was more than a recognition of personal scientific merits; it also accounted for a tradition founded by Prandtl and exported to new centers by Prandtl's pupils. By the 1930s, Göttingen was no longer the only center of modern fluid dynamics. Theodore von Kármán spread the gospel of his teacher first to Aachen and then to America, where he made the Guggenheim Aeronautical Laboratory of the California Institute of Technology (GALCIT) in Pasadena a center of world renown. Another branch was flourishing in Switzerland, where Jacob Ackeret directed a new institute for aerodynamics at the Technical University (Eidgenössische Technische Hochschule, or ETH) in Zurich.

The example of these new centers illustrates how local research traditions spread and grow beyond cultural and national borders – and partially fuse with other local traditions and practices. Although fluid dynamics in Aachen, Pasadena and Zurich, as in other new centers emerging in the 1930s, became institutionalized as an engineering discipline, interest in fundamental prob-

*The Dawn of Fluid Dynamics: A Discipline between Science and Technology.* Michael Eckert
Copyright © 2006 WILEY-VCH Verlag GmbH & Co. KGaA, Weinheim
ISBN: 3-527-40513-5

lems did not fade away. Practical problems often became an incentive for new efforts to study fundamental riddles. Kármán acquired the roots for this research mentality in Göttingen; he once wrote to Ackeret that "both you and I were exposed to the inspiring influence of Ludwig Prandtl, whom I consider as a great master in combining simple mathematical formulation and a clear physical picture in solving problems important for technical applications" [543]. Both Kármán and Ackeret felt the same tendency to combine science and engineering in the Göttingen tradition and to educate their students in Aachen, Pasadena, and Zurich in this spirit. Beyond this common research mentality, there is also more tangible evidence for the spread of Göttingen traditions: most new centers also introduced the experimental techniques employed in Göttingen, particularly the Göttingen closed-circuit-type wind tunnels. Like the fundamental mentality, research techniques and practices did not spread by themselves but were actively propagated by Prandtl's pupils. This fact was documented in one chapter of Kármán's autobiography, titled "Trailing the wind around the world" [85, p. 184].

## 9.1
## Aachen

The first offshoot of Prandtl's school grew at the Technical University Aachen. The seeds of Aachen's research in aerodynamics were planted in 1909 by the professors of machine construction and mechanics, Hugo Junkers and Hans Reissner, whose joint efforts also resulted in the foundation of an institute for aerodynamics in 1913. When Kármán succeeded Reissner as professor of mechanics in Aachen in the same year, he did not have to start from scratch, but the First World War prevented a smooth expansion. The institutes personnel was called to arms. Kármán himself, as a native Hungarian, was drafted into the Austro-Hungarian army. His only aerodynamic research from this period was a theoretical paper co-authored with his Aachen assistant, Erich Trefftz, about how Joukowsky's method of conformal mapping could be extended to account for a broader variety of wing profiles [544]. When he returned to Aachen after the war in November 1919, his laboratory was in a "deplorable condition," as he recalled many years later. "The French and Belgian troops, who were quartered in the university, hadn't bothered to do much cleaning before they left. Empty bottles were strewn on the floor. Paint was peeling from the walls. In the towing tank, which we had built to test models of hydroplanes, I found the remains of two drowned cats. I came away from my first inspection not feeling too kindly disposed toward the victorious Allied soldiers" [85, p. 96].

After these unpromising beginnings, it was initially the glider obsession from which Kármán and his students drew the motivation to proceed with research in aerodynamics. The Aachen Aeronautical Association (Flugwissenschaftliche Vereinigung), founded by Kármán and his students, was among the first academic glider clubs, which then emerged at many technical universities in Germany. Kármán's assistant, Werner Klemperer, won Rhön contests with aerodynamically refined gliders (see Chapter 6) and thus brought international renown to the Technical University Aachen and Kármán's institute for aerodynamics. Among the more theoretically minded colleagues, the "Kármán–Pohlhausen method" for the calculation of boundary layers (see Chapter 5) signaled that this institute was budding as a new center of fluid dynamics. It also followed the Göttingen model by publishing its research results in a regular series, the *Abhandlungen aus dem Aerodynamischen Institut an der Technischen Hochschule Aachen*. In 1925, the old wind tunnel built in 1912 on the institute's roof was enlarged and modified into a Göttingen-type closed-circuit tunnel (see Fig. 9.1). Three years later, the institute was expanded with a new building, so that a broad scope of experimental research in fluid dynamics could be undertaken. In addition to the aerodynamic wind tunnel investigations, precautions were taken to perform experiments on gas dynamics and hydraulics with a 30-m-long towing tank and a 40 $m^3$ windkessel. In 1929, Kármán invited his national and international colleagues to Aachen for a conference in order to celebrate the institute's expansion and underline his claim that from that point on, Göttingen would not be the only place in Germany where modern fluid dynamics flourished [545]. In the following year Kármán succeeded in establishing a new chair for Applied Mathematics and Fluid Mechanics, to which another Prandtl pupil, Carl Wieselsberger, was called. With Kármán, Ludwig Hopf, and Wieselsberger, the Technical University Aachen boasted Germany's most distinguished experts in fluid dynamics next to Göttingen [305].

Internationally, both Kármán and Wieselsberger were most coveted consultants for the design of wind tunnels. Before he was called to Aachen, Wieselsberger spent several years in Japan advising the Japanese government in matters of aeronautical research. "I am hired by the university and the navy, he wrote to Kármán in 1925 from Amimura, but he asked him to keep this information confidential "because according to the Peace Treaty a Reich's German is not allowed to be employed at one of the winning powers for military purposes. Here, I build wind tunnels, a propeller testing facility, a water towing tank including the required measurement instruments, and I have to deliver talks from time to time." He confided that the Japanese also wished to invite Kármán as a consultant [547]. Kármán asked his university for a one-year leave of absence in 1926/27 in order to travel to Pasadena, where the plans for GALCIT took shape, and then to Japan, where he served in Kobe as an advisor for the design of wind tunnels of the Göttingen closed-circuit-type at the Kawanishi Works, Japan's most important airplane manufacturer. Wiesels-

**Fig. 9.1** The institute for aerodynamics of the Technical University Aachen with its new Göttingen-type wind tunnel on the roof [546].

berger later benefited from his contacts with Japanese industries when he provided the new Aachen wind tunnel with novel measuring devices he had first introduced in Japan [548].

With regard to research planning in aeronautics in Germany, too, Aachen became a new center next to Berlin and Göttingen. In 1928, Prandtl and Kármán established the German Research Council for Aeronautics (Deutscher Forschungsrat für Luftfahrt), an advisory board for the department of aeronautics in the Reich's Ministry of Traffic administrated by Adolf Baeumker [120, pp. 146–149]. Kármán and Baeumker came to know one another through this council so well that they kept in contact even after Kármán's emigration to America in 1933. In his autobiography Kármán recalled how Baeumker, even after he became the head of aeronautical research in the new Air Ministry under Göring, "kept in touch with me. On one of my return trips to Germany he even took me on a tour of the air facilities. This was in 1934, during the early Hitler period. Baeumker let me know that since the situation for non-Aryans at the universities was, 'to say the least, delicate,' Göring's office had suggested that I join the Air Ministry as a consultant. It is known that Göring once said: 'Who is or is not a Jew is up to me to decide.' I had a good laugh and remained in Pasadena" [85, p. 146]. Although Kármán's recollections should not be read as a primary historical account, the friendship Baeumker felt for Kármán is also evident from their preserved correspondence. "We will always regard you as one of these friendly-minded year-long collaborators," Baeumker closed a long letter to Kármán in December 1934 [549]. Kármán responded warmly and hoped for an opportunity to visit him in Germany in the summer of 1935. However, Baeumker cannot have overlooked Kármán's sarcasm when in November 1935, he read about how Kármán reacted to a talk presented by a visiting colleague about the progress of German aeronautics: "He also reported about the subterranean hangars and the three thousand rooms in the Air Ministry. I see that I have to visit Germany soon in order to watch all these wonders. I am particularly curious to learn where the three thousand brains for the three thousand rooms will come from" [550].

Kármán, an avowed enemy of the Nazis, was already ready to leave Germany before 1933. In 1929, the California Institute of Technology (CalTech) offered him the directorship of the newly founded GALCIT, which Kármán accepted first as a half-time position. The Technical University Aachen agreed to the arrangement that Kármán lecture during the summer semesters in Aachen and spend the winter terms at CalTech. In order to keep him in Germany, the Science Ministry also offered Kármán a professorship in Berlin. Even the Nazi purge of the universities in April 1933 did not affect Kármán like other Jewish professors. Otto Blumenthal, professor of mathematics at Technical University Aachen, was dismissed despite his status as a "front fighter" of the First World War, which was granted as an exception clause for the purge. Hopf, who usually replaced Kármán during his leaves of absence, was dismissed, despite his expert knowledge in aerodynamics. Kármán, in contrast, was granted an extended leave of absence during the purge. In January 1934, however, the Reich Science Ministry denied a further extension and ultimately demanded his return. But in May 1933, Kármán had already applied for the American citizenship and resigned from his Aachen position [494, pp. 130–132], [551].

The Nazi seizure of power, therefore, severely damaged the rise of Aachen as a new center of modern fluid dynamics in Germany. Aachen's prominence in aerodynamics and applied mathematics was gone with Kármán, Hopf, and Blumenthal. Hopf was replaced by Wilhelm Müller, an ardent anti-Semite who could hardly compete with his predecessor in fluid dynamics. He became most infamous as Johannes Stark's ally in their struggle against modern theoretical physics as a Jewish bluff. Prandtl testified that Müller's work was "mathematically in order" but "uninteresting because he consequently avoids all non-linearities and only tackles problems where one succeeds with classical mathematics" [552]. The occasion of this evaluation was Müller's call as Sommerfeld's successor to Munich in 1939, a blatant scandal which mobilized the death-blow against the so-called Aryan physics fanatics [488, 553, 554]. In Aachen, however, Müller's call to Munich was welcomed because it made the chair vacant for another Prandtl pupil, Fritz Schultz-Grunow [305].

**Fig. 9.2** The enormous windkessel in the institute for aerodynamics of the Technical University Aachen could be evacuated for transient airflow at supersonic speed [546].

Although the Aachen Institute for Aerodynamics was less prominent than it was during the 1920s under Kármán's charismatic direction, particularly regarding fundamental theoretical work, Wieselsberger was able to maintain a high standard of experimental research. Practically all published Aachen reports after 1933 addressed wind tunnel investigations. A particular focus was experimental high-speed aerodynamics. At the International Volta Congress in 1935 in Rome, Wieselsberger reported on the first Aachen supersonic measurements. He alerted the assembled experts to a phenomenon caused by condensation of water vapor which become known as "spirit" in supersonic wind tunnels. After that point, special precautions were taken to reduce the humidity in supersonic wind tunnels to a minimum [555, p. 558], [305, p. 257]. In 1937, at a meeting of the Lilienthal Society, Wieselsberger presented a review of Aachen's supersonic facility. It consisted of two tubes, equipped with Laval nozzles and test chambers of 10 cm × 10 cm, and 20 cm × 20 cm, respectively, through which air was sucked by evacuating the enormous windkessel of the institute (see Fig. 9.2). It was possible to create supersonic air flows at a speed of 3.16 Mach for periods of 6 s in the larger test chamber, and 25 s in the smaller test chamber. Wieselsberger and his assistant, Rudolph Hermann, made a particular effort to develop new measurement techniques: the forces upon the models in the test chamber were measured by a novel method based on electromagnetic induction rather than mechanically as in traditional wind tunnels; optical observations were enabled by a refined and sophisticated Schlieren technology [556]. Hermann moved in 1937 to the Army's secret rocket development center, Peenemünde, where he directed a new aerodynamic institute for supersonic tests on models of rockets. Preliminary tests of the stability of rockets had also been performed in the Aachen supersonic facility. At Peenemünde, both Hermann himself and the employed measurement techniques were manifestations of a direct continuation of Aachen's high-speed aerodynamics [557].

Shortly before the beginning of the Second World War, the work reports of the Aachen Institute for Aerodynamics communicated measurements on the drag of propellers in addition to investigations of supersonic phenomena, which were important, as the report revealed, "for the problem how to reduce the final speed of modern airplanes in dives," and on thermal problems like the thawing of frozen oil coolers, performed at the request of the Airforce Testing Facility Rechlin. Most of the institute's research was probably kept secret, because the annual work report of 1938/39 contains only three publications [558].

## 9.2
## Pasadena

The California Institute of Technology (CalTech) rose to prominence when Robert A. Millikan was called to Pasadena in 1921 as new university president. Millikan was known for his far-reaching ambitions both as a physicist and as a science manager. He put CalTech on the map as a top university by inviting the world's most renowned scientists for guest lectures and by hiring internationally distinguished scientists to new chairs. With theoretical physicist Paul Epstein, a pupil of Sommerfeld's, Millikan brought modern atomic physics to CalTech in the early 1920s, and with Kármán, he pursued the same strategy a few years later in order to lure the best available aerodynamicist from Europe to Pasadena. In Kármán's case, the transfer was enabled by the lavish Daniel Guggenheim Fund, which provided a number of American universities with the means to catch up with Europe in aeronautical research. By 1926, Millikan had negotiated a plan with Daniel and Harry Guggenheim for the foundation of a new aerodynamic institute at CalTech [551,559].

The new "Guggenheim Aero Laboratory at the California Institute of Technology" (GALCIT), as it was first called, should be equipped with a large wind tunnel and meet the highest standards of teaching and research in aeronautical science. The following six goals were formulated: CalTech's theorists – the mathematicians Harry Bateman, E.T. Bell, and theoretical physicist Paul Epstein—were charged with presenting theoretical courses in "aero-

and hydrodynamics"; practical exercises should be organized in collaboration with Californian airplane manufacturer, the Douglas Company; the institute should also pursue a "comprehensive program of research on airplane and motor design"; as a practical engineering task, a novel airplane concept (the "stagger-decalage, tailless airplane") should be further developed; a number of research fellowships in aeronautics should be established; and the institute's wind tunnel should be used in addition to regular model testing for measurements of "full size experimental gliders and power planes for free flight" [560, pp. 1–2]. In order to bring such an ambitious plan to fruition, Millikan at first planned "to get Prandtl for a short time." G.I. Taylor, too, was named as a possible candidate, but he ranked under Prandtl and Kármán on Millikan's list who regarded Prandtl's institute as a role model for GALCIT. Epstein was a friend of Kármán's and often served as Millikan's advisor when it came to European candidates; he strongly recommended that Kármán be invited instead of Prandtl. There were also more practical reasons which spoke against Prandtl: he was older than Kármán; his avowed commitment to the German cause could make negotiations with American industrial and military contractors difficult, while no such concerns were to be expected with Kármán; and Prandtl appeared less approacheable and too academic. In the end, after discussions with Epstein, Millikan concluded "that in view of Prandtl's advanced age and his somewhat impractical personality he would be far less useful to us than v. Kármán." Harry Guggenheim shared Millikan's view that "Dr. Prandtl of course stands alone in the aeronautical world," but he also agreed with him "that for a practical visit such as you have in mind and which will fit in very nicely with our plans, Professor v. Kármán is the right man" [551, pp. 94, 99].

Epstein was asked to forward an invitation to his friend in Aachen. In September 1926 Kármán traveled to Pasadena for further negotiations. From the very beginning, Millikan planned to hire Kármán as a consultant and future director of the planned GALCIT, but he did not yet offer him a permanent position. They first arranged an exchange program between Aachen and Pasadena: Epstein would teach one semester each year in Aachen, while Kármán would come for one semester to Pasadena. In 1928/29 Kármán combined this arrangement with a voyage around the world. For the first three years after 1926, the bulk of the work involved with building GALCIT was performed by Robert Millikan's son, Clark Millikan, who soon made himself a name as one of America's outstanding aeronautical scientists. Only in 1929 did the plan to call Kármán permanently to Pasadena and to entrust him with the directorship of GALCIT get put into action. By that time, Kármán was a widely coveted aerodynamical consultant and candidate for new chairs. Stanford University, for example, also offered to hire Kármán. CalTech, in the meantime, planned to establish, together with the University of Akron in Ohio, a common Daniel Guggenheim Airship Institute, also to be directed by Kármán. To make the offer more tempting, Harry Guggenheim and Robert Millikan reassured Kármán that it would only occasionally require his personal presence in Akron, but nevertheless would have an advantageous effect on his salary. Kármán's total income in America would be three times as high as in Germany. Kármán did not immediately accept, but he used this offer in order to negotiate with the Prussian Science Ministry about the expansion of his Aachen institute. Wieselsberger's call to Aachen was one result of these talks. Kármán arranged with the CalTech president to assume his directorship on a half-time basis. Harry Guggenheim remarked in a letter to Millikan, "obviously your friend goes into the water inch by inch." Kármán's decision to move permanently to the USA, however, was largely influenced by private considerations, because he would no go without his mother, who was only "gradually getting used to the idea of Pasadena," as Kármán wrote to a friend in December 1929 [551, p. 131].

As we have seen, Kármán only resigned from his Aachen position in 1934. When the Nazi Science Ministry did not seem to extend Kármán's leave of absence from Aachen beyond the winter semester 1933/34, he could no longer defer the decision to make his American sojourn permanent. "I got a short letter from Berlin suggesting that I take up my activities over there in the fall," Kármán responded a letter from Prandtl in August 1933, choosing English as his language for the first time in their correspondence, I do not think I will do this; I find my situation here quite satisfactory. The German academic life has some advantages, for instance a definitively better beer than here, but I think you will agree with me that this is not sufficient reason for me to neglect the disadvantages" [551, p. 133]. In his autobiography Kármán quotes from the original letter with which he declared his resignation to the official in the Berlin Ministry: "I hope that you will be able to do for German science in the next years as much as you accomplished in this year for foreign science" [85, p. 146].

By that time, GALCIT was already fully active. The new institute building was completed in 1928. With a ground surface of 1,000 m$^2$ and five storeys, it was one of the university's largest buildings. It housed a wind tunnel, whose design was changed upon Kármán's advice from the original Eiffel-type into a closed-circuit Göttingen-type (see Fig. 9.3). Its dimensions were comparable to the largest wind tunnels of the epoch, with a maximal diameter of almost 7 m (20 feet) and a narrowest diameter of more than 3 m (10 feet). The building was designed so that the tunnel occupied three of the five storeys; on the first floor were the workshops for the model construction and the observation room; the tunnel walls were made from concrete and carefully insulated from the building so that the vibrations of the tunnel had no impact on the institute's rooms. The inner tunnel wall was painted with an asphalt emulsion in order to keep the airflow smooth and free from dust. The guiding vanes were made from steel and formed according to a design by Betz for the Göttingen wind tunnel. With the exception of a novel closed measuring room adjusted to the circular cross-section of the tunnel, it was obvious that Göttingen was also the model for GALCIT with regard to the tunnel design [561].

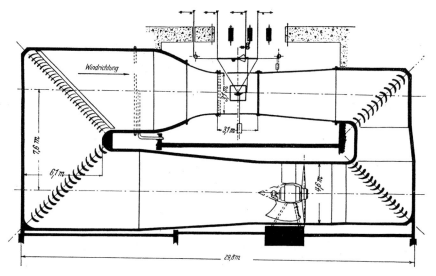

**Fig. 9.3** GALCIT's wind tunnel followed the role model of Göttingen-type wind tunnels [562, p. 283].

Göttingen traditions entered Pasadena in many ways. As formulated by Millikan in 1926, one of the six main items on GALCIT's agenda addressed research fellowships. In 1930, Kármán used the fellowship as an opportunity to offer two of Prandtl's pupils, Walter Tollmien and Rudolf Seiferth, the opportunity for extended research sojourns in Pasadena. Prandtl even asked Kármán to keep them as long as possible, because at the height of the economic crisis, they were better off in Kármán's institute than in his own institute in Göttingen. Tollmien stayed until 1933. "He was of invaluable assistance to graduate students and researchers at GALCIT," Kármán later wrote about Tollmien's visit. During the early years the focus in Pasadena was on practice rather than theory, and Kármán felt there was a need for more fundamental research when he took over the directorship in 1930. Tollmien was the right man at the right time to present to his students an "integrated approach to the technical problems"—a characteristic of Prandtl's school in Göttingen—"i.e., consider the physical ideas behind the mathematical operations and always visualize the relation between theoretical and experimental research (...) it appeared to me that we needed a person, at least in the first period, who would help to transplant some elements of the Göttingen approach to the new institution" [563].

Seiferth's sojourn, too, was long remembered. He had forged a reputation as an expert on wind tunnels, and Kármán hoped Seiferth could transmit a feel for the practical aspects of the Göttingen tradition to his students. This he did – although not necessarily in the way that Kármán had wished: "Seiferth told me," Kármán recalled in his autobiography, "that in Göttingen he and his colleagues had developed a model out of plaster of Paris, and that it was easier to form and much better than wood in low velocity tests. He wanted to try to make such a model for us at CalTech. (...) When he was finished, I invited the class to watch the wind-tunnel experiment with the first model made of plaster of Paris. I could not have created a worse demonstration if I had tried. As the class crowded around the tunnel, the first gust of wind smashed the plaster of Paris model into tiny bits. Particles were scattered all over the machinery. It took us two days to clean it out. Seiferth, the expert, of course had forgotten to tell me that he himself had had no experience in making wind-tunnel models. In Germany this kind of work was done by glassblowers, cabinetmakers, and other skilled mechanics, not by professors" [85, p. 154].

By the mid 1930s, CalTech ranked among the leading technical universities in the USA, with aeronautical science as its prime asset. Among the seven American Schools for Aeronautics funded by Guggenheim, GALCIT had the highest score in terms of publications. By 1939, GALCIT's scientific productivity was expressed by 122 publications and 51 unpublished dissertations; during the first decade after its official opening in 1929, a total of 162 engineering students had completed their study there, among them 73 with an M.S. degree in mechanical engineering, 68 with an M.S. degree in aeronautical engineering, and 21 with a Ph.D. in Aeronautics. 31 Navy and Army officers were trained in special one-year courses as aeronautical engineers. Eighty-seven students, among them also many soldiers, finished their study with an M.S. degree in meteorology, a discipline with close ties to fluid dynamics. That both mechanical engineering and meteorology were counted among GALCIT's specialties shows that Kármán considerably expanded Millikan's initial program and pursued aeronautical research and teaching in a very broad sense: "Theoretical and experimental investigations were directed both toward the solution of fundamental basic problems in Elasticity and Fluid Mechanics, and toward results that could be applied immediately to aircraft design," GALCIT's history comments on the wide range of topics [560, pp. 6–7]. Meteorology had at first been a sub-discipline of earth physics in the geology department, but in 1933 it was integrated into aeronautical engineering because of its importance for aviation [564].

The wide scope of themes appears unusual for an institute whose primary focus was to train aeronautical engineers, but Kármán's agenda during his first years as director of GALCIT was less dictated by comparisons with other

American engineering schools than by his own experiences from Göttingen and Aachen. He also brought with him the competitive spirit in which he attempted to catch up with or outstrip his role model at Prandl's institute. GALCIT is "one of the centers of the modern 'turbulence research,' stimulated by friendly competition with other centers"—the institutional history mirrors Kármán's personal rivalry with Prandtl [560, p. 7]. Besides turbulence, research in fluid dynamics included investigations of boundary layer phenomena, gas dynamics, and a host of other themes. As in Göttingen and in Aachen, basic and applied research proceeded hand in hand, as the list of GALCIT's research themes illustrates. They ranged from lubrication studies for the support of the new 5-m-telescope at the Mount Palomar Observatory to the design of steam turbine blades for General Electric. Within few years, GALCIT became a top resource for governmental, military, industrial, and academic contractors for research in fluid dynamics. In 1938, for example, the Soil Conservation Service of the Department of Agriculture charged Kármán with an investigation of soil erosion, a problem of particular importance for the southwestern parts of the USA where sand storms resulted in grave losses of farmland. By planting rows of trees, one hoped to reduce the damage, but the efficiency of such shelterbelts was disputed and required further investigation. Kármán's student Frank J. Malina reviewed the available evidence of this problem in a report on "Experimental Methods for the Study of the Fundamentals of the Wind Erosion Process Under Controlled Conditions," and Kármán designed a movable outdoor wind tunnel in which sand grains of variable size could be blown over different agricultural surfaces and the resulting deposit of sand could be measured [560, p. 23].

Despite the wide scope of research themes the GALCIT's primary orientation was on aerodynamics related to aeronautics, with a focus on the Californian aviation industry. GALCIT played a similar role for the West Coast firms as the NACA's Langley Laboratory did for the aviation industry on the East Coast or the Göttingen AVA for German firms. Investigations for an industrial contractor usually took one, two or three weeks; an aviation firm would first make a scale model of the planned airplane with a maximal span of 3 m in the institute's model workshop and subject it to tests in the GALCIT's wind tunnel; the data were summarized in a report, compared with the corresponding data derived from theory, and sent to the manufacturer together with suggestions on how to improve the design. The performance of these tests was entrusted to advanced students, who thereby not only became familiar with aerodynamics as applied to practical problems but also became acquainted with aviation firms. After their graduation they often became hired by former GALCIT contractors. William Rees Sears, for example, who completed his Ph.D. under Kármán's supervision in 1938, was employed as chief aerodynamicist at Northrop Aircraft Inc., then a new aviation manufacturer with close ties to the GALCIT aerodynamicists [565]. For industrialists in Aachen like Georg Talbot or Hugo Junkers, Kármán was already an approachable academic – but these were humble beginnings compared to his relationships with governmental, military, and industrial circles in America. In the 1930s, airplane manufacturers in the USA and elsewhere usually did not yet posses their own wind tunnels, and thus, GALCIT's role as an external testing facility for the Californian aircraft industry can hardly be overstated. "Every major military or commercial airplane produced in Southern California within the past several years has been designed on the basis of wind tunnel tests carried out at GALCIT," the institute's history proudly reported in 1939: "Well known examples of such planes are the Douglas DC-2, DC-3, DC-4, and DC-5, the Douglas B-18 and B-23 Bombers, the new Douglas (Northrop) Attack Bombers built for the French and U.S. governments, the Lockheed Model 12 and 14 Transports and P-38 high-motored Pursuit, the Consolidated 4-engine Patrol Boat and the new Model 31 long range flying Boat, the North American Observation, Basic Combat and Bomber planes, the Boeing B-17 Bomber, the Vultee Attack Bomber, the Hughes Racer, the Curtiss-Wright Model 20 Transport, etc. etc. At present the tunnel is operating 75 hours per week with two complete shifts of workers and has an operating staff of about 20 persons. Its facilities are completely engaged 6 months in advance" [560, p. 20].

## 9.3
## Zurich

In small countries like Switzerland, the circumstances for the emergence of aeronautical engineering with its concomitant growth of fluid dynamics as a new engineering discipline were different from those in the larger industrial countries, where this growth happened largely in the wake of the boom of aviation after the First World War. Until the mid-1930s, Switzerland had no wind tunnel, let alone a national aeronautical research center. But in view of the rapid developments in the , , , , and other countries, Swiss politicians, industrialists, and academics did not want to be left aloof and made mere bystanders of their neighbors' booming aeronautical capacities. As in other countries, military interests were most predominant. In 1928 the Department of War Technology (Kriegstechnische Abteilung) of the Swiss War Ministry (Eidgenössisches Militärdepartment) and the Minister of Education (Schulratspresident) decided to establish the foundation for a national Swiss aviation. The ETH Zurich, renowned as one of the world's leading technical universities, was chosen as the site of a national aerodynamical testing establishment. However, it took a few more years before this decision resulted in tangible measures. In March 1930, representatives of the War Ministry, the Ministry for Mail and Railway, and the ETH convened in order to decide the next steps. They suggested that the ETH's laboratory for mechanical engineering (Maschinenlaboratorium), which was already slated for a major renovation, assume the role of the central Swiss aerodynamic research laboratory. Jacob Ackeret, who had recently returned from Prandtl's institute and appointed professor at the ETH, was to be entrusted with the directorship of this laboratory. Ackeret pointed out that the laboratory would need "first of all a wind tunnel"; furthermore, the new institute should be equipped with "a facility for very high air speeds" so that "important problems of ballistics" could be investigated. "In this regard the ETH can gain a lead over other centers abroad," Ackeret advised the ministerial representatives [566].

Ackeret was also to chair a new Scientific Commission for Aeronautics (Wissenschaftlichen Komission für das Flugwesen, or WKF) which was to coordinate the various Swiss civil and military interests in flight. For the time being, however, this commission produced no results "because of personal divergences among the leading men," as the minutes reported in September 1932 [567]. But this did not prevent the construction of the new laboratory at the ETH. The institute for aerodynamics, which accounted for the larger part of the laboratory, was furnished largely according to Ackeret's wishes. Although Ackeret's own teacher, Aurel Stodola, did not agree with his former student's plan of a novel closed-circuit supersonic wind tunnel, a design which had not materialized in even the most advanced aeronautical research laboratories abroad, Ackeret's plan was given the blessing of the authorities [568]. His institute should first serve the interest of the ETH for teaching and research, Ackeret explained in a budget report in 1934; research for government agencies and private users came second. Although the main focus was on "aeronautical research," Ackeret explicitly emphasized investigations "on fluid dynamics related to mechanical engineering, physics, and meteorology"; as an example, he mentioned "tests of wind pressure for building and railways." [569].

Ackeret had worked for several years after his return from Göttingen as an engineer at the Escher-Wyss AG, the leading Swiss turbine manufacturer, before he became the director of the ETH's new institute of aerodynamics. He always kept strong ties with his former employer and other Swiss machine manufacturers. Escher-Wyss, for example, designed a novel axial blower for the wind tunnel of Ackeret's institute and supported the first experiments "by providing materials and personnel at their own costs" [570]. The supersonic wind tunnel was also equipped with a newly developed turbine from Brown Boveri and Cie. (BBC); Escher-Wyss could not yet offer a comparable turbine, and therefore, Ackeret trusted his former employer's competitor with this project. BBC's "axial turbo blower" could suck in large volumes of air and compress the air in several stages to the required extent so that the tunnel could be operated in a closed circuit. Ackeret used his industrial contacts to competing firms so that he was always able to furnish his institute with the newest available technology [571–573].

Ackeret's institute immediately ranked among the best aerodynamic research centers of the world when it became operational in 1934. The large wind tunnel had a cross-section of 3 m × 2.1 m and allowed tests at an airspeed of 90 m/s (324 km/h). The supersonic closed-circuit wind tunnel was unparalleled. "This wind tunnel is at present the only one of this kind in the world," the German journal *Luftwissen* announced the news of its construction in January 1934, "it will be used not only for aerodynamic tests but also for the investigation of steam and gas flow through turbines." In view of a maximal air speed higher than Mach two, it was correctly assumed that "it will be particularly valuable for ballistic investigations and rocket tests" [505, p. 24]. In Italy the supersonic tunnel attracted such interest that Ackeret was hired as a consultant for the construction of an almost identical copy at the central aeronautical research facility, Guidonia, in Rome [507, pp. 537–542], [574]. In 1935 Ackeret was particularly proud that he could add to his congratulatory letter for Prandtl's 60th birthday the news that his institute was now operational. That both Zurich wind tunnels followed the Göttingen closed-circuit principle was hardly worth mentioning because at that time this principle had won recognition all around the world [561, 575].

There are more parallels between the new aerodynamic center in Zurich and the Göttingen model. Since 1934, a series of communications, the *Mitteilungen aus dem Institut für Aerodynamik an der ETH*, was published, following the example of the *Ergebnisse der Aerodynamischen Versuchsanstalt zu Göttingen* and a similar series from the aerodynamic institute in Aachen. Already the first issues showed that the fluid dynamicists in Zurich were also following a broad scope of research themes, ranging from investigations about wings close to the ground (G. Dätwyler, 1934) or axial blowers seen from the perspective of airfoil theory (C. Keller, 1934) to investigations of boundary layer suction (A. Gerber, 1938) and the application of gas dynamical methods to free-surface water flow (E. Preiswerk, 1938). If we include Ackeret's own publications from the 1930s, the thematic variety becomes even broader: three papers deal with wind pressure against chimneys and gas containers; five articles are concerned with the design of turbines, particularly steam turbines; six with cavitation; and several papers are dedicated to the equipment of his institute, particularly the measurement technique involved with the wind tunnel tests [576]. Although it is not possible to overlook the parallels between Zurich's and Göttingen's research programs (for example, wind pressure experiments were on the agenda of Prandtl's institute at the same time – see Chapter 7), Ackeret's institute did not simply follow the Göttingen model but developed its own characteristic profile. It had a stronger emphasis on mechanical engineering than other aerodynamical research centers, which is perhaps not astonishing in view of the institute's affiliation with the laboratory of mechanical engineering and the tradition founded by Stodola. Prandtl teased Ackeret in 1939 that "you apparently apostatized aerodynamics related to aeronautics" [577]. Besides Ackeret's contribution to the Prandtl Festschrift in the ZAMM in 1935, where he tackled the fundamental problem of vortex generation, many of Ackeret's papers appeared in journals like the *Schweizerische Bauzeitung* or the *Escher-Wyss-Mitteilungen* and dealt with applied engineering problems. But if we examine the period beyond the 1930s, when the construction of his institute no longer required Ackeret's exclusive attention, it becomes apparent that aeronautical research had not faded away from the agenda. By and large, Ackeret displayed a similar breadth of research interests as other Prandtl pupils: "I always felt that our ways of thinking are similar," Kármán reminded Ackeret of their common roots in Prandtl's institute when he congratulated Ackeret on his 60th birthday [576, p. 55].

Kármán was not merely flattering. He had displayed great interest in Ackeret's institute as early as in 1935 and expressed his intent to send students to Europe for further study: "Both Millikan (who visited your institution last year) and I found that the best place for them to go would be your department in Zurich," he wrote in a letter to Ackeret. "We were both very much impressed, not only by the laboratory but also by the manner in which you are conducting your work." Kármán envisioned a student exchange between his and Ackeret's institutes, so that both universities would share the costs [578]. In the end, the plan failed for financial reasons [579], but the mere intent is evidence for his high esteem of Zurich as a new center of modern fluid dynamics next to Göttingen and Aachen. On the other hand, Pasadena was also regarded as a most attractive center from the European perspective. Gottfried Dätwyler, for example, a student of Ackeret's, spent several years at GALCIT at his own expense after he had finished his dissertation in Zurich in 1933 [580]. There were also close relationships between Zurich and Prandtl's pupils in Germany. Ackeret's colleague and friend during his employment at Escher-Wyss, Fritz Schultz-Grunow, went to Göttingen for further studies and became Prandtl's closest collaborator for research on turbulence, before he was called to Aachen [581]. By the end of the 1930s, many of Prandtl's former pupils were themselves professors and spread the gospel of their greatly respected teacher to new generations of engineers both in Germany and abroad.

A particular occasion to cultivate the relationship among Prandtl's pupils came in 1935 with the Volta Congress on high-speed aerodynamics in Rome. Ackeret used this opportunity to present to his colleagues the concept of the new supersonic wind tunnel (see Fig. 9.4) which had become operational shortly before [507, pp. 487–537]. Prandtl met there with his most outstanding former students, all of them representatives of the elite of fluid dynamics: Ackeret, Busemann, Kármán, and Wieselsberger. The enthusiasm about the progress of their field of research was stronger than the irritation which some of them must have felt about the propaganda role this conference played for Italy's Fascist government. "Mussolini chose the opportunity to announce the invasion in Ethiopia," Kármán recalled. "This bold announcement stirred some spirited discussion among my colleagues – especially in view of Mussolini's use of high-speed bombing planes against the poorly equipped Abyssinians." Although it was obvious how pertinent high-speed aerodynamics – the theme of this conference – was for this and future wars, the collective excitement about the newest experimental and theoretical results was stronger than the dissenting views on Mussolini's politics. Kármán, certainly no friend of Fascism, was fully aware of the propaganda role which was attributed to him and his colleagues, but accepted its call without hesitation [85, pp. 218–222]. Ackeret, for his part, also had no fears of contact with Nazi Germany. Invitations to the annual meetings of the Lilienthal Society offered him a welcome opportunity to meet his former Göttingen colleagues and friends. In 1939, such invitations were no longer regarded as apolitical abroad, as we have seen (see Chapter 8): when Prandtl offered Ackeret an invitation that year to the meeting to be held in the Reich's capital of Berlin, he also forwarded an argument of how Ackeret could respond to criticism of Swiss compatriots: "After the through and through peaceful tendency of yesterday's declaration of our Führer and Reich's Chancellor," Prandtl wrote on 29 April 1939 to Zurich, "there should hardly arise political problems for you if you participate in a German conference by hinting at your former close collaboration with us" [582].

Even during the Second World War, the contact between Ackeret and his German colleagues was not interrupted. In 1941, Ackeret, together with two researchers at his institute and a representative from each of Escher-Wyss and BBC, traveled to the AVA in Göttingen, where great interest was expressed to acquire the newest technology of supersonic wind tunnels. After this visit, Ackeret wrote in a report to his authorities: "Besides the Göttingen gentlemen, in particular Prof. Prandtl and Prof. Betz, there were gentlemen present from the Reich's Air Ministry and the German Testing Facility Adlershof; in fact, only people who are directly involved with the special questions. It may be said that on the part of the Germans very rich material was presented with great candor; for example, extended test series with artillery and infantry shells and new types of aircraft. The material which we presented was largely composed from results which are now prepared for publication or which has already recently been presented in Basle. By and large we can say with gratification that our work was very esteemed, which is also evident from the fact that contracts for one or two supersonic wind tunnels from BBC are now very likely" [583]. In December 1942, Ackeret traveled again to Germany in order to present a talk on gas turbines before the German Academy of Aero-

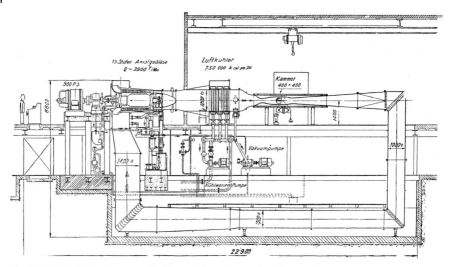

**Fig. 9.4** At the Volta Congress in Rome Ackeret presented the novel scheme of the Zurich closed-circuit supersonic wind tunnel. The air was circulated by a 900 horsepower axial blower produced by the Swiss turbine manufacturer BBC [507, p. 529].

nautical Research. Ackeret played a similar role for the development of fluid dynamics in Switzerland as Prandtl in Germany and Kármán in the USA – particularly with regard to the military applications and as an advisor in the political, industrial and military arena. Although it is beyond the scope of this book it seems apt to mention that in the Second World War, Prandtl served as the chair of a high-ranking advisory board ("Forschungsführung") for the Air Ministry [120, pp. 246–260]; Kármán became famous as General "Hap" Arnold's advisor for the foundation of the research establishments of the U.S. Air Force [584]; Ackeret was busy as an advisor for the Swiss War Ministry (Eidgenössisches Militärdepartment) [585, 586].

# 10
# Fluid Dynamics on the Eve of the Second World War

The emergence of new centers in Aachen, Pasadena, and Zurich, directed by Prandtl's pupils Kármán, Wieselsberger, and Ackeret, shows how Prandtl's school began to spread its influence outside Göttingen in the course of the 1930s. We can easily find examples of other Prandtl pupils who became professors at other technical universities or at new aeronautical research facilities, like Busemann, who became the director of an institute for gas dynamics at the Aeronautical Research Establishment Hermann Göring (Luftfahrtforschungsanstalt Hermann Göring, LFA). The extent of Prandtl and his school's influence on the course of fluid dynamics becomes even more apparent if we examine, in addition to the emergence of new institutions, the burst of new theoretical and practical applications, which resulted in quite a different outlook for this discipline compared to its beginnings three or four decades earlier. On the eve of the Second World War, fluid dynamics was an engineering science in which applied mathematicians and practical engineers found ample opportunity to display their virtuosity with ever new applied theories and practical applications. This is not to say that the proverbial schism between theory and practice had been completely resolved. A closer look at three major subject areas—airfoil theory, turbulence, and gas dynamics – will illustrate how closely science and practice in fluid dynamics had approached one another, and yet remained separate with distinct interests and diverging tendencies.

## 10.1
### Airfoil Theory

"An engineer needs successful design materials that enable him to oversee the effect of his measures quickly and safely." This is how, in 1936, Betz introduced an article on the "Tasks and Prospects of Theory in Fluid Dynamics." Ideally, theoretical research would make the required knowledge available in the form of diagrams and formulae, from which everything an engineer

*The Dawn of Fluid Dynamics: A Discipline between Science and Technology.* Michael Eckert
Copyright © 2006 WILEY-VCH Verlag GmbH & Co. KGaA, Weinheim
ISBN: 3-527-40513-5

would need for his design could be obtained. Depending on the specific subject matter, however, this ideal goal was more or less within reach. To describe how close theory came to practical application, Betz defined four levels. On the lowest level, theory was not advanced enough to offer solutions for all practical cases. He named turbulence as an example, where theory was still unable to offer practical methods for engineering. Not only mathematical virtuosity but also "sensitive guessing of the physical processes" were required in order to make use of the theory, and the prospects of success were uncertain. On the second level, the physical mechanism was well-understood in theory, but the mathematical development was not yet advanced enough. Progress in this phase required "the work of highly talented people," but also depended to some extent on chance. An example of such progress was propeller theory in the 1920s, a field to which Betz himself had made pioneering contributions. On the third level, the solution of a problem was possible both from the underlying physics and the mathematical analysis, but the latter involved such a tremendous amount of effort that it was still beyond the reach of practical design; airfoil theory was on that level. Only on the fourth level physics, mathematics and the procedures of analysis were all in place so that a practitioner could use them; Betz placed the theory of the induced drag in this category [587].

Betz had airplane design in mind as he reflected on the use of fluid dynamics theory for practical applications. His favorite example was airfoil theory, which he placed "at about level three or four" with regard to its utility for airplane design. Prandtl's airfoil theory, however, was based on ideal fluid theory, an approximation that was only useful for the calculation of lift and induced drag. Many practical problems, such as the theoretical determination of the angle of attack at which the air flow detaches from the wing such that it stalls, were beyond its reach. Here, Prandtl's boundary layer theory offered the prospect "to find the laws which make it possible to pursue the processes by theory at least approximately (level one and two)," Betz argued. "With the airfoil, one has the impression that not much is lacking before we can determine theoretically all we need to know." Conceptually, airfoil theory was composed of two parts: profile theory, which could be treated mathematically as a two-dimensional problem; and the spatial part of the theory, which addressed all questions related to the finite span and the planform of a wing. Both had been the subject of considerable mathematical effort and resulted in sophisticated methods and procedures, but could not surmount the final hurdle to become practical tools for the industrial airplane designer (Chapter 6). Betz rated the various profile theories as level-three theories, with a good chance of proceeding to the fourth level. Regarding the three-dimensional problem, two groups of problems were discerned: first, determining the planform and the induced drag from a given distribution of lift over the span; and second, the more difficult inverse problem of calculating the distribution of lift over the span for a given planform. The first problem had already reached level four after the First World War; the second approached the third level with Irmgard Lotz's theory by the early 1930s [587].

Betz had good reasons for why he regarded profile theories close to the fourth level. For thin profiles Munk, Glauert and others had developed the method of singularities far enough such that the flow over an almost arbitrary profile could be calculated from potential theory by substituting the profile with a suitable distribution of sources and sinks. Thicker profiles were treated using the method of conformal mapping, either according to the original Kutta–Joukowsky method or a modification introduced by Kármán and Trefftz [544]; the latter avoided the problem of an infinitely sharp rear edge as assumed with Joukowsky profiles. However, both the Joukowsky and Kármán–Trefftz profiles assumed a circular mean line, which involved the unfavorable consequence that the location of minimal pressure moved backwards with increasing angle of attack. Practical wing shapes avoided this with a stronger curvature in the front part compared to the rear part of the profile. In 1937, Betz and his collaborator Fritz Keune generalized the Kármán–Trefftz procedure by superposing an S-shaped line to the circular mean line, so that the theoretical profile shapes became further adjusted to the needs of practical wing design (see Fig. 10.1). However, the calculation required some effort: "By the use of supplementary constructions which address recurring steps, the construction of such profiles and the calculation of the pressure distribution can be accomplished in about one day," the authors commented on the practical use of this procedure [588].

In a survey on the state of profile research from summer 1939 a researcher at the AVA listed 52 publications, yet he was unable to "sufficiently answer all pertinent questions" of this specialty. He limited his study to simple profiles, excluding investigations on profiles with flaps and other details that complicated theory but were nevertheless very important for the practical design of wings. Regarding the procedures based on conformal mapping, he concluded that it was possible now to determine the distribution of pressure around "arbitrarily shaped profiles with sufficient precision." The published catalogs of Göttingen and NACA profiles provided enough data to check the theoretical results with wind tunnel data. However, skin friction and wind tunnel turbulence required further tests before the experimental results could be compared with theory. A comparison of test results from Munk's Variable-Density Tunnel (VDT) with corresponding measurements from the NACA's Full-Scale Tunnel (FST) showed, for example, that the profile drag coefficients as measured in the VDT were systematically higher than those from the FST. Nevertheless, the reviewer concluded his survey of profile investigations with a positive assessment: tangible progress resulted from comparisons of results obtained in different wind tunnels [589].

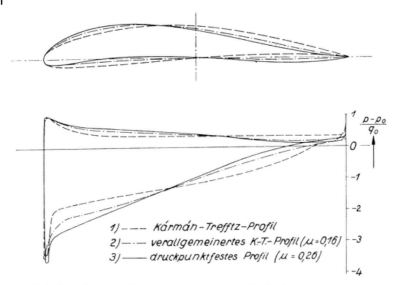

**Fig. 10.1** Betz–Keune profiles account for pressure distributions as required for practical wing design [588, p. I-46].

From the perspective of the practical airplane designer, the greatest benefit could be gleaned from theory if it was able to guide systematic profile investigations. This was the opinion of Kurt Tank, for example, who had gathered accounts of practical design at two aircraft factories, Messerschmitt and Focke-Wulf, and who counted himself among the new breed of "scientifically minded airplane designers" [590]. The boom of aeronautics in the wake of the Nazi war preparation resulted in the establishment of research laboratories at the various airplane factories themselves, in addition to the central aeronautical research facilities like the AVA and the institutes at the technical universities. In 1941, Tank introduced an issue of *Luftfahrt-Forschung* with reflections about the role of industrial research in the rapid progress of German airplane production. "The speed with which the development of German airplane production has been pushed forward was possible because the industry could do their own research to a very large extent." The "hard constraint" of finding solutions to very special problems by certain deadlines implied different research strategies than at the central aeronautical research laboratories, where systematical research prevailed. "Both types are necessary and will be fruitful for the development as a whole," Tank argued, "if a vivid mutual exchange of theoretical and practical research results takes place as much as possible" [591].

Compared with profile theory, other parts of airfoil theory were less advanced. The director of the Institute for Flight Research and Aeronautics at the AVA wrote in 1937 that "up to now about 70 methods" had been developed for the problem of determining the lift distribution over the span of a wing with a given planform, but all of them "fairly intricate, and not every design bureau has the required experience and the necessary instruments" [592].

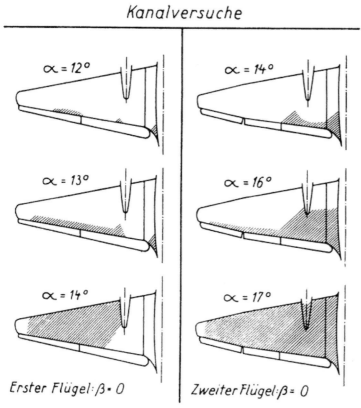

Kanalversuche

Erster Flügel: β = 0    Zweiter Flügel: β = 0

**Fig. 10.2** Comparison of Ju-86 wing models in wind tunnel tests with respect to tip-over controllability. The shaded areas indicate flow separation [593, p. 166].

An aerodynamicist at Junkers, August Quick, illustrated with the case of the wing of the Ju-86 why it was so desirable to know the theoretical distribution of lift in advance. An early test version of the Ju-86 showed a tendency to tip over sideways quite rapidly. The cause for this behavior was that the flow separated sooner from the wing along the outer parts rather than closer to the fuselage (see Fig. 10.2). In this situation, the designer was confronted with the task of improving the controllability, such that with an increasing angle of attack, flow separation would proceed along the span from the fuselage outwards rather than the other way around. Without prior knowledge of the distribution of lift, it was impossible to predict such wing behavior, and a large number of wind tunnel testing was necessary in order to determine the optimal profile and wing planform by trial and error. Even at a firm like Junkers, which by the end of the 1930s already had a long tradition of wind tunnel testing, one found it "enjoyable that in this regard progress may be expected for the next time" [593].

Aerodynamicists from the academic and central research laboratories welcomed the information on the firms' own research results, such as in the case of the Ju-86 wing design, because it offered them a new look at the practical problems involved with aeronautical engineering. Previously they obtained only complete wing designs for wind tunnel tests without further information on the specific goals that motivated one development or another, as the director of the Berlin DVL once complained [594]. As a consequence of closer contact between theorists and practitioners, airfoil theory became more practical by the end of the 1930s. In 1938, Hans Multhopp, a theoretical aerodynamicist from the AVA, published a method to determine the lift distribution over the span of a wing, which "already stood the test both in industry and at various German research establishments" [595]. Multhopp's method represented the final stage of a development, which had begun with Betz's dissertation in 1919 and which aimed at a solution to the lift distribution problem in terms of Prandtl's "lifting line" concept. At the core of the problem was the integral equation from Prandtl's theory, which expressed the relationship between lift and angle of attack. Reviews like Durand's encyclopedia reported numerous attempts to solve this problem [596,597], but Multhopp's method was the first one which stood the test of practical utility in industry. It transformed the integral equation into a system of equations, corresponding to a set of pre-determined points along the "lifting line," which was solved iteratively. It was crucial for practical use that the coefficients of the system of equations only depended on the total number of pre-determined points, so that certain data could be calculated in advance and presented in the form of tables for the final calculation of a specific wing. If the number of points was chosen as 7, for example, the calculation of the lift distribution could be executed in half an hour; with 15 points it took two hours, and with 31 four to five hours [598]. If the lift distribution had to be determined only for small angles of attack (i.e., as long as lift varies linearly with the angle of attack), the effort could be further reduced [599].

Multhopp's method considerably narrowed the gap between theory and practice in the design of airplanes. "Today, such calculations, particularly along Multhopp's procedure, have also become standard in industry," Betz wrote in a review in 1941 [403, p. 44]. Multhopp himself pursued his career at Focke-Wulf, where he focused, for example, on calculating the lift of the fuselage of airplanes [600]. Other former theorists of the AVA also became employed in design offices of airplane manufacturers: Heinrich Helmbold, Friedrich Keune, and Oskar Schrenk at Heinkel, and Hans Winter at Messerschmitt. Their publications after 1939 illustrate how much theory these "scientifically minded airplane designers" brought to industry. In 1940, for example, Schrenk published an approximation procedure which made it possible to calculate the distribution of lift "within a few minutes" and which could be used when it was more important to obtain data upon which a first design could be founded rather than to reach high degrees of precision for later design phases. A comparison of these data with the more precise data obtained from Multhopp's method showed only minor differences [601]. In 1941, Helmbold published formulae with which it was possible to account for variations of a wing's planform and twist in a simple manner [602].

The lifting line concept, however, failed in a number of practically important cases. Short wings, for example, could not be treated with this theory. In the late 1930s Prandtl published a new airfoil theory, based on earlier work by his pupils, Walter Ackermann and Walter Birnbaum—the "lifting surface" concept [603]. Once more, Kármán was his rival. A year before Prandtl's first paper on the lifting surface concept was published in the ZAMM, Kármán had published a similar theory in the same journal [604]. Although Prandtl was able to present more developed theoretical results for the lift of spherical and square plates, for example – cases that were beyond the applicability of the lifting line theories – the new concept still had a long way to go before it reached a level of practical utility for wing design comparable to those theories based on the lifting line concept. Nevertheless, it was important from a practical perspective, because it served to prove the applicability of the lifting line concept. Similarly, airfoil theory could not account for wing flutter and other phenomena that implied non-stationary flow. Here, too, Birnbaum and others had paved the way as early as the 1920s, but these theories were limited to idealizations (such as the two-dimensional case of plane flow); furthermore, wind tunnel experiments were more difficult for non-stationary flow problems, such that theoretical results could not easily be compared to experimental data [605–607].

Apart from these cases, however, airfoil theory according to the lifting line concept yielded satisfactory solutions in most cases involving the calculation of lift. Multhopp's method placed airfoil theory on the fourth level of Betz's scale of practical utility. In view of the unrealistic assumption of ideal flow, upon which Prandtl's airfoil theory was based, the far-reaching applicability of this theory is astonishing. Even the drag of a wing was not entirely beyond its reach. The induced drag or wingtip resistance due to the wing's finite span resulted from the lifting line theory without additional assumptions. Only profile drag, due to skin friction and eddy formation in the wake of each wing section, required additional knowledge involving assumptions regarding the boundary layer and vortex-shedding. The development of approximate solutions for the drag problem, however, required more than practical methods to solve the boundary layer equations. Whenever skin friction and vortex-shedding played a role, the aeronautical engineers became aware that a complete mastery of theoretical wing design was hindered by the unsolved riddle of turbulence. In contrast to procedures like Multhopp's method, where applied mathematics bridged the gap between theory and practice, turbulence was still to a large extent a physical problem.

## 10.2
## Turbulence

During the 1930s, the drag problem was as a major challenge for airplane design. The quest for high speed involved systematic research into all possible sources of drag. Apart from the wing itself, other parts of an airplane were subjected to scrutiny with regard to the possibility of drag reduction. The period between the First and Second World Wars is regarded in the history of airplane design as the epoch of "streamlining" and "drag clean-up" [42, Chapter 8], [233, Chapter 5]—a period of "reinventing the airplane" [608, p. 41]. As a consequence of streamlining, airplane design was subject to additional constraints. In 1936, Clark Millikan remarked before the Lilienthal Society in his talk on aerodynamical research and its consequence for airplane design that for this reason "almost all new American airplanes have surprisingly similar shapes" [562]. In Germany, too, the drag problem was a recurring theme throughout this period. The DVL published a scheme for estimating how much the actual drag of an airplane exceeded the minimal theoretical drag. The ratio between minimal and actual drag was designated as the "aerodynamical efficiency number," from which one could estimate to what extent the speed could be further increased by additional "aerodynamical refinement" [609]. In 1938, Ernst Heinkel called the further increase of speed, possibly beyond the speed of sound, the "crucial problem" of airplane design in Germany and argued that in the pursuit of this goal, overall progress will be fostered most efficiently. Among the various drag contributions, skin friction was most significant. He estimated that for a high-speed airplane, 50 percent of the total drag was caused by skin friction. Its reduction involved an improvement of the airplane's surface. He communicated test results according to which polished surfaces and the use of rivets with immersed heads reduced the drag of an airplane's surface by 27 percent, compared to surfaces with normal rivets and conventional paint [610].

Solving the skin friction problem involved the solution of the boundary layer equations for both laminar and turbulent flow (see Fig. 10.3). Betz placed turbulence on the lowest level on his scale of practical utility, as we have seen, where "sensitive guessing of the physical processes" was required in order to make use of the theory, and the prospects of success were uncertain. Turbu-

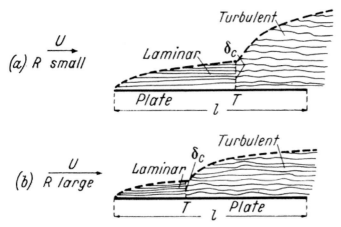

(a) R small

(b) R large

**Fig. 10.3** The theoretical determination of skin friction requires knowledge of the boundary layer, in particular on the ranges of the laminar and turbulent boundary layers. At higher Reynolds numbers, the transition point from laminar to turbulent boundary layer flow moves upstream [612, p. 60].

lent skin friction, however, had found such extensive treatment in the semi-empirical approaches developed by Prandtl, Kármán, Nikuradse, Schlichting, Tollmien, and others since the 1920s (see Chapter 5) that Betz exempted this case from his general evaluation of the turbulence problem and placed boundary layer turbulence on the third and certain special cases even on the fourth level [587]. The DVL's "aerodynamical efficiency number" for high-speed airplanes, for example, contained a coefficient for skin friction based on theoretical work published by Prandtl and Schlichting in 1934 [611]. However, their formula for turbulent skin friction applied to the case of a flat plate rather than wings with different profiles. Upstream from the leading edge along the upper side of a profile the pressure decreases to a minimum from where it rises again as the flow approaches the rear end of the wing. At some point along this passage the laminar boundary layer becomes turbulent; as the flow in the turbulent boundary layer experiences an increasing pressure towards the rear end of the wing, it detaches from the surfaces. In other words, to calculate the drag of curved surfaces, it was crucial to know how the boundary layer behaved under varying pressure.

In 1930 Prandtl suggested this problem to Eugen Gruschwitz as the theme of a doctoral dissertation. Gruschwitz built a special wind tunnel (see Fig. 10.4) in which the pressure in the air flow along a flat plate could be varied in a controlled manner. In order to compare the measured data with theoretical results, Gruschwitz derived an approximate procedure to determine the turbulent boundary layer under the influence of given pressure gradients [613]. The result was in Betz's judgment a "physically unfounded but usable prac-

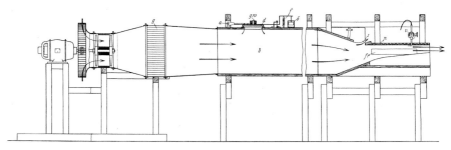

**Fig. 10.4** Gruschwitz's wind tunnel for measuring turbulent boundary layers [613, p. 324].

tical calculation procedure" for the turbulent boundary layer: "Gruschwitz's method is relatively simple, and there is now the strange situation that with this method the calculation of the turbulent friction layer, which in reality is much more intricate, becomes easier than the calculation of the laminar friction layer according to the Pohlhausen method" [403, p. 48]. The Pohlhausen method (see Chapter 5) was also made the subject of further study. In 1940, Betz's fellow researcher at the AVA derived a graphical method for the determination of laminar boundary layers [614]. With these procedures, laminar and turbulent boundary layers could be calculated for curved surfaces, and it became possible to derive formulae for the drag due to skin friction in each case; however, it was not possible to determine without additional information the point of transition where the laminar boundary layer became turbulent. That deficiency was solved empirically after evaluating Gruschwitz's diagrams [609].

The theory of skin friction of wings shows how far academic turbulence research approached the practical problems involved with airplane design. This convergence was enabled by supplementing empirical results derived from wind tunnel experiments. However, the unsolved riddle of wind tunnel turbulence made such empirical foundations of dubious value. Measurements of skin friction in a wind tunnel were critically dependent on the amount of turbulence present in the air stream of the specific tunnel. Kármán and Clark Millikan illustrated this dependence in 1934 with a diagram (see Fig. 10.5) in which they compared the drag coefficients of a sphere determined at different levels of wind tunnel turbulence (controlled by inserting a grid at various positions upstream the test model).

"No theory is advanced enough that the distribution of mean velocity and of the turbulence level can be predicted or calculated in such cases," Kármán concluded in 1937 at the end of a lecture on turbulence before the Royal Aeronautical Society in London. Turbulence should not be regarded "merely as an interesting chapter of mathematical physics," because "if we meet a practical question in aerodynamic design which we are unable to answer, the reason

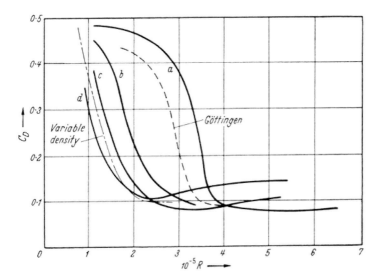

**Fig. 10.5** The drag curve of a sphere ($c_D$ versus Reynolds number) depends on wind tunnel turbulence. The *solid curves* refer to the GALCIT wind tunnel (a) without a grid, (b) with a grid at 48 inches, (c) with a grid at 20.5 inches, and (d) with a grid at 10.5 inches upstream from the sphere; for comparison, equivalent results from the NACA's VDT and the Göttingen wind tunnel are shown as *dashed curves* [612, p. 57].

that we are unable to give a definitive answer is almost certainly that it involves turbulence." But he also expressed optimism: "I believe that in spite of the complex mathematical and physical aspects of the problem of turbulence, the scientist is justified in saying to the practical engineer: Tua res agitur (your case is on trial)" [615]. Kármán's optimism was sparked by G.I. Taylor's recent attempt to solve the turbulence problem with a new theory [616]. Although the title of Taylor's treatise, "Statistical Theory of Turbulence," did not at first sight raise high expectations for a practitioner engaged in aeronautical engineering, the theory was of immediate pertinence for aviation because, as Kármán explained at a conference of the Institute of Aeronautical Sciences, "it is concerned with the important problem of wind-tunnel turbulence and its results could be compared directly with experimental work" [617]. For Kármán himself Taylor's work served as an enticement to focus on turbulence again [618]. Another practically minded fluid dynamicist, Hugh Dryden, took the new theory as an opportunity to pursue new experimental work on turbulence at the National Bureau of Standards. At the request of the NACA, Dryden and his colleagues made a new effort to investigate wind tunnel turbulence with an improved hot-wire technique [619]. The reliability of wind tunnel data hinged on solid knowledge about wind tunnel turbulence. No aeronautical testing facility or industrial airplane designer could afford to ignore the problem of how wind tunnel turbulence affected the measured data for long. In Göttingen, too, Taylor's theory was received with great interest: "The recent statistical theory of turbulence which mainly was developed in England and America, has procured valuable results on these relations," a colleague of Prandtl's remarked in 1941 in a summary review for *Luftfahrt-forschung*—a journal which would hardly have had room at that time to review such a theory if it were not pertinent for practical aeronautics [620].

Disregarding the specific type of wind tunnel, the stream of air usually arrives in the test chamber after it is "calmed" by a honeycomb or a grid, where major flow inhomogeneities are leveled out. The passage through the meshes causes new vortical motion, but this intentionally imposed turbulence decreases downstream such that a largely homogeneous stream of air arrives at the model in the test chamber. From a theoretical perspective, it is important that the turbulence caused by the meshes of the honeycomb or grid be largely isotropic, i.e., the mean deviation from the speed of the airflow is the same in all directions. By correlating the deviations to one another at points separated by various distances, one obtains so-called correlation coefficients which could be directly measured by the hot-wire technique. Taylor's theory predicted a "scale of turbulence" (a length related to the mesh size) and another characteristic length, defined by the dissipation of turbulent energy (i.e., the kinetic energy derived from the deviations from the mean velocity of the air stream). Dimensional analysis revealed a relationship between both characteristic lengths, which predicted a linear decrease of turbulence with increasing distance downstream from the grid – a result that could be verified experimentally for a wide range of air speeds and mesh sizes [616, p. 314].

The physical concept underlying Taylor's "statistical theory" was akin to Prandtl's idea of a "mixing length." As early as 1915, in the context of a study on turbulent mixing in the atmosphere, Taylor had suggested focusing the mathematical analysis on the *mean* vortical motion rather than studying the hopelessly complex individual motion of single vortices; his model was kinetic gas theory, where one also does not consider the collision of individual molecules but only the mean motion that results from many such collisions. Taylor attempted to describe the turbulent transfer of energy in a fluid as a diffusion process [621]. In 1921, Taylor further pursued the parallel between a statistical description of gases and turbulent diffusion, aiming at a calculation of the correlation function of the velocities fluid particle assumes at a given position at different times. The extent by which the correlation decreased with increasing time served him as a quantitative measure of turbulence [622].

Prandtl's mixing length hypothesis (Chapter 5) was motivated by similar ideas – except for one important difference: Taylor imagined that turbulent mixing in a fluid resulted from diffusion of vorticity, while Prandtl's concept envisioned a diffusion of momentum. In 1932, Taylor published a paper on turbulent heat transport in which he further stressed vorticity as the dominant exchange mechanism; a recent wind tunnel experiment at the National

Physical Laboratory, in which the heat exchange in the wake of a heated cylinder exposed to a stream of air was measured, seemed to confirm Taylor's assertion [623]. Prandtl, however, regarded the experimental result only as evidence "that there are two different forms of turbulence, one belonging to the fluid motions along walls and the other belonging to a mixture of free jets." For the former, he insisted that turbulence involved a transfer of momentum as described in his mixing length theory; in free flow Taylor's concept might be correct, he agreed: "In the meantime we have experimentally studied the flow of a cold jet of air through a warm room, and found that in this case, the turbulence is of your kind." Prandtl had not known of Taylor's "old papers of 1915 and 1922," he regretted in a postscript, "I think that if I had known about these papers, I would have found the way to turbulence earlier" [624].

Taylor also had not taken notice of Prandtl's turbulence theory for a long time, as he admitted in his paper in 1932 [623, p. 254]. Although they met during Prandtl's visit to England in 1927 and corresponded with one another since that time, their mutual relationship was at first prone to misunderstandings. Taylor wrote in English and in a handwriting that caused Prandtl some pain to decipher. Only when they became better known to one another in the course of the International Congress for Applied Mechanics in 1934 in Cambridge did their relationship turn from distant respect into collegial friendship. Taylor invited Prandtl to live in his house, which was an "enjoyable surprise" for Prandtl. "If we now become engaged again in scientific debates it will work better than in 1927," he wrote to Taylor, "because since then I have been in America for three months and therefore understand English better than then" [625]. After the Congress they kept in close contact and coordinated their research on turbulence. Prandtl remarked in the annual report of his institute for the year 1935: "The experimental investigations about the properties of turbulent fluctuations have been pursued and resulted in a collaboration with Professor G.I. Taylor in Cambridge (England), who develops this field theoretically" [626]. In the same year, Taylor presented a first result of this division of labor between Cambridge and Göttingen in the form of an article to the Festschrift for Prandtl's 60th birthday. Although it dealt with a specific case, the turbulence in the convergent part of wind tunnel nozzles, the paper may be regarded as a direct prelude to Taylor's burgeoning statistical theory. This contribution shows, to a better extent than the subsequent publications, how closely theorizing was based on experimental experience. Taylor's study addressed the specific problem of how the air in the contracting passage of a wind tunnel upstream from the measuring chamber is calmed and how the remaining fluctuations of velocity in the air stream can be assessed theoretically and compared with experimental data [627].

The correspondence between Prandtl and Taylor provides further evidence of how heavily Taylor's statistical theory relied on wind tunnel experiments in Göttingen and elsewhere. Taylor was particularly interested to calculate the "rate of decay of turbulent energy in a wind stream," he explained to Prandtl; among the experimental data with which he compared his theoretical results were "Schlichting's measurements of the spreading of a wake," obtained in 1930 in Prandtl's institute [284, 628]. Prandtl sent Taylor a more recent experimental study motivated by his interest in meteorology, in which a student of Prandtl's analyzed with a refined hot-wire technique the correlation of velocity fluctuations in a wind tunnel whose walls were kept at different temperatures, so that turbulence was influenced by a temperature gradient transverse to the direction of flow [629]. "Your paper with Dr. Reichardt shows me once again how difficult it is to suggest to people research work that is worth doing and yet has not been tried at Göttingen," Taylor responded [630]. Kármán may have had the same thought when Prandtl reported to him on "beautiful progress" made with these experiments [631]. The old rivalry with his former teacher, which was a recurring feeling for Kármán, especially when the turbulence riddle was concerned, enticed him to new efforts. He invited Dryden's collaborator Arnold Kuethe from the National Bureau of Standards, an expert of the hot-wire measurement technique, to the GALCIT. Subsequently, Kuethe and Kármán's assistant, Frank Wattendorf, investigated the velocity fluctuations in GALCIT's wind tunnel using this technique [560]. When Taylor informed Prandtl in April 1935 about the progress of his theory, he added the recent measurements from Pasadena to those from Göttingen as a confirmation of his results: "It gives, of course, exactly the data which I wanted to for comparison with my theory of energy dissipation," Taylor triumphed, "the agreement is very good indeed" [630].

Although Taylor's theory involved sophisticated mathematical and physical arguments and appeared by its title rather as a contribution to theoretical physics than to applied fluid dynamics, it primarily addressed the engineering community: "The study of turbulence is not only of direct interest to aeronautics, but is indispensable for the advance of the sciences of meteorology and oceanography. But the chief incentive to explore such problems in recent years has been provided by the study of the practical problems of flight." In this manner turbulence was presented in 1938, for example, as a pertinent specialty for practical engineering in the two-volume textbook *Modern Developments in Fluid Dynamics*. This treatise became a standard text for several decades. Composed by a panel of the British Aeronautical Research Committee and edited by Sydney Goldstein, it was addressed to a broad audience of theorists and practitioners – to all "who are primarily interested in the physical and engineering aspects of fluid flow, as well as those who are engaged on its mathematical study" [632]. Kármán, too, elaborated and propagated the new theory in wider circles, from the academic members of the American National Academy of Sciences to the practical-minded readership of the reports of The Royal Aeronautical Society in England or L'Aerotecnica in Italy [615, 617, 618, 633].

At the Fifth International Congress for Applied Mechanics in Cambridge, Massachusetts, USA, turbulence ranked high on the agenda. Kármán, who served as one of the main organizers, invited Taylor to present a general lecture on turbulence. "I have told him of our plan to have an afternoon discussion of turbulence in which the various workers on the problem might come together and report progress," Kármán informed Prandtl. "Professor Taylor thought it most appropriate for you to preside and to lead the discussion" [634]. Prandtl accepted the role as chair of the symposium on turbulence; he, therefore, became more an arbiter than a pleader for his own cause. He made no effort to present the mixing-length approach to the turbulence problem – based on the exchange of momentum – as superior to others, particularly to those of Taylor and Kármán. "We are forced to discern different kinds of turbulence for which the rules of calculation are different," Prandtl argued. "In addition to the older kinds, wall turbulence and free turbulence, there has been added in the meantime the turbulence of stratified flows and recently the isotropic turbulence, which decays in time." He mentioned work in progress at his institute, such as investigations of wall flows from smooth to rough surfaces and about the spectral distribution of turbulent velocity fluctuations [635]. Besides Prandtl, Taylor, and Kármán, the participants of the turbulence symposium comprised the elite of theoretical and experimental research on turbulence. More than 20 talks were held on various aspects of turbulence, with applications from aeronautics to meteorology [636].

Prandtl's own contribution to the advancement of turbulence theory in the 1930s is less visible from his published record. His Collected Papers contain six publications in the section "Turbulence and Vortex Formation" for the years from 1925 to 1929, seven for the subsequent five-year period, but just one with only 5 printed pages between 1935 and 1942. It would be wrong, however, to infer a fading interest in turbulence research on the part of Prandtl from this record; one sees from the work of Prandtl's pupils (Hans Reichardt, Fritz Schultz-Grunow, Henry Görtler) that turbulence remained a major research topic throughout the 1930s and beyond. Although Prandtl did not himself contribute to a new theory, he attempted to define new goals of turbulence research and to consolidate those areas where theory had been confirmed by experiment. This tendency may be clearly seen in his *Führer durch die Strömungslehre* which appeared as a first edition in 1942 and became a standard textbook for the German-speaking countries. Originally conceived as an updated edition of the *Abriss der Strömungslehre*, which appeared in two editions in 1931 and 1935, the 1942 edition turned into a thoroughly revised new textbook. In the preface Prandtl explicitly referred to the new sections on turbulence. However, in contrast to the English textbook counterpart, *Modern*

*Developments in Fluid Dynamics*, the theoretical derivations of the new statistical turbulence theory were reduced to a minimum and, where unavoidable, were relegated to small-print passages which "may conveniently be skipped at first reading" according to the preface. Beyond such differences, however, Prandtl's *Führer durch die Strömungslehre* shared an emphasis on practical applications [637, preface and pp. 119–121].

## 10.3
## Gas Dynamics

In contrast to turbulence, which addressed both practical and theoretical interests, gas dynamics was a concern of engineers without noticeable ramifications in pure science. Gas dynamics found practical application "in the flow in steam turbines and similar machines, as well as in the motion of projectiles and fast airplanes, and furthermore with fast rotating propellers," such was Prandtl's portrayal of the practical importance of gas dynamics, a specialty which could simply be characterized as "fluid dynamics for high speeds" from the vantage point of the late 1930s [637, p. 233]. In terms of the physical principles, "high speeds" meant that the gas could no longer be assumed as incompressible. This change of premise involved modifications of the flow equations; although compressibility effects involved new phenomena which could not easily be accounted for by the theoretical and experimental methods of incompressible fluid dynamics, there were no fundamental physical riddles comparable to those in turbulence research. Gas dynamics in the 1930s called for the development of sophisticated mathematical methods and experimental techniques rather than the discovery of new physical mechanisms. Nevertheless, the relationship between theory and practice in gas dynamics was different from other aerodynamic specialties like airfoil theory, for example, where the underlying physics was also solved in principle: first, because technical applications and, consequently, practical interests were more multifarious and far exceeded the realm of aeronautics; second, because there was a lack of experience with trans- and supersonic phenomena so that despite a basic knowledge of the involved physical processes, the further development of gas dynamics had to await the availability of highly sophisticated experimental facilities such as supersonic wind tunnels and Schlieren optics.

It is particularly interesting to consider the role of theory in supersonic flight. Compared to the early years of aviation around 1900, the relationship between theory and experiment in supersonic flight was reversed. Prandtl's airfoil theory had arrived at an explanation of the basics of lift and induced drag for finite wings only after airplanes were already in the sky for more than a decade. But the principles of lift and drag in the supersonic velocity range were largely known long before 1947, the year in which the first airplane flew faster than sound. The first approaches were proposed in the 1920s with Prandtl, Ackeret, and Glauert's "rule" (see Chapter 7). These theories addressed the case of plane flow only (i.e., the wing of infinite span), but the analog of Prandtl's lifting line theory for a finite wing span followed soon after; the principles of a finite wing theory for supersonic flight were published in 1936 by Prandtl and Schlichting [638, 639]. Theory, therefore, preceded practical experience with supersonic flight. New phenomena that would be encountered beyond the sound barrier, in particular, the new kind of air resistance due to the creation of shock waves ("wave drag"), were known in principle from these theories. This does not mean, however, that theory had that much to offer practical engineering before supersonic flight became a reality. The development of a successful engineering design for the transonic speed range involved more than the derivation of basic principles. This did not happen before airplanes were really flying at such speeds [640].

The special problems of high speed fluid dynamics first became clearly explained in 1935 at the Volta Congress in Rome. Although gas dynamics had been dealt with in earlier review articles (see Chapter 7), the 1935 Congress assembled for the first time a group of international researchers concerned with this field for an exchange of ideas and experiences. A provisional list of main lecturers, which the organizer Arturo Crocco, a general and renowned aerodynamicist, circulated in preparation of the meeting, contained these names and topics [641]:

"1. Prof. L. Prandtl (Göttingen): A survey of the theory of current incompressible fluids.

2. Prof. J. Ackeret (Zurich): Aerodynamic lift at supersonic velocities.

3. Prof. E. Pistolesi (Pisa): Aerodynamic lift at speeds approaching that of sound.

4. Prof. Th. v. Kármán (Pasadena): The problem of resistance in compressible fluids.

5. Prof. M. Panetti (Torino): Problems with experimental techniques at high speeds.

6. Prof. A. Busemann (Dresden): High speed wind tunnels.

7. Prof. G.I. Taylor (Cambridge): Well-established results of high-speed researchers.

8. Prof. G. Douglas (Farnborough): Results of research on high-speed airscrews."

The list of speakers and themes was further modified, but this draft shows more clearly than the final program what problems "The high velocity in aviation," as the theme of the Volta Congress was chosen, had to tackle [507]. Prandtl's talk was originally to close "the persisting gap" between theories of lift for velocities below and above the speed of sound. However, Prandtl responded that there was no new theory for the intermediate range of velocities, so the gap further persisted [642]. Busemann chose the theory of lift for supersonic velocities as his theme – in contrast to Crocco's proposal. He was obviously urged by Baeumker to focus "more on the scientific side" because others, like Ackeret, "are far ahead of us regarding the practical side." Baeumker's involvement behind the scene indicates that the theme of the Volta Congress was not without political ramifications. "Here, too, I recommend for various reasons to be guarded," he wrote to Prandtl with respect to Busemann's presentation. "Perhaps you can give him a little hint in Rome" [643]. In his talk at the Volta Congress, Busemann derived from a simple theoretical argument "that the effective Mach numbers may be reduced by arranging the wings in a dihedral form," in other words, that V-shaped wings are more advantageous for supersonic flight than the traditional straight wing arrangement. "The arrow-like wing arrangement is more advantageous," Busemann argued, "because the actions of the pressure are fully active in the direction of the lift, while they act in the direction of the resistance only with the component parallel to the direction of flight" [507, p. 343].

The example of the swept-back wing shows how short the path from theory to practice could be even in the absence of prior experience and elaborations of airfoil theory for supersonic speeds. Busemann's talk was reprinted in Germany in the journal *Luftfahrtforschung* [644]. The airplane manufacturer Messerschmitt showed an interest and awarded a contract to the AVA for wind tunnel tests on swept-back wings; it turned out that this wing arrangement is already advantageous at high velocities below the sound barrier [645]. Outside Germany, however, Busemann's suggestion went by unnoticed – although the proceedings of the Volta Congress were published and Kármán himself had chaired the session in which Busemann presented his talk. In retrospect, he wondered why he paid no attention: "My direction of effort at this time was not in design, but in developing supersonic theory," Kármán argued in his autobiography about his research interests in the mid-1930s [85, p. 219].

In Germany, after the Rome Congress, Busemann was called to be the director of an institute for gas dynamics at the new Aeronautical Research Establishment Hermann Göring (Luftfahrtforschungsanstalt Hermann Göring, or LFA) at Völkenrode near Brunswick, a secret facility directed by Prandtl's pupil Hermann Blenk [646]. Kármán was taken by complete surprise when he visited the Völkenrode facility after the war as head of an American intelligence unit ("Operation Lusty"): "The whole thing was incredible. Over a thousand people worked there, yet not a whisper of this institute had reached the ears of the Allies" [85, p. 274]. If Kármán's recollection is correct, this reveals a blatant failure of Allied intelligence, because it would have sufficed to study what was published on gas dynamics in Germany before the war to infer that Völkenrode would become a major center for advanced aeronautical research. Although the bulk of the LFA's work was kept secret, a number of theoretical papers from Busemann's institute was presented at meetings of the Lilienthal Society and published in the *Schriften der deutschen Akademie für Luftfahrtforschung* or the ZAMM, and therefore, this research activity was not entirely concealed from observers abroad. (See the list of publications in [646] and, for example, Busemann's outline of [647]).

Nevertheless, it is obvious that from the vantage point of the Volta Congress in 1935 that the concern about supersonic flight appeared in a different light than in retrospective accounts from the period following the Second World War. Other presentations attracted more interest because they addressed current rather than future problems, including Kármán's talk on the resistance in compressible flow, which was of immediate relevance for ballistics. Together with his pupil, Norton B. Moore, Kármán had already published a first paper on this subject in 1932 [648]. His Rome talk was an elaboration of the same theme [507, pp. 222–276]. Ackeret's presentation on wind tunnels for high speeds also addressed the interests of the assembled researchers because there was an urgent need to compare theoretical results of gas dynamics with experimental data. Ackeret could offer images and plans from his institute in Zurich, where he operated the first closed-circuit supersonic wind tunnel with a continuous air stream in the world [507, pp. 460–537]. Luigi Crocco, the general's son, presented the concept of the supersonic wind tunnel at Guidonia based on the blueprint of the Zurich tunnel [507, pp. 542–562]. The primary research goal with these tunnels was ballistics, not supersonic flight, as Ackeret pointed out, although he also envisioned aeronautical investigations. He thought of rotating propeller blades, whose tips came close to the speed of sound, and military airplanes "which reach extraordinary high speeds in nose dives." For all of these applications, compressibility effects played an important role. Theory had been successfully developed during the past years to cope with these effects, but phenomena such as heat conduction were still neglected. "For the time being, only experimental research can close this gap," Ackeret argued [507, pp. 488–490].

A major problem for the application of theoretical gas dynamics to ballistics was the extension of two-dimensional flow theory to the three-dimensional configurations pertinent for projectiles. A first attempt to this end was Busemann's calculation of pressures against "conical tips moving at supersonic velocity," published in 1929 [420]. Conical supersonic flow was not only important for ballistics; beyond the sound barrier, wingtips would also become the source of conical flow, as Busemann mentioned in his talk at the Volta Congress [507, pp. 345–356]. In a memorial issue of *Luftfahrtforschung* at the occasion of Wieselsberger's death in 1941, Busemann published an article on conical supersonic flow, which demonstrates the effort dedicated to this special problem: The resulting flow was obtained graphically by the method of characteristics; in addition, two researchers at Busemann's institute, working in a special "computer office," integrated the differential equations for conical flow via the Runge–Kutta method – with the result that the graphical solution was verified and the accuracy of the result further improved. The aim was to integrate the dynamic gas flow equations with such a fine mesh that a complete survey of all possible conical angles and Mach numbers could be obtained. The result was a catalog of solutions for supersonic conical flow, which became useful not only in order to account for the shock waves around the tips of projectiles but also, for example, in supersonic wind tunnels, where conical shock waves emerged from the nozzle wall [649, 650].

Theoretical gas dynamics also found applications in new areas far from ballistics and aviation. A striking cross-over relation emerged between hydraulics and gas dynamics. Due to a formal analogy between the differential equations of two-dimensional gas flow and the flow of water in a shallow horizontal channel, certain quantities of both cases could be compared to one another: so-called hydraulic jumps, i.e., sudden changes of the water level at flow speeds faster than the speed of (gravitational) water waves, correspond to shock waves in gas flow. The analogy between water depth and pressure change is not perfect (because it would require a gas having a ratio of specific heats $c_p/c_v = 2$, which does not exist) but yields qualitatively analogous results, so that a number of supersonic phenomena may be conveniently studied in water tables. Ackeret made this hydraulic analogy, as it became known, the subject of a doctoral dissertation at his institute [652, 653]. Kármán, too, made some effort to study this analogy in collaboration with colleagues from the hydraulics department at CalTech [654].

Gas dynamics was also important for turbo-machinery – a technology whose emergence accelerated at the turn of the century, as we have seen (see Chapter 7). In the 1930s, gas turbines and axial compressors provided new challenges. Both in Ackeret's institute at the ETH Zurich and in the Göttingen AVA, this technology was regarded with great interest – not only as new targets of fluid dynamics research but in particular because axial compressors were used as blowers in supersonic wind tunnels. In the course of the "turbojet revolution," axial compressors evolved into turbojet fan engines [651]. The transition from propeller-driven to jet-powered aircraft opened a new era of flight – and new modes of science-technology interactions. However, these are beyond the scope of our focus here.

# 11
# Epilogue

On 1 March 1939—a few months before Hitler's Army invaded Poland and thus started the Second World War – at the annual meeting of the German Academy of Aeronautical Research, Göring congratulated its members for their contribution to make "the German Air Force superior to the Air Force of every other country." Göring explicitly praised "the exemplary collaboration of aeronautical science and technology in the past years which substantially contributed to the rebuilding of our political situation." At the same occasion, Prandtl was awarded with the Hermann-Göring-medal, the Academy's highest distinction, "in view of his extraordinary merits for the scientific foundations of fluid dynamics," as the lauding argument declared [513, pp. 131–134]. Beyond the rituals of Nazi propaganda, with which such events were celebrated and communicated to the broader public (see Chapter 8), both the award and the emphasis of collaborative scientific and technological efforts indicate that fluid dynamics had reached a climax. No other science had flourished under the Nazi regime to such an extent, and few other scientists found themselves so highly esteemed by a regime, which was regarded as rather hostile towards science in general.

Outside Germany, too, fluid dynamics was considered as a science that opened the door to ever new applications, from hydraulics to aeronautics. Fluid dynamics benefited from the high expectations with which patrons in government and philanthropic foundations, at the armed forces and in industry, regarded this science. The International Congresses for Applied Mechanics, which became a forum for the community of fluid dynamicists all over the world, clearly illustrated this appreciation. Between the "zeroth" Innsbruck meeting in 1922 and the Fifth Congress for Applied Mechanics in 1938 in Cambridge, Massachusetts, fluid dynamics had acquired considerable weight as a fledgling discipline, with dozens of spokespersons from all over the world who represented hundreds of practitioners in academic, industrial, and governmental research institutes. The community of international fluid dynamicists was a mix of practical engineers and academic scientists whose insti-

*The Dawn of Fluid Dynamics: A Discipline between Science and Technology.* Michael Eckert
Copyright © 2006 WILEY-VCH Verlag GmbH & Co. KGaA, Weinheim
ISBN: 3-527-40513-5

tutional affiliation varied from country to country and whose specializations ranged from applied mathematics to mechanical engineering. Despite this heterogeneous composition, however, they regarded Prandtl as the founding father of their discipline. G.I. Taylor addressed him as "our chief" who deserved a Nobel prize [542].

The heterogeneous mix of practical engineering and academic science calls for further reflection. Prandtl raised no hopes to receive the Nobel prize because he regarded fluid dynamics as part of mechanics, a discipline "which nowadays is no longer regarded as part of physics, but stands as a self-contained field between mathematics and engineering science" [655]. Prandtl's master pupil and rival, Kármán, who was still regarded in 1920 as a candidate for a professorship in theoretical physics, displayed a similar attitude. On 27 December 1939, for example, he presented the 15th Josiah Willard Gibbs Lecture in Columbus, Ohio, on the question of how "The Engineer Grapples with Nonlinear Problems." Kármán left no doubt that his attitude was closer to that of practical engineers than abstract scientists. He distanced himself and his presentation from what his audience might associate with the name of Gibbs, whose main interest he described as "certainly centered on basic conceptions of mathematical physics," while Kármán's own interest was focused on "practical applications of applied mathematics" [656]. It is impossible to overlook the aversion of fluid dynamicists like Kármán, Prandtl, or Taylor against abstract science. They favored intuition and visual thinking over the more formal attitude of "pure" mathematics; rather, they focused on the tangible problems mechanics offered so generously in contrast to the remote problems to which theoretical physics sometimes paid so much effort. That is not to say that they abhorred theory. Turbulence is a blatant example to the contrary. But even when intricate problems such as those in turbulence theory were involved, visual thinking prevailed over abstract theorizing. Prandtl confessed to a preference of this approach, for example, when he explained in 1947 to an audience of physicists: "When faced with problems in mechanics, I slowly became accustomed to 'see' forces and accelerations in the equations or to 'sense' them by tactile feeling" [68].

But personal recollections often tend to embellish and mystify scientific discoveries and technological innovations, such that the complex processes behind them are portrayed as mere individual acts of creativity. We find similar characterizations elsewhere: "I try to identify myself with the atoms," Linus Pauling described his approach in structural chemistry. "I ask what I would do if I were a carbon atom or a sodium atom under these circumstances" [658, p. 706]. There should be no doubt that "feeling and intuition" are important aspects of a researcher's creativity, but for an evaluation of fluid dynamics as a discipline between practical engineering and academic science, we need other criteria.

For the past decades, historians of science and technology took great pains to consider the territory between science and technology. In 1985, the editor of *Technology and Culture* reviewed some of these efforts in his book, *Technology's Storytellers*, without arriving at a general framework for evaluating scientific or technological developments. He quoted an article in this journal on "The science–technology relationship as a historiographic problem," in which the inherent problems behind the science–technology distinction were expounded with the result that all efforts to pinpoint that distinction with historical case-studies were doomed to failure. "Indeed, such inquiries can be, and perhaps should be, conducted under complete avoidance of the terms 'science' and 'technology,'" the author of this study (Otto Mayr) suggested. "Instead, we should recognize that the concepts of science and technology themselves are subject to historical change; that different epochs and cultures had different names for them, interpreted their relationship differently, and as a result, took different practical actions" [659], [660, p. 95].

Since then, a number of studies have shed further light on the problematic relationship. The old linear model, conceived by Vannevar Bush shortly after the Second World War, which portrayed basic science as the pacesetter of technological progress and which seemed at first sight plausible in view of the host of new technologies obtained as a result of "scientific" war projects like radar and the atom bomb, was demystified as a strategic postwar paradigm of U.S. science policy in the Cold War: "This sort of dynamic linear-model thinking gave rise to the Department of Defense's categories for R & D, which soon accounted for the major share of postwar federal spending on research." The linear model falsely assumed a one-way flow from scientific discovery to technological innovation and that science is exogenous to technology. "The annals of science suggest that this premise has always been false to the history of science and technology," one study argued [661, pp. 11 and 20]. In the meantime, a broad consensus among historians of science and technology has been reached about the demise of simplistic models of the science–technology relationship. A new rhetoric supersedes the older discourse. Contemporary analysts of science and technology address their subject matter as "technoscience" [662], "mode-2-science" [663], or more recently and with particular regard of research policy, as a "triple-helix" [664], indicating a transcendence of boundaries between formerly distinct (or imagined as distinct) scientific/technological, social, and political realms. The consensus about the demise of the linear model, however, does not imply a new synthesis. Recent science studies raise little hope that the disputed question will find an answer to which most analysts would agree [665–668].

Even in the absence of a generally accepted framework, however, it is possible to draw from the preceding narrative on the emergence of modern fluid dynamics some lessons concerning the science–technology debate. It has been argued, for example, that the "mysterious harmony between science and technology" may be explained as an inherent tendency of science to be oriented towards externally determined goals. The thesis of this "finalization" concept was that "science reacts to external control institutions," such that "external goals become an integral part of its cognitive content, and the development of theories and methods becomes modified accordingly." The concept is of particular concern here because its proponents chose fluid dynamics as an example for finalization. In a nutshell, they argued that classical hydrodynamical theory was complete by the mid-nineteenth century with the formulation of the Navier–Stokes equations; what followed only served external goals, i.e., the adaptation towards technical applications. Prandtl's boundary layer and the mixing length theories were regarded as manifestations of the transition from fundamental theory towards special applications. Airfoil theory and the theory of fluid resistance were ranked among the latter – fully directed towards technological ends, in total contrast to the general, but still useless, basic hydrodynamical theory of the nineteenth century from which they emerged. Boundary layer theory, like every other theory that mediates the transition from the fundamental to the applied, was called "basic application theory." Theories at this intermediate level are not by themselves applied to technology and may appear as basic or fundamental science. They are celebrated as scientific breakthroughs because they render more fundamental principles practical. "These are the cognitive prerequisites for finalization: it is the external goals, here in the realm of technology, which determine the further development of theories. Scientists are constantly on the heel of technological development. The 'research frontier' is always shifting in a direction where technical progress makes the application of theory interesting" [669, pp. 72, 96, 113–114].

Although there was widespread attention paid to the debate in the 1970s, the finalization thesis gradually vanished from the stage of science studies in the 1980s after critical reviews. The concept was largely dismissed on epistemological grounds [670]. I do not attempt to revive a defunct thesis, but beyond its epistemological deficiencies (for example concerning the claim that hydrodynamical theory was "complete" after the formulation of the Navier–Stokes equations), it serves as a useful starting point for further reflection. In view of the historical development of fluid dynamics in the first decades of the twentieth century, as described in this book, there can be little doubt about the role of technical goals as a sort of direction-finder for the scientific and technological research front of fluid dynamics, and this major tenet of the finalization concept should not be dismissed just because the overall argument is flawed (or too simplistic) for other reasons.

The overall orientation towards external goals is not incompatible with inherent "scientific" goals emerging in the pursuit of a certain path of research, and a good deal of fluid dynamics may even appear as "knowledge for its own sake." Turbulence research is a good example. Isolated from its context, the statistical theory of isotropic turbulence appears as pure fundamental physics. "It was Taylor's fortunate idea to simplify the problem by the consideration of a uniform and isotropic field of turbulent fluctuation," Kármán remarked in a review in 1948. That such a condition is satisfied "approximately in the wind tunnel" was mentioned just in passing in order to indicate how the theory could be subjected to experimental tests. Otherwise, the review reported on the recent theories of Kolmogorov, Heisenberg, and Weizsäcker, names not usually associated with technology. Only the form of its publication – in the *Journal of Marine Research*—reminded a reader that this review was a report on progress in the pursuit of goals set by technological demands [671].

In a study of the research activities at Prandtl's institute in the Second World War the notion of "epistemic things" was used to characterize this type of research, which appears fundamental despite its underlying technological goals. Epistemic things were originally defined in a different context [672] but were adapted as "the problem-generating, future-oriented elements of research." On the agenda for Prandtl's Kaiser Wilhelm Institute for Fluid Dynamics in the Second World War were epistemic things like "the 'boundary layers' of aerodynamic flow around bodies or the phenomena of 'turbulence' and 'cavitation.' All of them dealt with a more or less precisely defined set of research topics, which involved more unsolved than understood aspects; in particular, they were all related both to questions concerning the 'fundamental laws' of fluid mechanics and the design of a host of technological objects (airplanes, wind tunnels, projectiles,...). We are concerned here with a branch of research whose epistemic things have been from the very beginning in touch with useful military technology" [511, p. 41]. Kármán expressed the same in less sophisticated language when he discerned in 1943 a host of fluid mechanics problems suggested "by needs of our war effort. It is a dire necessity to tackle these problems, whether or not they are interesting from the general point of view of science and engineering, and whether they are relatively easy or so complex that there is little hope for their fundamental solution. However, it is a comforting idea to the engineer and scientist that most of the problems connected with needs of warfare are also fascinating from the viewpoint of the progress of science and art, and have attracted and fascinated a great many noble minds in the past" [673, p. 205].

If fluid dynamics, by and large, is goal-oriented, why not simply designate it as applied science? Most of its new research results were presented under the label of applied mechanics or mathematics, such as at the International Congresses of Applied Mechanics, or in the issues of ZAMM. Labeling for conferences and journals, however, serves other purposes and should not relieve us from the effort of analyzing the peculiar practices of research and the nature of knowledge from the perspective of social studies of science and technology. The question of whether technology in general may be regarded as applied science has often been debated in philosophical, sociological, and historiographic papers. Most analysts tend to refute the applied science model and insist that "technological knowledge is irreducibly distinct" and that "science cannot claim the role as technology's sole source of knowledge," as reviewed in *Technology's Storytellers* [660, pp. 95–103].

The question, then, is what characterizes technological knowledge as unique and distinct from other forms of knowledge; or more precisely, coined for the epoch considered in this narrative – the age of Prandtl—What is the difference between engineering and science? "The predominance of graphic description of airfoil characteristics in NACA reports seems to reflect the power of nonverbal thought in the engineering mind. Unlike scientists who tend to think in mathematical or verbal terms, engineers work principally from learned mechanical alphabets, models, and curves," the historian of NACA's Langley laboratory distinguished engineering from scientific knowledge, alluding to reflections of Eugene S. Ferguson on the "mind's eye" as the most important organ of an engineer for successful theorizing in technology [233, chapter 4], [674], [675]. The visual and intuitive approach has also been contrasted against more abstract theorizing in other cases such as the development of gyro-technology. "I feel that the mechanical and graphical side is very enlightening and has true cultural and educational value to a very much wider group than the more abstruse and purely mathematical aspect," gyroscope inventor Elmar Sperry is quoted as saying in a historical study of this technology [676, p. 131]. Walter Vincenti, a historian of aeronautics and former NACA engineer, presented a variety of cases for this peculiar mode of engineering thought and discerned *design* as the decisive criterion for analyzing engineering knowledge [677]. Pursuing this argument, John D. Anderson, Jr., a distinguished textbook author and historian of aerodynamics, titled a recent article "The evolution of aerodynamics in the twentieth century: engineering or science?" and employed this definition: "Science: A study of the physical nature of the world and universe, where the desired end product is simply the acquisition of new knowledge for its own sake. Engineering: The art of applying an autonomous form of knowledge for the purpose of designing and constructing an artifice to meet some recognized need. Engineering Science: The acquisition of new knowledge for the specific purpose of qualitatively or quantitatively enhancing the process of designing and constructing an artifice." In conclusion, he argued that our modern understanding of aerodynamics was achieved "through an intellectual process that blended the disciplines of science, engineering science, and pure engineering" [678].

But if the "mysterious harmony between science and technology" noticed by finalization analysts, aviation historians, and many others may really be described as a blending of various forms of knowledge, where does it come from? Historians of technology resorted to biology in order to account for the dynamics with which science and technology evolved hand in hand. "Transposed to technology, the concept of co-evolution implies that the development of one set of devices may be linked intimately to the development of other devices within the same macrosystem, and that the two sets of devices may exert powerful, mutual selective pressure on each other." This concept of "technological co-evolution" was originally conceived in order to explain how within larger technological systems a given technology, say steam engines, becomes linked to others, say power transmission. With the additional concept of "presumptive anomaly" it was further developed into a model of scientific and technological co-evolution. Presumptive anomaly occurs "when assumptions derived from science indicate either that under some future conditions the conventional system will fail (or function badly) or that a radically different system will do a much better job." Within this conceptual framework, Edward W. Constant described the "turbojet revolution" as a preeminent example in which developments that were originally quite disparate came together and resulted in a radical technological change [651, pp. 14–15].

It is not accidental that the analogy of biological evolution was found attractive by historians of technology, particularly those concerned with the airplane, the "defining technology" of the twentieth century [608]. Vincenti, for example, used the "variation-selection model" inferred from his aeronautical case-studies in order to explain other cases, like the spectacular wind-induced collapse of the Tacoma Narrows bridge as an example of "Darwinian selection" of technological evolution [164]. Anderson recommended that the "evolution of aerodynamics" be used as a role model for studies of scientific engineering in the twentieth century: "Perhaps, within the scope of the history of technology, aerodynamics is one of the best examples of such blending." [678]. This view finds ample endorsement in the preceding narrative. The advancement of airplane design and airfoil theory in the First World War, for example, may well be regarded as an evolutionary process – with Prandtl's Göttingen facility as playing a major role as the variation-selection mechanism (Chapter 3). Theory, however, was a beneficiary rather than a pacesetter of this process, or, as aviation historian Richard K. Smith once remarked, "The airplane did more for science than science ever did for the airplane" [678, p. 256]. Problems from the technology of steam turbines and the need to account for compressibility effects in the airflow of fast-rotating propellers contributed to the rise of gas dynamics (Chapter 7) and are appropriately described within Constant's framework of "presumptive anomaly" and "technological co-evolution." The development of airfoil theory between the

First and Second World Wars, from the first formulation of the "lifting line" concept into a multitude of procedures appropriate for practical wing design (Chapter 10), is also appropriately described as an evolutionary process. Here, too, the airplane shaped the "variation-selection" processes.

Nevertheless, the growth of modern fluid dynamics should not be regarded as a mere corollary of the evolution of aerodynamics in the twentieth century. Although it is true that fluid dynamics emerged as a modern engineering discipline mainly in the wake of the airplane, this "defining technology" cannot account for the entire scope of specialties beyond aerodynamics, such as hydraulics, ballistics, or meteorology. Although "technological co-evolution" is a powerful concept and includes other engineering specialties besides aerodynamics, it cannot explain the full extent of fluid dynamics. In particular, it is not clear how this discipline is related to physics. We have to recall the history of fluid dynamics before the age of Prandtl, and examine its development in the second half of the twentieth century, in order to better appreciate its multifaceted history.

From the Renaissance to the seventeenth century, the flow of air and water was a fundamental problem, both in practice and in theory. For Leonardo da Vinci, Galileo Galilei, Isaac Newton and others with whom we associate the rise of the modern era, or the Scientific Revolution, problems of fluid flow were inseparably connected with the abstractions that gave rise to classical mechanics – the pillars of classical physics. Newton himself failed to account for fluid motion. Only in the post-Newtonian age did it become possible to add "ideal flow" to the discussion. We celebrate Johann and Daniel Bernoulli, Leonard Euler, and others for this breakthrough. However, in view of the non-ideal reality, they were also blamed for this idealization, because it paved the way of two diverging avenues. For Euler and the Bernoullis, hydraulics and hydrodynamics were still synonymous; for their successors in the nineteenth century, these terms became expressions for an ever-increasing gap between practice and theory.

In this book, we have described to what extent the gap was bridged in the age of Prandtl. Surprisingly, physics showed little interest in fluid dynamics. This is most clearly illustrated in Prandtl's own sphere of activity, Göttingen. Although his first professorship in Göttingen was designated as technical physics, it was separate from the physics institute. In 1907 Prandtl's professorship was renamed from technical physics to applied mechanics; in the 1920s the Society for Applied Mathematics and Mechanics (GAMM, founded in 1922 and chaired by Prandtl as its first president) became his true professional community. Kurt Magnus, an assistant of Prandtl's successor at the institute of applied mechanics, described Prandtl's affiliations in Göttingen as a "kind of shell model," with the Kaiser Wilhelm Institute of Fluid Dynamics as the inner-most shell, followed by Betz's AVA as the second shell; next came the colleagues from the institute of applied mechanics; the other colleagues from Göttingen university were arranged in the periphery. Little direct contact existed with the Göttingen physicists, Max Born, James Franck, Robert Wiechard Pohl, and their assistants and students. "However," Magnus concluded with some irony, "the outstanding importance of Prandtl did not remain concealed from the physicists. In 1947 they offered him an honorary membership to the German Physical Society" [679, p. 290].

Most of Prandtl's students made a career in engineering rather than physics. Internationally, too, fluid dynamics during the first half of the twentieth century was closer to engineering and applied mathematics than physics. It will be the subject of another book to extend the history of fluid dynamics beyond the age of Prandtl, when the interest of physics in fluid dynamics finally awakened. The foundation of new journals corroborates this process: in 1958, for example, the American Institute of Physics (AIP) announced the foundation of *Physics of Fluids*, a journal devoted to a broad spectrum of physical subdisciplines: "The scope of these fields of physics includes magneto-fluid dynamics, ionized fluid and plasma physics, shock and detonation phenomena, hypersonic physics, rarefied gases and upper atmosphere phenomena, liquid state physics and superfluidity, as well as certain basic aspects of physics of fluids bordering geophysics, astrophysics, biophysics, and other fields of science" [680]. Other fluid flow problems – in particular, turbulence and instabilities – gave rise to an avalanche of studies about nonlinear dynamics and complex systems, often included since the 1980s under the label of chaos theory. When a renowned expert of nonlinear dynamics was asked in 2001 where he saw the largest deficits in the past development of physics, he answered: "We have neglected, for example, applied mathematics or the physics of fluids including hydrodynamics" [681].

The renewed interest of physicists in the second half of the twentieth century does not mean that fluid dynamics moved from engineering towards science. Fluid dynamics remained a discipline of utmost importance in many engineering branches. Engineering fluid mechanics did not cease to further evolve as a discipline for practitioners. "The science of fluid mechanics is developing at a rapid rate," we are informed, and yet many problems "still rely heavily on empiricism," like the "flow of multiphase mixtures such as solids in a liquid (slurries)" or "oil recovery operations" [682, p. 7]. At the same time, physicists began to become better acquainted with fluid dynamics from the vantage point of their own discipline. For a physicist, fluid dynamics is of interest not only because of its technical applications, argued the author of the textbook *Fluid Dynamics for Physicists* in 1995, but also "because most other subjects in the physics curriculum are almost exclusively concerned with *linear* processes, whereas fluid dynamics leads one into the *non-linear* domain. And, lastly, because there are so many curious and beautiful natural phenomena, visible every day in the world about us, which a physicist with no knowledge of fluid mechanics is unable to appreciate to the full" [683, p. 1].

Is fluid dynamics at the beginning of the twenty-first century still the same twin-discipline as it was a hundred years ago? Is what was then labeled as hydrodynamics versus hydraulics, reappearing as *physical* versus *engineering* fluid dynamics? Are we witnessing merely the old schism between theory and practice in a new guise?

If we are entitled to extrapolate from the decades before the Second World War the ensuing course of fluid dynamics, we have little reason to conclude that the present science–technology dichotomy is still the same old gap between theory and practice. Know-how in fluid dynamics, whether it is labeled as engineering or physical, transformed technologies of war and peace to such an extent that there is little resemblance with the situation at the end of the nineteenth century. No other discipline was closer to the heart of the military–industrial–academic complex of the Cold War than fluid dynamics. Although physicists and engineers regard it from their own perspectives with their own specific and sometimes diverging interests, this divergence is different from the gap between theory and practice a hundred years ago. However, only a closer historical inspection of fluid dynamics beyond the age of Prandtl can tell us how.

# Appendix

## Abbreviations

| | |
|---|---|
| AEG | Allgemeine Elektricitaets-Gesellschaft (Corporate name). |
| AIP | American Institute of Physics. |
| APK | Artillerie-Prüfungs-Kommission. |
| AVA | Aerodynamische Versuchsanstalt. |
| BBC | Brown Boveri and Cie. (Corporate name). |
| CalTech | California Institute of Technology. |
| CWTK | Collected Works of Theodore von Kármán, 5 volumes. London: Butterworths, 1956. |
| DFG | Deutsche Forschungsgemeinschaft. |
| DLR | Deutsche Gesellschaft für Luft- und Raumfahrt. |
| DMA | Deutsches Museum, Archive, München. |
| DMV | Deutsche Mathematiker-Vereinigung. |
| DPG | Deutsche Physikalische Gesellschaft. |
| DVL | Deutsche Versuchsanstalt für Luftfahrt. |
| ETHA | Zürich, Archiv der Eidgenössischen Technischen Hochschule (ETH). |
| ETH | Eidgenössische Technische Hochschule Zürich |
| FLZ | Flugzeugmeisterei. |
| FST | Full Scale Tunnel. |
| GALCIT | Guggenheim Aeronautical Laboratory of the California Institute of Technology. |
| GAMM | Gesellschaft für angewandte Mathematik und Mechanik. |
| GOAR | Göttingen Archive of the Deutsche Gesellschaft für Luft- und Raumfahrt (DLR). |
| HSPS | Historical Studies in the Physical Sciences. |
| IEB | International Education Board. |

*The Dawn of Fluid Dynamics: A Discipline between Science and Technology.* Michael Eckert
Copyright © 2006 WILEY-VCH Verlag GmbH & Co. KGaA, Weinheim
ISBN: 3-527-40513-5

| | |
|---|---|
| IRC | International Research Council. |
| IUTAM | International Union of Theoretical and Applied Mechanics. |
| JSTG | Jahrbuch der Schiffbau-Technischen Gesellschaft. |
| KWG | Kaiser-Wilhelm-Gesellschaft zur Förderung der Wissenschaften. |
| KWI | Kaiser Wilhelm Institut. |
| LFA | Luftfahrtforschungsanstalt. |
| LMAL | Langley Memorial Aeronautical Laboratory. |
| LPGA | Ludwig Prandtl Gesammelte Abhandlungen, 3 volumes. Walter Tollmien, Hermann Schlichting, Henry Görtler (eds.) Berlin: Springer, 1961. |
| MAN | Maschinenfabrik Augsburg-Nürnberg AG. |
| MIT | Massachusetts Institute of Technology. |
| MPGA | Archiv zur Geschichte der Max-Planck-Gesellschaft, Berlin. |
| MVA | Modellversuchsanstalt. |
| NACA | National Advisory Committee for Aeronautics. |
| NACP | National Archives II, College Park, Maryland. |
| NaPhil | National Archives, Philadelphia. |
| NPL | National Physical Laboratory (Great Britain). |
| NRC | National Research Council. |
| NSUB | Niedersächsische Staats- und Universitätsbibliothek, Göttingen. |
| RLM | Reichsluftfahrtministerium. |
| RVM | Reichsverkehrsministerium. |
| RWTH | Rheinisch-Westfälische Technische Hochschule Aachen. |
| Sef | Segelflugzeugbau G.m.b.H. Aachen. |
| SPGIT | The Scientific Papers of Sir Geoffrey Ingram Taylor. Volume II: Meteorology, Oceanography and Turbulence. G.K. Batchelor (ed.) Cambridge: Cambridge University Press, 1960. |
| SVK | Seeflugzeug-Versuchskommando. |
| TB | Technische Berichte, edited by the Flugzeugmeisterei der Inspektion der Fliegertruppen, Berlin-Charlottenburg. |
| TKC | Theodore von Kármán-Collection, California Institute of Technology Archives. There are microfiche copies available at the Smithsonian Air and Space Museum in Washington, DC, and the Institute for Philosophy at the University of Berne, Switzerland. |
| VDI | Verein Deutscher Ingenieure. |
| VDT | Variable Density Tunnel. |
| WAF | Wissenschaftliche Auskunftei für Flugwesen. |
| WGL | Wissenschaftliche Gesellschaft für Luftfahrt. |
| WKF | Wissenschaftlichen Komission für das Flugwesen. |
| ZAMM | Zeitschrift für Angewandte Mathematik und Mechanik. |
| ZFM | Zeitschrift für Flugtechnik und Motorluftschifffahrt. |
| ZVDI | Zeitschrift des Vereins Deutscher Ingenieure. |

# References

1 Sydney Goldstein: Fluid Mechanics in the First Half of This Century. In: Annual Review of Fluid Mechanics, 1, 1969, 1–28.

2 Arnold Sommerfeld: Neuere Untersuchungen zur Hydraulik. In: Verhandlungen der Gesellschaft deutscher Naturforscher und Ärzte, 72, 1900, 56.

3 Itiro Tani: History of Boundary-Layer Theory. In: Annual Review of Fluid Mechanics, 9, 1977, 87–111.

4 Marshall Clagett: The Science of Mechanics in the Middle Ages. Madison: University of Wisconsin Press, 1959.

5 René Dugas: A History of Mechanics. New York: Dover, 1988 (first published 1955).

6 Galileo Galilei: Dialogues Concerning Two New Sciences. New York: Dover, 1950. Unaltered republication of the translation by Henry Crew and Alfonso de Salvio first published in 1914, based on Galileo's original Italian text published in 1638.

7 J. MacLachlan: Galileo's experiment with pendulums: Real and imaginary. In: Annals of Science, 33, 1976, 173–185.

8 Ronald Naylor: Galileo: Real Experiment and Didactic Demonstration. In: Isis, 67, 1976, 398–419.

9 Michael Segre: The Role of Experiment in Galileo's Physics. In: Archive for History of Exact Sciences, 23, 1980, 227–252.

10 Michael Segre: Stieg Galilei auf den Schiefen Turm? In: Kultur und Technik, 12(3), 1988, 166–172.

11 Herman Erlichson: Galilei and the High Tower Experiments. In: Centaurus, 36, 1993, 33–45.

12 Thomas B. Settle: Galileo and Early Experimentation. In: R. Aris, T. Davis, R.H. Stuewer (eds.): Springs of Scientific Creativity. Minneapolis: University of Minnesota Press, 1983, 3–20.

13 I. Bernard Cohen: The Birth of a New Physics. London: Penguin Books, 1992.

14 Theodore von Kármán: Aerodynamics. Selected Topics in the Light of Their Historical Development. Ithaca: Cornell University Press, 1954.

15 Michael Segre: In the Wake of Galileo. New Brunswick: Rutgers University Press, 1991.

16 E.J. Aiton: The Vortex Theory of Planetary Motions. London: MacDonald, 1972.

17 Clifford A. Truesdell: An Idiot's Fugitive Essays on Science. New York: Springer, 1984.

18 R. Giacomelli, E. Pistolesi: Historical Sketch. In: William Frederick Durand (ed.): Aerodynamic Theory, Volume I. Berlin: Springer, 1934, 305–394.

19 Isaac Newton: The Mathematical Principles of Natural Philosophy. Translated into English by Andrew Motte in 1729. London: Dawsons of Pall Mall, 1968.

20 Istvan Szabo: Geschichte der mechanischen Prinzipien und ihrer wichtigsten Anwendungen. Basel: Birkhäuser, 1979.

21 Niccolò Tartaglia: Nova scientia. Venice, 1537.

22 Edme Mariotte: Traité du Mouvement des Eaux. Paris, 1718.

23 A. Rupert Hall: Ballistics in the Seventeenth Century. Cambridge: Cambridge University Press, 1952.

24 P. Charbonnier: Essais sur l'Histoire de la Ballistique. Paris: Imprimerie Nationale, 1928.

25 Jürgen Renn, Peter Damerow, Simone Rieger: Hunting the White Elephant: When and How did Galileo Discover the Law of Fall? In: Science in Context, 13, 2000, 299–419.

26 Michael Segre: Torricelli's Correspondence on Ballistics. In: Annals of Science, 40, 1983, 489–499.

27 J. Laurence Pritchard: The Dawn of Aerodynamics. In: Journal of the Royal Aeronautical Society, 61, 1957, 149–180.

28 Michael Eckert: Euler and the Fountains of Sanssouci. In: Archive for History of Exact Sciences, 56, 2002, 451–468.

29 Michael Eckert: Hydraulics for Royal Gardens: Water Art as a Challenge for 18th Century Science – and 20th Century Physics Teaching. In: Science and Education. To be published in 2005.

30 Clifford A. Truesdell: Essays in the History of Mechanics. Berlin: Springer, 1968.

**31** Robert Fox: The Rise and Fall of Laplacian Physics. In: Historical Studies in the Physical Sciences, 4, 1974, 89–136.

**32** Eda Kranakis: Hybrid Careers and the Interaction of Science and Technology. In: Peter Kroes, Martijn Bakker (eds.): Technological Development and Science in the Industrial Age. New Perspectives on the Science–Technology Relationship. Dordrecht, Boston, London: Kluwer, 1992, 177–204.

**33** Maria Yamalidou: Molecular Ideas in Hydrodynamics. In: Annals of Science, 55, 1998, 369–400.

**34** William Thomson (Lord Kelvin): The Scientific Work of Sir George Stokes. Obituary Notice. In: Nature, 12 February 1903, 337–338.

**35** Georg Gabriel Stokes: Mathematical and Physical Papers, 5 vols. New York: Johnson Reprint Corporation, 1966.

**36** Salvatore P. Sutera, Richard Skalak: The History of Poiseuille's Law. In: Annual Reviews of Fluid Mechanics, 25, 1993, 1–19.

**37** Ostwald's Klassiker der Exakten Wissenschaften, no. 237, 1933.

**38** Gustav Lejeune Dirichlet: Über die Bewegung eines festen Körpers in einem incompressiblen flüssigen Medium. In: Bericht über die Verhandlungen der Königlich Preussischen Akademie der Wissenschaften, 1852, 12–17.

**39** Olivier Darrigol: From Organ Pipes to Atmospheric Motions: Helmholtz on Fluid Mechanics. In: Historical Studies in the Physical and Biological Sciences, 29(1), 1998, 1–51.

**40** Hermann von Helmholtz: Ueber Integrale der hydrodynamischen Gleichungen, welche der Wirbelbewegung entsprechen. In: Journal für die reine und angewandte Mathematik, 55, 1858, 25–55.

**41** Hermann von Helmholtz: Ueber diskontinuirliche Flüssigkeitsbewegungen. Berliner Berichte, 1868, 215–228.

**42** John D. Anderson, Jr.: A History of Aerodynamics. Cambridge: Cambridge University Press, 1997.

**43** P.G. Drazin, W.H. Reid: Hydrodynamic Stability. Cambridge: Cambridge University Press, 1981.

**44** Crosbie Smith, Norton Wise: Energy and Empire. A Biographical Study of Lord Kelvin. Cambridge: Cambridge University Press, 1989.

**45** Daniel M. Siegel: Thomson, Maxwell, and the Universal Ether in Victorian Physics. In: G.N. Cantor, M.J.S. Hodge (eds.): Conceptions of Ether. Studies in the History of Ether Theories 1740–1900. Cambridge: Cambridge University Press, 1981, 239–268.

**46** Robert H. Silliman: William Thomson: Smoke Rings and Nineteenth-Century Atomism. In: Isis, 54, 1963, 461–474.

**47** Moritz Epple: Die Entstehung der Knotentheorie. Kontexte und Konstruktionen einer modernen mathematischen Theorie. Vieweg: Braunschweig/Wiesbaden, 1999.

**48** Osborne Reynolds: On the Action of Rain to Calm the Sea. On Vortex Motion. Reprinted in Osborne Reynolds: Papers on Mathematical and Physical Subjects, Vol. 1: 1869–1882. Cambridge: Cambridge University Press, 1900, 86–88, 184–191.

**49** Osborne Reynolds: An Experimental Investigation of the Circumstances Which Determine Whether the Motion of Water Shall Be Direct or Sinuous, and of the Law of Resistance in Parallel Channels. In: The Philosophical Transactions of the Royal Society, 1883. Reprinted in Osborne Reynolds: Papers on Mathematical and Physical Subjects, Vol. 2: 1881–1900. Cambridge: Cambridge University Press, 1901, 51–105.

**50** N. Rott: Note on the History of the Reynolds Number. In: Annual Reviews of Fluid Mechanics, 22, 1990, 1–11.

**51** Olivier Darrigol: Turbulence in 19th-century hydrodynamics. In: Historical Studies in the Physical and Biological Sciences, 32(2), 2002, 207–262.

**52** J.D. Jackson: Osborne Reynolds: scientist, engineer and pioneer. In: Proceedings of the Royal Society London, A 451, 1995, 49–86.

**53** Enzo O. Macagno: Historico-critical Review of Dimensional Analysis. In: Journal of the Franklin Institute, 292, 1971, 391–402.

**54** Osborne Reynolds: On the Two Manners of Motion of Water. In: Proceedings of the Royal Institution of Great Britain, 1884.

Reprinted in Osborne Reynolds: Papers on Mathematical and Physical Subjects, Vol. 2: 1881–1900. Cambridge: Cambridge University Press, 1901, 153–162.

55 Wilhelm Wien: Lehrbuch der Hydrodynamik. Leipzig: Hirzel, 1900.

56 P. Forchheimer: Hydraulik. In: Enzyklopädie der mathematischen Wissenschaften, 4(20), 1906, 324–472.

57 Moritz Rühlmann: Hydromechanik oder die Technische Mechanik flüssiger Körper. Hannover: Hahn'sche Buchhandlung, 1880.

58 Hunter Rouse, Simon Ince: History of Hydraulics. Ann Arbor: Iowa Institute of Hydraulic Research, 1957.

59 Carl Cranz: Compendium der theoretischen äusseren Ballistik. Leipzig: Teubner, 1896.

60 Friedrich Ritter von Loessl: Die Luftwiderstandsgesetze, der Fall durch die Luft und der Vogelflug. Wien: Hölder, 1896.

61 Sebastian Finsterwalder: Aerodynamik. In: Enzyclopädie der mathematischen Wissenschaften, 4(17), 1902, 149–184.

62 Robert Marc Friedman: Appropriating the Weather. Vilhem Bjerknes and the Construction of Modern Meteorology. Ithaca, London: Cornell University Press, 1989.

63 Hermann Schlichting: Grenzschicht-Theorie, 8th edition. Karlsruhe: Verlag G. Braun, 1982.

64 Ludwig Prandtl: Über Flüssigkeitsbewegung bei sehr kleiner Reibung. In: Verhandlungen des III. Internationalen Mathematiker-Kongresses, Heidelberg 1904, Leipzig: Teubner, 1905, 484–491. Reprinted in Ludwig Prandtl Gesammelte Abhandlungen, 2, 575–584.

65 Karl-Heinz Manegold: Universität, Technische Hochschule und Industrie: Ein Beitrag zur Emanzipation der Technik im 19. Jahrhundert unter besonderer Berücksichtigung der Bestrebungen Felix Kleins. Berlin: Duncker und Humblot, 1970.

66 Gottfried Richenhagen: Carl Runge (1856–1927): Von der reinen Mathematik zur Numerik. Göttingen: Vandenhoeck und Ruprecht, 1985.

67 Julius C. Rotta: Die Aerodynamische Versuchsanstalt in Göttingen, ein Werk Ludwig Prandtls. Ihre Geschichte von den Anfängen bis 1925. Göttingen: Vandenhoeck und Ruprecht, 1990.

68 Ludwig Prandtl: Mein Weg zu hydrodynamischen Theorien. In: Physikalische Blätter, 4, 1948, 89–92. Reprinted in Ludwig Prandtl Gesammelte Abhandlungen, 3, 1604–1608.

69 Karin Reich: Die Rolle Arnold Sommerfelds bei der Diskussion um die Vektorrechnung, dargestellt anhand der Quellen im Nachlass des Mathematikers Rudolf Mehmke. In: Joseph W. Dauben: History of Mathematics: States of the Art. San Diego: Academic Press, 1996, 319–341.

70 Arnold Sommerfeld: Zu L. Prandtls 60. Geburtstag am 4. Februar 1935. In: Zeitschrift für Angewandte Mathematik und Mechanik, 15, 1935, 1–2.

71 J. Georgi: Professor Dr. Fritz Ahlborn, ein vergessener Pionier der Strömungsforschung. In: Abhandlungen und Verhandlungen des Naturwissenschaftlichen Vereins in Hamburg, N. F. II, 1957, 5–18.

72 Friedrich Ahlborn: Hydrodynamische Experimentaluntersuchungen. In: Jahrbuch der Schiffbautechnischen Gesellschaft, 1903, 3–34.

73 Correspondence from Ahlborn to Prandtl, 13 September 1909. GOAR 3684.

74 Friedrich Ahlborn: Hydrodynamische Experimentaluntersuchungen. In: Physikalische Zeitschrift, 11, 1910, 201–206.

75 Draft in Prandtl's handwriting, to be signed by Hartmann (astronomy), Klein, Prandtl, Riecke (experimental physics), Simon (applied electricity), and Voigt (theoretical physics), 20 December 1909. GOAR 3684.

76 Correspondence from Ahlborn to Prandtl, 24 December 1909. GOAR 3684.

77 Friedrich Ahlborn: Die Widerstandsvorgänge im Wasser an Platten und Schiffskörpern. Die Entstehung von Wellen. In: Jahrbuch der Schiffbautechnischen Gesellschaft, 1909, 3–65.

78 Friedrich Ahlborn: Die Wirbelbildung im Widerstandsmechanismus des Wassers. In: Jahrbuch der Schiffbautechnischen Gesellschaft, 6, 1905, 67–81.

79 Ludwig Prandtl Gesammelte Abhandlungen, 3 volumes, edited by Walter Tollmien, Hermann Schlichting, Henry Görtler. Berlin: Springer, 1961.

80 Correspondence from Blasius to Prandtl, 13 August 1907. GOAR 3684.

81 Heinrich Blasius: Grenzschichten in Flüssigkeiten bei kleiner Reibung. Dissertation, Universität Göttingen, 1907. Published in Zeitschrift für Mathematik und Physik, 56, 1908, 1–37.

82 Ernst Boltze: Grenzschichten an Rotationskörpern in Flüssigkeiten mit kleiner Reibung. Dissertation, Universität Göttingen, 1909.

83 Karl Hiemenz: Die Grenzschicht an einem in den gleichförmigen Flüssigkeitsstrom eingetauchten geraden Kreiszylinder. Dissertation, Universität Göttingen, 1911. Published in Dinglers Polytechnisches Journal, 326, 1911, 321–324, 344–348, 357–362, 372–376, 391–393, 407–410.

84 Wilhelm Kutta: Beitrag zur näherungsweisen Integration totaler Differentialgleichungen. In: Zeitschrift für Mathematik und Physik, 46, 1901, 435.

85 Theodore von Kármán: The Wind and Beyond. Boston, Toronto: Little, Brown and Company, 1967. (Deutsche Übersetzung: Die Wirbelstrasse. Mein Leben für die Luftfahrt. Hamburg: Hoffmann und Campe, 1968.)

86 Theodore von Kármán: Über den Mechanismus des Widerstandes, den ein bewegter Körper in einer Flüssigkeit erfährt. In: Nachrichten der Königl. Gesellschaft der Wissenschaften, mathematisch-physikalische Klasse, 1911, 509–517 and 1912, 547–556.

87 Theodore von Kármán, H. Rubach: Über den Mechanismus des Flüssigkeits- und Luftwiderstandes. In: Physikalische Zeitschrift 13, 1912, 49–59.

88 Ludwig Prandtl: Ergebnisse und Ziele der Göttinger Modellversuchsanstalt (vorgetragen am 4. November 1911 in der Versammlung von Vertretern der Flugwissenschaften in Göttingen). In: Zeitschrift für Flugtechnik und Motorluftschiffahrt, 3, 1912, 33–36.

89 Emil Wiechert, Ludwig Prandtl: Vorschläge für Arbeiten des Unterausschusses für dynamische Fragen der Motorluftschiff-Studiengesellschaft. In: Jahrbuch der Motorluftschiff-Studiengesellschaft, 1906–1907, 71–72. Reprinted in Ludwig Prandtl Gesammelte Abhandlungen, 3, 1204–1205.

90 Ludwig Prandtl: Die Bedeutung von Modellversuchen für die Luftschiffahrt und Flugtechnik und die Einrichtungen für solche Versuche in Göttingen. In: Zeitschrift des Vereins Deutscher Ingenieure, 53, 1909, 1711–1719. Reprinted in Ludwig Prandtl Gesammelte Abhandlungen, 3, 1212–1233.

91 Georg Fuhrmann: Widerstands- und Druckmessungen an Ballonmodellen. In: Zeitschrift für Flugtechnik und Motorluftschiffahrt, 1, 1910, 129–130; 161–163; 2, 1911, 165–166.

92 Ludwig Prandtl: Bericht über die Tätigkeit der Göttinger Modellversuchsanstalt. In: Jahrbuch der Motorluftschiff-Studiengesellschaft, 1910–1911, 43–50. Reprinted in Ludwig Prandtl Gesammelte Abhandlungen, 3, 1239–1247.

93 Georg Fuhrmann: Theoretische und experimentelle Untersuchungen an Ballonmodellen. Dissertation, Universität Göttingen, 1912. Reprinted in: Jahrbuch der Motorluftschiff-Studiengesellschaft, 1911–1912, 64–123.

94 Claudine Fontanon: De L'Air Au "Plus Lourd que l'Air." In: Cahiers de Science et Vie, 1996, 56–72.

95 Otto Föppl: Ergebnisse der aerodynamischen Versuchsanstalt von Eiffel, verglichen mit den Göttinger Resultaten. In: Zeitschrift für Flugtechnik und Motorluftschiffahrt, 3, 1912, 118–121.

96 Gustave Eiffel: Sur la Résistance des Sphéres dans l'Air en Mouvement. In: Comptes Rendues, 155, 1912, 1597–1599.

97 Ludwig Prandtl: Bericht über die Tätigkeit der Göttinger Modellversuchsanstalt. In: Jahrbuch der Motorluftschiff-Studiengesellschaft, 1912–1913, 73–81. Reprinted in Ludwig Prandtl Gesammelte Abhandlungen 3, 1263–1270.

98 Correspondence from Eiffel to Prandtl, 28 September 1913 and 28 October 1913. GOAR 3684.

99 Carl Runge: Über die Berechtigung von aerodynamischen Modellversuchen. In: Zeitschrift für Flugtechnik und Motorluftschiffahrt, 4, 1913, 241–243.

100 Ludwig Prandtl: Der Luftwiderstand von Kugeln. In: Nachrichten der Gesellschaft der Wissenschaften zu Göttingen, Mathematisch-physikalische Klasse, 1914,

177–190. Reprinted in Ludwig Prandtl Gesammelte Abhandlungen, 2, 597–608.

**101** Carl Wieselsberger: Der Luftwiderstand von Kugeln. In: Zeitschrift für Flugtechnik und Motorluftschiffahrt, 5, 1914, 140–145.

**102** Wilhelm Martin Kutta: Über eine mit den Grundlagen des Flugproblems in Beziehung stehende zweidimensionale Strömung. In: Sitzungsberichte der Kgl. Bayerischen Akademie der Wissenschaften, Mathematisch-physikalische Klasse, 1910, 1–58.

**103** Nikolai Joukowsky: Über die Konturen der Tragflächen der Drachenflieger. In: Zeitschrift für Flugtechnik und Motorluftschiffahrt, 1, 1910, 281–284.

**104** Otto Föppl: Auftrieb und Widerstand eines Höhensteuers, das hinter der Tragfläche angeordnet ist. In: Zeitschrift für Flugtechnik und Motorluftschiffahrt, 2, 1911, 182–184.

**105** Albert Betz: Auftrieb und Widerstand eines Doppeldeckers. In: Zeitschrift für Flugtechnik und Motorluftschiffahrt, 4, 1913, 1–3.

**106** Ludwig Prandtl: Flüssigkeitsbewegung. In: Handwörterbuch der Naturwissenschaften, 4, 1913, 101–140. Reprinted in Ludwig Prandtl Gesammelte Abhandlungen, 3, 1421–1485.

**107** Albert Betz: Untersuchungen von Tragflächen mit verwundenen und nach rückwärts gerichteten Enden. In: Zeitschrift für Flugtechnik und Motorluftschiffahrt, 5, 1914, 237–239.

**108** Carl Wieselsberger: Beitrag zur Erklärung des Winkelfluges einiger Zugvögel. In: Zeitschrift für Flugtechnik und Motorluftschiffahrt, 5, 1914, 225–229.

**109** Ludwig Prandtl: Tragflügeltheorie, I and II. Nachrichten der Gesellschaft der Wissenschaften zu Göttingen, Mathematisch-physikalische Klasse, 1918 and 1919, 107–137. Reprinted in Ludwig Prandtl Gesammelte Abhandlungen, 1, 322–372.

**110** Correspondence from Prandtl to Wieghardt, 13 December 1921. NSUB, Cod. Ms. L. Prandtl, 1:78.

**111** Helmuth Trischler: Die neue Räumlichkeit des Krieges: Wissenschaft und Technik im Ersten Weltkrieg. In: Berichte zur Wissenschaftsgeschichte, 19, 1996, 95–103.

**112** Erich Truckenbrodt: Aerodynamik. In: Ludwig Bölkow (ed.): Ein Jahrhundert Flugzeuge. Düsseldorf: VDI-Verlag, 1990, 52–77.

**113** Werner Schwipps: Schwerer als Luft. Die Frühzeit der Flugtechnik in Deutschland. Koblenz: Bernard und Graefe Verlag, 1984.

**114** Roger E. Bilstein: Flight in America. From the Wrights to the Astronauts. Baltimore and London: Johns Hopkins University Press, 1994.

**115** John H. Morrow, Jr.: German Air Power in World War I. Lincoln: University of Nebraska Press, 1982.

**116** Martin Kunz: Das Flugzeug als Waffe – Der Erste Weltkrieg als Experimentierfeld des Luftkriegs. In: Museum für Verkehr und Technik Berlin (ed.): Hundert Jahre deutsche Luftfahrt. Lilienthal und seine Erben. Gütersloh: Bertelsmann, 1991, 39–59.

**117** John H. Morrow, Jr.: The Great War in the Air. Military Aviation from 1909 to 1921. Washington: Smithsonian Institution Press, 1993.

**118** Volker Koos: Flugerprobung bis zum Ende des Ersten Weltkrieges. In: Heinrich Beauvais et al. (eds.): Flugerprobungsstellen bis 1945. Bonn: Bernard und Graefe Verlag, 1998.

**119** Technische Berichte, Vol. 1. Flugzeugmeisterei der Inspektion der Fliegertruppen (ed.). Charlottenburg, 1917.

**120** Helmuth Trischler: Luft- und Raumfahrtforschung in Deutschland 1900–1970. Politische Geschichte einer Wissenschaft. Frankfurt: Campus, 1992.

**121** Akten "betreffend Errichtung eines Kaiser-Wilhelm-Instituts für Aerodynamik und Hydrodynamik in Göttingen, Angefangen 1911. Geschlossen 1916." MPGA, Abt. I, Rep. 1A , No. 1466.

**122** Ludwig Prandtl: Die Modellversuchsanstalt in Göttingen. In: Zeitschrift für Flugtechnik und Motorluftschiffahrt, 11, 1920, 84ÂÛ87. Reprinted in Ludwig Prandtl Gesammelte Abhandlungen, 3, 1271–1293.

**123** Correspondence from Prandtl to the Ministry of War, 13 September 1918. MPGA, Abt. III, Rep. 61, No. 2107.

**124** Correspondence from Prandtl to APK, 29 October 1918. MPGA, Abt. III, Rep. 61, No. 2107.

**125** DMV, confidential survey, winter 1917/18; Correspondence from Prandtl to Gutzmer, 20 February 1918. GOAR 3664.

**126** Correspondence from Hoff to Prandtl, 16 May 1917 and 21 June 1917. GOAR 1354.

**127** Correspondence from Prandtl to Hoff, 9 July 1917. GOAR 1353.

**128** Correspondence from Hoff to Pfalz Flugzeugwerke G.m.b.H., 6 July 1917, copied to Prandtl. GOAR 1354.

**129** Correspondence from Prandtl to Mannesmann Waffen- und Munitions-Werke, 23 February 1918, in response to the firm's request on 19 February 1918. GOAR 1360.

**130** Correspondence from Hoff to Prandtl, January 1917. GOAR 1353.

**131** Max Munk: Bericht über Luftwiderstandsmessungen von Streben. TB, 1, 85–97.

**132** Correspondence from MVA to Luftschiffbau Zeppelin, 16 February 1917. GOAR 1371.

**133** Max Munk: Weitere Widerstandsmessungen an Streben. TB, 2, 15–17 and Tables 11 and 13.

**134** Karl Pohlhausen: Widerstandsmessungen an Seilen und Profildrähten. TB, 2, 13 and Table 10.

**135** Max Munk: Die Messungen an Flügelmodellen in der Göttinger Anstalt. TB, 1, 135–147.

**136** Max Munk, Erich Hückel: Weitere Göttinger Flügelprofiluntersuchungen, TB, 2, 407–450.

**137** Günter Schmitt: Hugo Junkers. Ein Leben für die Technik. Planegg: Aviatik Verlag, 1991.

**138** H. Kumbruch: Änlichkeitsversuche an Flügelprofilen. In: Zeitschrift für Flugtechnik und Motorluftschiffahrt, 10, 1919, 95–109.

**139** Albert Betz: Einfluss der Spannweite und Flächenbelastung auf die Luftkräfte von Tragflächen. TB, 1, 98–102.

**140** Max Munk: Modellmessungen an drei Tragflächen von verschiedener Spannweite. TB, 1, 203.

**141** Einzelflügel mit zugespitzten Enden. TB, 2, 397–399.

**142** Max Munk: Beitrag zur Aerodynamik der Flugzeugtragorgane. TB, 2, 187–273.

**143** Max Munk: Isoperimetrische Aufgaben aus der Theorie des Fluges. Dissertation, Universität Göttingen, 1919.

**144** Albert Betz: Beiträge zur Tragflügeltheorie mit besonderer Berücksichtigung des einfachen rechteckigen Flügels. Dissertation, Universität Göttingen, 1919.

**145** Correspondence from Munk to Prandtl, 18 December 1912. Göttingen, NSUB, Teilnachlass Prandtl, Acc. Mss. 1999.2, Cod. Ms. L. Prandtl 5 (Briefwechsel mit Bewerbern um Assistentenstellen).

**146** Correspondence from Munk to Prandtl, 26 February 1915. Göttingen, NSUB, Teilnachlass Prandtl, Acc. Mss. 1999.2, Cod. Ms. L. Prandtl 5 (Briefwechsel mit Bewerbern um Assistentenstellen).

**147** Correspondence from Prandtl to Munk, 9 March 1915, and subsequent correspondence in Göttingen, NSUB, Teilnachlass Prandtl, Acc. Mss. 1999.2, Cod. Ms. L. Prandtl 5 (Briefwechsel mit Bewerbern um Assistentenstellen).

**148** Correspondence from Munk to Prandtl, 7 April 1918; correspondence from Prandtl to Munk, 10 April 1918; correspondence from Munk to Prandtl, 12 April 1918. GOAR 2647.

**149** Max Munk: Beitrag zur Aerodynamik der Flugzeugtragorgane. Dissertation, TH Hannover, 1919. Printed in Göttingen: Dieterichsche Universitäts-Buchdruckerei.

**150** Max Munk: Isoperimetrische Aufgaben aus der Theorie des Fluges. Dissertation, Universität Göttingen, 1919. Printed in Göttingen: Dieterichsche Universitäts-Buchdruckerei.

**151** Correspondence from Prandtl to Munk, 23 April 1918; correspondence from Munk to Prandtl, 30 April 1918; correspondence from Prandtl to Munk, 4 May 1918. GOAR 2647.

**152** Ludwig Prandtl: Votum zur Dissertation Munks, 18 June 1918. GOAR 2647.

**153** A.H.G. Fokker, Bruce Gould: Der fliegende Holländer. Das Leben des Fliegers und Flugzeugkonstrukteurs A.H.G. Fokker. Zürich: Rascher, 1933.

**154** A.R. Weyl: Fokker: The Creative Years. London: Putnam, 1965.

**155** Ernst Heinkel: Stürmisches Leben. Stuttgart: Mundus-Verlag, 1953.

**156** "Angaben zur Lebensgeschichte von Professor Junkers und zur Geschichte der Junkers-Werke. Beitrag von Oberingenieur Steudel, zusammengestellt von Dr. M. Conzelmann," 6 December 1938. DMA, Junkers, Sammlung Pulst, No. 18.

**157** Philipp von Doepp: Stroemungstechnische Entwicklung bei Junkers. In: Die Junkers-Lehrschau. Dessau: Junkers, 1939, 6–20.

**158** Junkers's Diary, October 1915 to May 1917. DMA, Nachlass Junkers (NL 21), Notizbücher 008.

**159** "Angaben zur Lebensgeschichte von Professor Hugo Junkers, Beitrag von Professor Dr. Ludwig Prandtl, zusammengestellt von Dr. Margarete Conzelmann," 5 June 1948. DMA, Nachlass Conzelmann (NL 123), no. 11.

**160** Max Munk: Spannweite und Luftwiderstand. TB, 1, 199–202.

**161** Correspondence from Deutsche Flugzeug-Werke Leipzig to Prandtl, 17 May 1918; correspondence from Prandtl to Deutsche Flugzeug-Werke Leipzig, 22 May 1918. GOAR 1360. (This exchange occurred in response to Albert Betz: Einfluss der Spannweite und Flächenbelastung auf die Luftkräfte von Tragflächen. TB, 1, 98–102.)

**162** Ludwig Prandtl et al.: Ergebnisse der Aerodynamischen Versuchsanstalt zu Göttingen, I.–IV. Lieferung. München: Oldenbourg, 1921–1935.

**163** Richard Fuchs, Ludwig Hopf: Aerodynamik. Berlin: Richard Carl Schmidt und Co., 1922.

**164** Walter G. Vincenti: Real-world Variation-Selection in the Evolution of Technological Form: Historical Examples. In: John Ziman (ed.): Technological Innovation as an Evolutionary Process. Cambridge: Cambridge University Press, 2000, 174–189.

**165** "Summary of the Political Situation for the Month of February, 1922," Service Report, no. 3009, 6 March 1922. U.S. Military Intelligence Reports. Germany. 1919–1941. National Archives, Washington, RG 155. (Microfilm in Munich, Bayerische Staatsbibliothek, Sign. Film R 84.16–1919/41,1)

**166** Brigitte Schroeder-Gudehus: Deutsche Wissenschaft und internationale Zusammenarbeit 1914–1928. Ein Beitrag zum Studium kultureller Beziehungen in politischen Krisenzeiten. Genf: Dumaret and Golay, 1966.

**167** Grover Cleveland Loening: Progress in Aerodynamics. Studying the Aeroplane in the Laboratory. Scientific American, 14 October 1911, 340–341.

**168** A. Lawrence Rotch: Aerial Engineering. In: Science, 35, 1912, 41–46.

**169** William F. Trimble: Jerome C. Hunsaker and the Rise of American Aeronautics. Washington, London: Smithsonian Institution Press, 2002.

**170** Alex Roland: Model Research. The National Advisory Committee for Aeronautics 1915–1958. Volume 1. Washington, DC: NASA, 1985. (NASA SP-4103).

**171** I.B. Holley, Jr.: Ideas and Weapons. Exploitation of the Aerial Weapon by the United States During World War I; A Study in the Relationship of Technological Advance, Military Doctrine, and the Development of Weapons. Washington: Office of Air Force History, 1983.

**172** Report of the National Research Council. In: Report of the National Academy of Sciences for the Year 1917. Washington: Government Printing Office, 1918.

**173** Roy MacLeod: Secrets among Friends: The Research Information Service and the "Special Relationship" in Allied Scientific Information and Intelligence, 1916–1918. Minerva, 37, 1999, 201–233.

**174** Report of the National Research Council. In: Report of the National Academy of Sciences for the Year 1918. Washington: Government Printing Office, 1919.

**175** Minutes of Regular Meeting of Executive Committee of the NACA, 6 September 1918, 17 February 1919, 10 April 1919 and 24 April 1919. NACP, RG 255, Entry 7, A1, Box 1.

**176** Minutes of Regular Meeting of Executive Committee of the NACA, 20 June 1919. NACP, RG 255, Entry 7, A1, Box 1.

**177** Corrrespondence from Ames to Knight, 30 July 1919. NACP, RG 255, Entry 1, A1, Box 248, (51–6G) Paris Office, Miscellaneous, 1919–1920.

**178** Correspondence from Knight to NACA, 2 July 1919. NACP, RG 255, Entry 1, A1, Box 248, (51–6G) Paris Office, Miscellaneous, 1919–1920.

179 Minutes of Regular Meeting of Executive Committee of the NACA, 12 September 1919. NACP, RG 255, Entry 7, A1, Box 1: 1915–1920.

180 Correspondence from Knight to Ames, 18 August 1919. NACP, RG 255, Entry 1, A1, Box 248, (51–6G) Paris Office, Miscellaneous, 1919–1920.

181 Correspondence from Knight to Prandtl, 15 November 1919. MPGA, Abt. III, Rep. 61, No. 836.

182 Correspondence from Prandtl to Knight, 1 December 1919. MPGA, Abt. III, Rep. 61, No. 836.

183 Correspondence from Knight to Prandtl, 26 January 1920. MPGA, Abt. III, Rep. 61, No. 836.

184 Correspondence from Prandtl to Knight, 4 March 1920. MPGA, Abt. III, Rep. 61, No. 836.

185 Daniel J. Kevles: Into Hostile Political Camps: The Reorganization of International Science in World War I. Isis, 62, 1971, 47–60.

186 Correspondence from Stratton (Secretary of the NACA) to Knight, 27 February 1920. NACP, RG 255, Entry 1, A1, Box 248, (51–6G) Paris Office, General Correspondence, 1915–1942.

187 Minutes of Regular Meeting of Executive Committee, 1. April 1920. NACP, RG 255, Entry 7, A1, Box 1; Correspondence from Lewis to Kimball, 24 June 1920. NACP, RG 255, Biographical Files, Box 28, Folder: George W. Lewis.

188 Correspondence from Lewis to Knight, 17 June 1920. NACP, RG 255, Entry 1, A1, Box 248.

189 Correspondence from Hunsaker to Prandtl, 16 June 1920. MPGA, Abt. III, Rep. 61, No. 724.

190 Minutes of Regular Meeting of Executive Committee of the NACA, 20 September 1920. NACP, RG 255, Entry 7, A1, Box 1: July - December 1920.

191 Correspondence from Prandtl to Hunsaker, 16 July 1920. MPGA, Abt. III, Rep. 61, No. 724.

192 Correspondence from NACA to Knight, 8 May 1920. NACP, RG 255, Entry 1, A1, Box 248, (51–6G) Paris Office, General Correspondence, 1915–1942.

193 Correspondence from Hunsaker to Prandtl, 29 January 1921. MPGA, Abt. III, Rep. 61, No. 724.

194 Edward P. Warner: Report on German Wind Tunnels and Apparatus. September 1920. NACP, RG 255, Entry 18: Reports on European Aviation, 1920–1951. Box 1: 1920–1923.

195 Ludwig Prandtl: Theory of Lifting Surfaces, Parts I and II. Technical Notes. NACA, Reports Nos. 9 and 10. July and August 1920.

196 Ludwig Prandtl: Göttingen Wind Tunnel for Testing Aircraft Models. Technical Notes. NACA, Report no. 66. November 1920. (Based on L. Prandtl: Die Modellversuchsanstalt in Göttingen. Berichte und Abhandlungen der Wissenschaftlichen Gesellschaft für Luftfahrt, 1, 1920, 51–59; reprinted in Ludwig Prandtl Gesammelte Abhandlungen, 3, 1271–1293).

197 Minutes of Semiannual Meeting of the NACA, 21 April 1921; Minutes of Regular Meeting of Executive Committee of the NACA, 12 May 1921. NACP, RG 255, Entry 7, A1, Box 1; correspondence from Ide to Victory (Secretary of the NACA), 1 May 1921; correspondence from Lewis to Ide, 9 July 1921; correspondence from Ide to Lewis and Ames, 15 March 1923. NACP, RG 255, Entry 1, A1, Box 250; Minutes of Annual Meeting of the NACA, 18 October 1923. NACP, RG 255, Entry 7, A1, Box 1.

198 Correspondence from Knight to Prandtl, 30 December 1920. MPGA, Abt. III, Rep. 61, No. 836. Reichsverkehrsminister to Prandtl, Karman, Hopf, v. Mises, Reissner, Eberhardt (Darmstadt), Kutzbach, Pröll, Baumann, Huppert, 18 February 1921. GOAR 1372.

199 Ide to NACA, 17 October 1921. NACP, RG 255, Entry 18: Reports on European Aviation, 1920–1951. Box 1: 1920–1923; correspondence from Ide to Lewis, 19 September 1921; correspondence from Lewis to Ide, 5 October 1921. NACP, RG 255, Entry 1, A1, Box 250, (51–6G) Paris Office, Miscellaneous, 1921–1923.

200 Reinhard Siegmund-Schultze: Rockefeller and the Internationalization of Mathematics Between the Two World Wars. Basel: Birkhäuser, 2001.

201 Takehiko Hashimoto: The Wind Tunnel and the Emergence of Aeronautical Re-

search in Britain. In: Peter Galison, Alex Roland (eds.): Atmospheric Flight in the Twentieth Century. Dordrecht: Kluwer, 2000, 223–239.

**202** William Knight: Standardization and Aerodynamics. Technical Notes, NACA Report No. 134 (March 1923). (This is a reprint of the series of articles in Aerial Age from June 1920 to December 1922).

**203** Correspondence from Knight to NACA, 27 September 1919. NACP, RG 255, Entry 1, A1, Box 248, (51–6G) Paris Office, Miscellaneous, 1919–1920.

**204** Correspondence from Warner to Ames, 4 November 1919. NACP, RG 255, Entry 1, A1, Box 248, (51–6G) Paris Office, Miscellaneous, 1919–1920.

**205** Correspondence from Prandtl to Knight, 3 August 1921. MPGA, Abt. III, Rep. 61, No. 837.

**206** Correspondence from Knight to Prandtl, 10 January 1922. MPGA, Abt. III, Rep. 61, No. 838.

**207** D.L. Bacon, E.G. Reid: The Resistance of Spheres in Wind Tunnels and in Air. NACA-Report No. 185, 1924.

**208** E.G. Reid: Standardization Tests of NACA No. 1 Wind Tunnel. NACA-Report No. 195, 1925.

**209** E.G. Reid: Interference Tests on an NACA Pitot Tube. NACA-Report No. 199, 1925.

**210** Montgomery Knight: Wind Tunnel Standardization Disk Drag. NACA-Report No. 253, 1926.

**211** Correspondence from Prandtl to Knight, 26 February 1923. MPGA, Abt. III, Rep. 61, No. 839.

**212** Correspondence from Kármán to Levi-Civita, 12 April 1922. TKC 18.8.

**213** Theodore von Kármán, Tullio Levi-Civita (eds.): Vorträge aus dem Gebiete der Hydro- und Aerodynamik (Innsbruck 1922). Berlin: Springer, 1924.

**214** Correspondence from Burgers to Kármán, 15 May 1923. TKC 4.21.

**215** J.L. van Ingen: Ludwig Prandtl and Early Fluid Dynamics in the Netherlands. In: Gerd E.A. Meier (ed.): Ludwig Prandtl, ein Führer in der Strömungslehre. Braunschweig/Wiesbaden: Vieweg, 2000, 219–241.

**216** Correspondence from Burgers to Prandtl, 22 October 1923. MPGA, Abt. III, Rep. 61, No. 210.

**217** Correspondence from Prandtl to Burgers, 30 October 1923. MPGA, Abt. III, Rep. 61, No. 210.

**218** Correspondence from Kármán to Prandtl, 7 December 1923. MPGA, Abt. III, Rep. 61, No. 792.

**219** Correspondence from Burgers to Prandtl, 8 December 1923. MPGA, Abt. III, Rep. 61, No. 210.

**220** Correspondence from Prandtl to Burgers, 15 December 1923. MPGA, Abt. III, Rep. 61, No. 210.

**221** Zeitschrift für Angewandte Mathematik und Mechanik, 4, 1924, 272–276.

**222** S. Juhasz (ed.): IUTAM. A Short History. Berlin: Springer, 1988.

**223** Correspondence from Burgers to Prandtl, 3 March 1928. MPGA, Abt. III, Rep. 61, No. 210.

**224** Correspondence from Prandtl to Burgers, 29 March 1928. MPGA, Abt. III, Rep. 61, No. 210.

**225** Correspondence from Prandtl to Kármán, 15 June 1921. MPGA, Abt. III, Rep. 61, No. 792.

**226** H. Gericke: 50 Jahre GAMM. Berlin: Springer, 1972.

**227** Renate Tobies: Die "Gesellschaft für angewandte Mathematik und Mechanik" im Gefüge imperialistischer Wissenschaftsorganisation. Zeitschrift für Geschichte der Naturwissenschaften, Technik und Mathematik (NTM), 19, 1982, 16–26.

**228** Johanna Vogel-Prandtl: Ludwig Prandtl. Ein Lebensbild. Erinnerungen. Dokumente. Göttingen 1993. (Mitteilungen aus dem Max-Planck-Institut für Strömungsforschung, No. 107)

**229** Correspondence from Prandtl to Knight, 25 April 1923. MPGA, Abt. III, Rep. 61, No. 839.

**230** Correspondence from Bothezat to Lewis, 28 June 1920. NACP, RG 255, Biographical Files, Bothezat, Box No. 6, Folder: George de Bothezat.

**231** Minutes of Regular Meeting of Executive Committee, 11 November 1920. NACP, RG 255, Entry 7, A1: Minutes of the Executive Committee of NACA, 1915–1958, Box 1: 1915–1920, Folder 1–2 Minutes July-December 1920.

**232** Correspondence from Munk to Prandtl, 23 November 1920. MPGA, Abt. III, Rep. 61, No. 1110.

**233** James R. Hansen: Engineer in Charge. A History of the Langley Aeronautical Laboratory, 1917–1958. Washington: NASA, 1987 (NASA SP 4305).

**234** Correspondence from Norton to Lewis, 9 March 1921. NAPhil, RG 255, Entry 2, AV400–1 (1921–1944), Box 1025.

**235** Correspondence from Lewis to Norton, 18 April 1921. NAPhil, RG 255, Entry 2, AV400–1 (1921–1944), Box 1025.

**236** Correspondence from Norton to Lewis, 30 April 1921. NAPhil, RG 255, Entry 2, AV400–1 (1921–1944), Box 1025.

**237** Max Munk: On a New Type of Wind Tunnel. NACA, TN 60, 1921.

**238** Max Munk: General Theory of Thin Wing Sections. NACA, TR 142, 1922.

**239** M. Munk: Fundamentals of Fluid Dynamics for Aircraft Designers. New York: Ronald Press, 1929.

**240** Ira H. Abbott, Albert E. von Doenhoff: Theory of Wing Sections. New York: Dover, 1949.

**241** Correspondence from Norton to Ames, 6 October 1921. NAPhil, RG 255, Entry 2, AV400–1 (1921–1944), Box 1025.

**242** Correspondence from Munk to Prandtl, 25 December 1921. MPGA, Abt. III, Rep. 61, No. 1110.

**243** Robert T. Jones, in an interview with Walter Bonney, 24 September 1974, transcript p. 3. I am grateful to James Hansen for this information.

**244** Correspondence from Blasius to Prandtl, 31 May 1911. GOAR 3684.

**245** Heinrich Blasius: Das Aehnlichkeitsgesetz bei Reibungsvorgängen in Flüssigkeiten. Mitteilungen der Forschungen des VDI, 131, 1913, 1–39.

**246** Carl Wieselsberger, Albert Betz, Ludwig Prandtl: Ergebnisse der Aerodynamischen Versuchsanstalt zu Göttingen, 1. Lieferung. München: Oldenbourg, 1921.

**247** W.S. Diehl: Skin Frictional Resistance of Plane Surfaces in Air. Abstract of Recent German Tests. NACA, TN 102, July 1922.

**248** T.E. Stanton et al.: On the Conditions at the Boundary of a Fluid in Turbulent Motion. In: Proceedings of the Royal Society, London, 97A, 1920, 413.

**249** Johannes Martinus Burgers: Hitzdrahtmessungen. In: L. Schiller (ed.): Hydro- und Aerodynamik, 1. Teil: Strömungslehre und Allgemeine Versuchstechnik. Handbuch der Experimentalphysik, Band 4 (eds. W. Wien, F. Harms). Leipzig: Akademische Verlagsgesellschaft, 1931, 635–667.

**250** Johannes Martinus Burgers: The Motion of a Fluid in the Boundary Layer along a Plane Smooth Surface. In: C.B. Biezeno und J.M. Burgers (eds.): Proceedings of the First International Congress for Applied Mechanics. Delft: Techn. Boekhandel en Drukkerij J. Waltman Jr., 1924, 113–128.

**251** Ludwig Schiller: Experimentelle Untersuchungen zum Turbulenzproblem. ZAMM, 1, 1921, 436–444.

**252** Ludwig Schiller: Die Entwicklung der laminaren Geschwindigkeitsverteilung und ihre Bedeutung für Zähigkeitsmessungen. ZAMM, 2, 1922, 96–106.

**253** Ludwig Schiller: Über den Strömungswiderstand von Rohren verschiedenen Querschnitts und Rauhigkeitsgrades. ZAMM, 3, 1923, 2–13.

**254** Johann Nikuradse: Untersuchung über die Geschwindigkeitsverteilung in turbulenten Strömungen. Forschungsarbeiten auf dem Gebiet des Ingenieurwesens, Vol. 281, 1926.

**255** Fritz Dönch: Divergente und konvergente turbulente Strömungen mit kleinen Öffnungswinkeln. VDI Forschungsarbeiten, Vol. 282, 1926.

**256** Johann Nikuradse: Über turbulente Wasserströmungen in geraden Rohren bei sehr großen Reynoldschen Zahlen. Abhandlungen aus dem Aerodynamischen Institut der Technischen Hochschule Aachen, Vol. 8, 1928, 63–69.

**257** Johann Nikuradse: Gesetzmässigkeiten der turbulenten Strömung in glatten Rohren. VDI Forschungsarbeiten, Vol. 356, 1932.

**258** Ludwig Prandtl, Albert Betz (eds.): Ergebnisse der Aerodynamischen Versuchsanstalt zu Göttingen. 4. Lieferung. München: Oldenbourg, 1932.

**259** Julius C. Rotta: Ludwig Prandtl und die Turbulenz. In: Gerd E.A. Meier (ed.):

Ludwig Prandtl, ein Führer in der Strö-
mungslehre. Biographische Artikel zum
Werk Ludwig Prandtls. Braunschweig,
Wiesbaden: Vieweg, 2000, 53–123.

260 Loose sheets of paper, preserved in NSUB,
Teilnachlass Ludwig Prandtl, Acc. Mss.
1999.2, Cod. Ms. L. Prandtl 18.

261 Loose sheets of paper, preserved in NSUB,
Teilnachlass Ludwig Prandtl, Acc. Mss.
1999.2, Cod. Ms. L. Prandtl 19.

262 Manuscripts in MPGA, Abt. III, Rep. 61,
Nos. 2269, 2276, 2278, 2282, and 2285.

263 Fritz Noether: Das Turbulenzproblem.
ZAMM, 1, 1921, 125–138.

264 Ludwig Prandtl: Bemerkungen über die
Entstehung der Turbulenz. Physikalische
Zeitschrift, 23, 1922, 19–25. Reprinted in
Ludwig Prandtl Gesammelte Abhandlun-
gen, 2, 687–696.

265 Oskar Tietjens: Beiträge zur Entstehung
der Turbulenz. ZAMM, 5, 1925, 200–217.

266 Correspondence from Prandtl to Noether,
1 June 1922. MPGA, Abt. III, Rep. 61, No.
1155.

267 Correspondence from Noether to Prandtl,
8 June [1922]. MPGA, Abt. III, Rep. 61,
No. 1155.

268 Correspondence from Noether to Prandtl,
21 June 1922. MPGA, Abt. III, Rep. 61, No.
1155.

269 Correspondence from Prandtl to Kármán,
14 June 1921. MPGA, Abt. III, Rep. 61, No.
792.

270 Correspondence from Hopf to Prandtl, 26
May 1922. MPGA, Abt. III, Rep. 61, No.
704.

271 Werner Heisenberg: Über Stabilität und
Turbulenz von Flüssigkeitsströmen. Annalen der Physik, 74, 1924, 577–627.

272 Correspondence from Hopf to Prandtl, 28
May 1923. MPGA, Abt. III, Rep. 61, No.
704.

273 Correspondence from Prandtl to Heisenberg, 19 July 1923. MPGA, Abt. III, Rep.
61, No. 643.

274 Correspondence from Prandtl to Noether,
12 July 1926. MPGA, Abt. III, Rep. 61,
no.1155.

275 Fritz Noether: Zur asymptotischen Behandlung der stationären Lösungen im
Turbulenzproblem. ZAMM, 6, 1926,

232–243; Ludwig Prandtl: Zum Turbulenzproblem. ZAMM, 6, 1926, 339–340.
Further responses in ZAMM, 6, 1926, 428
and 497–498.

276 Correspondence from Prandtl to Hopf, 20
July 1926. MPGA, Abt. III, Rep. 61, No.
704.

277 Walter Tollmien: Über die Entstehung der
Turbulenz. Nachrichten der Gesellschaft
der Wissenschaften zu Göttingen,
Mathematisch-Physikalische Klasse, 1929,
21–44.

278 Ludwig Prandtl: Über die ausgebildete
Turbulenz. Verhandlungen des II. Internationalen Kongresses für Technische
Mechanik 1926. Zürich: Füssli, 1927, 62–
75. Reprinted in Ludwig Prandtl Gesammelte Abhandlungen, 2, 736–751.

279 Correspondence from Prandtl to Kármán,
10 October 1924. MPGA, Abt. III, Rep. 61,
No. 792.

280 Ludwig Prandtl: Bericht über Untersuchungen zur ausgebildeten Turbulenz.
ZAMM, 5, 1925, 136–139. Reprinted in
Ludwig Prandtl Gesammelte Abhandlungen, 2, 714–718.

281 Walter Tollmien: Berechnung turbulenter
Ausbreitungsvorgänge. ZAMM, 6, 1926,
468–478.

282 Albert Betz: Über turbulente Reibungss-
chichten an gekrümmten Wänden. In: A.
Gilles et al. (eds.): Vorträge aus dem Gebiete der Aerodynamik und verwandter
Gebiete (Aachen 1929). Berlin: Springer
1930, 10–18.

283 Johann Nikuradse: Untersuchungen über
die Strömungen des Wassers in kon-
vergenten und divergenten Kanälen.
Forschungsarbeiten auf dem Gebiet des
Ingenieurwesens, Vol. 289, 1929.

284 Hermann Schlichting: Über das ebene
Windschattenproblem. Ingenieur-Archiv,
1, 1930, 533–571.

285 Hermann Schlichting: An Account of
the Scientific Life of Ludwig Prandtl.
Zeitschrift für Flugwissenschaften, 23,
1975, 297–316.

286 Correspondence from Kármán to Prandtl,
12 February 1921. GOAR 3684.

287 Correspondence from Prandtl to Kármán,
16 February 1921. MPGA, Abt. III, Rep.
61, No. 792.

**288** Theodore von Kármán: Über laminare und turbulente Reibung. ZAMM, 1, 1921, 233–252. Reprinted in CWTK, 2, 70–97.

**289** Theodore von Kármán: Über die Oberflächenreibung von Flüssigkeiten. In: Theodore von Kármán and Tullio Levi-Civita: Vorträge aus dem Gebiet der Hydro- und Aerodynamik (Innsbruck 1922). Berlin: Julius Springer, 1924, 146–167. Reprinted in CWTK, 2, 133–152.

**290** Ludwig Prandtl, Albert Betz (eds.): Ergebnisse der Aerodynamischen Versuchsanstalt zu Göttingen. 3. Lieferung. München: Oldenbourg, 1927, 1–5.

**291** Karl Pohlhausen: Zur näherungsweisen Integration der Differentialgleichung der laminaren Grenzschicht. ZAMM, 1, 1921, 252–268.

**292** Ludwig Hopf: Turbulenz bei einem Flusse. Annalen der Physik, 32, 1910, 777–808.

**293** Ludwig Hopf: Die Messung der hydraulischen Rauhigkeit. ZAMM, 3, 1923, 329–339.

**294** Karl Fromm: Strömungswiderstand in rauhen Rohren. ZAMM, 3, 1923, 339–358.

**295** Correspondence from Prandtl to Hopf, 24 May 1922. MPGA, Abt. III, Rep. 61, No. 704.

**296** Correspondence from Prandtl to Kármán, 26 July 1928. MPGA, Abt. III, Rep. 61, No. 792. Prandtl alluded to these dissertations: M. Hansen: Die Geschwindigkeitsverteilung in der Grenzschicht an einer eingetauchten Platte; W. Fritsch: Der Einfluss der Wandrauhigkeit auf die turbulente Geschwindigkeitsverteilung in Rinnen. Abhandlungen aus dem Aerodynamischen Institut der Technischen Hochschule Aachen, Vol. 8, 1928.

**297** Ludwig Prandtl: Neuere Ergebnisse der Turbulenzforschung. Zeitschrift des VDI, 77, 1933, 105–114. Reprinted in in Ludwig Prandtl Gesammelte Abhandlungen, 2, 819–845. Prandtl presented these results on 6 May 1932 in a talk at the Technical University Prague, on 18 and 19 May 1932 at a conference at the Hamburger Schiffbau-Versuchsanstalt, and again on 16 October 1932 at a meeting of the VDI. See also J. Nikuradse: Strömungswiderstand in rauhen Rohren. ZAMM, 11, 1931, 409–411; and J. Nikuradse: Strömungsgesetze in rauhen Rohren. VDI-Forschungsheft 361. Berlin: VDI-Verlag, 1933.

**298** Theodore von Kármán: Mechanische Ähnlichkeit und Turbulenz. Nachrichten von der Gesellschaft der Wissenschaften zu Göttingen, Mathematisch-Physikalische Klasse, 1930, 58–76. Reprinted in CWTK, 2, 322–336.

**299** Correspondence from Kármán to Kempf, 26 September 1932. MPGA, Abt. III, Rep. 61, No. 793.

**300** Ludwig Prandtl: Erörterungsbeitrag zur Gruppe I. Reibungswiderstand. In: G. Kempf, E. Förster (eds.): Hydromechanische Probleme des Schiffsantriebs. Hamburg: Schiffbau-Versuchsanstalt, 1932, 87–91.

**301** Correspondence from Kármán to Prandtl, 26 September 1932. MPGA, Abt. III, Rep. 61, No. 793.

**302** Correspondence from Prandtl to Kármán, 29 September 1932. MPGA, Abt. III, Rep. 61, No. 793.

**303** Correspondence from Kármán to Prandtl, 9 December 1932. MPGA, Abt. III, Rep. 61, No. 793.

**304** Correspondence from Prandtl to Kármán, 19 December 1932. MPGA, Abt. III, Rep. 61, No. 793.

**305** E. Krause, U. Kalkmann: Prandtls Schüler in Aachen. In: Gerd E.A. Meier (ed.): Ludwig Prandtl, ein Führer in der Strömungslehre. Biographische Artikel zum Werk Ludwig Prandtls. Braunschweig, Wiesbaden: Vieweg, 2000, 243–276.

**306** W. Tollmien: Die Kármánsche Ähnlichkeitshypothese in der Turbulenztheorie und das ebene Windschattenproblem. Ingenieur-Archiv, 4, 1933, 1–15.

**307** Correspondence from Prandtl to Kármán, 11 August 1920. GOAR 3684.

**308** Correspondence from Prandtl to Kármán, 8 June 1920. GOAR 3684.

**309** Ludwig Schiller (ed.): Hydro- und Aerodynamik, 1. Teil: Strömungslehre und Allgemeine Versuchstechnik. Handbuch der Experimentalphysik, Vol. 4. Leipzig: Akademische Verlagsgesellschaft, 1931.

**310** Correspondence from Dryden to Prandtl, 5 March 1921. MPGA, Abt. III, Rep. 61, No. 361.

**311** Correspondence from Prandtl to Dryden, 20 April 1921. MPGA, Abt. III, Rep. 61, No. 361.

**312** W.L. LePage, J.T. Nichols: The Effect of Wind Tunnel Turbulence Upon the Forces Measured on Models. NACA, Technical Notes, No. 191, May 1924.

**313** H.L. Dryden, R.H. Heald: Investigation of Turbulence in Wind Tunnels by a Study of the Flow about Cylinders. NACA Report 231, 1926, 465–479.

**314** H.L. Dryden, A.M. Kuethe: The Measurement of Fluctuations of Air Speed by the Hot-Wire Anemometer. NACA Report No. 320, 1930, 359–382.

**315** H.L. Dryden, A.M. Kuethe: Effect of Turbulence in Wind-Tunnel Measurements. NACA Report 342, 1931, 147–170.

**316** Correspondence from Dryden to Prandtl, 30 March 1931. MPGA, Abt. III, Rep. 61, No. 361.

**317** John Stack: Tests in the variable density wind tunnel to investigate the effects of scale and turbulence on airfoil characteristics. NACA Technical Note No. 364, February 1931.

**318** Correspondence from Prandtl to Dryden, 18 April 1931. MPGA, Abt. III, Rep. 61, No. 361.

**319** Correspondence from Dryden to Prandtl, 8 May 1931. MPGA, Abt. III, Rep. 61, No. 361.

**320** David F. Channell: The History of Engineering Science. An Annotated Bibliography. New York, London: Garland, 1989.

**321** Ludwig Prandtl: Die Aerodynamische Versuchsanstalt und ihre Bedeutung für die Technik. Die Naturwissenschaften, 10, 1922, 169–176.

**322** Memorandum signed by H. Reissner, 15 January 1910. DMA, Junkers-Archiv, Flugzeugbau und Verwertung, 0101–T03.

**323** 50 Jahre Aerodynamisches Institut der Rhein.-Westf. Technischen Hochschule Aachen, 1913–1963. Historisches Archiv der RWTH, A 33/95.

**324** R. Otzen: Die technischen Hochschulen im Dienste der Flugtechnik. Jahrbuch der Wissenschaftlichen Gesellschaft für Luftfahrt, 3, 1915, 56–57.

**325** A. Pröll: Gedanken zur Frage des Hochschulunterrichtes im Luftfahrtwesen. ZFM, 13, 1922, 163–166.

**326** Henri Bouche: Aeronautic Instruction in Germany, NACA, TM 51, November 1921.

**327** ZFM, 16, 1925, 96.

**328** ZFM, 16, 1925, 145.

**329** ZFM, 20, 1929, 136, 172–173.

**330** ZFM, 22, 1931, 626.

**331** A. Pröll: Flugtechnik. Grundlagen des Kunstfluges. München, Berlin: Oldenbourg, 1919.

**332** ZFM, 11, 1920, 73–74.

**333** Ludwig Hopf, Richard Fuchs: Aerodynamik. Berlin: Richard Carl Schmidt und Co., 1922.

**334** ZFM, 14, 1923, 65.

**335** Richard von Mises: Fluglehre. Vorträge über Theorie und Berechnung der Flugzeuge in elementarer Darstellung, 3rd edition. Berlin: Springer, 1926.

**336** Hermann Glauert: The Elements of Aerofoil and Airscrew Theory. Cambridge: Cambridge University Press, 1926. This book was also translated into German: Die Grundlagen der Tragflügel- und Luftschraubentheorie. Berlin: Springer, 1929.

**337** E. Pfister: Grundlagen der Fluglehre, Teil I: Luftkräfte; Teil II (with V. Porger): Tragflügeltheorie. Berlin: C.J.E. Volkmann, 1927 and 1929.

**338** H. Schmidt: Aerodynamik des Fluges. Eine Einführung in die mathematische Tragflächentheorie. Berlin, Leipzig: De Gruyter, 1929.

**339** C. Eberhardt: Einführung in die theoretische Aerodynamik. München, Berlin: Oldenbourg, 1927.

**340** Ludwig Prandtl: Abriss der Strömungslehre. Braunschweig: Vieweg, 1931. The 2nd edition was published in 1935; the 3rd edition published in 1942 as Führer durch die Strömungslehre, which went through 11 editions until 2002.

**341** Lutz Budrass: Flugzeugindustrie und Luftrüstung in Deutschland 1918–1945. Düsseldorf: Droste, 1998.

**342** ZFM, 20, 1929, 173.

**343** A. von Parseval: Die Bedeutung des motorlosen Segelflugs. ZFM, 13, 1922, 280–281.

**344** Peter Riedel: Start in den Wind. Erlebte Rhöngeschichte 1911–1926. Stuttgart: Motorbuch, 1977.

**345** Gustav Lilienthal: Der geheimnisvolle Vorwärtszug. Untersuchungen über den Segelflug. München: Oldenbourg, 1913.

**346** Correspondence from Prandtl to Lilienthal, 25 November 1920. GOAR 1364. (For the previous correspondence from 1916, see GOAR 2747).

**347** Friedrich Ahlborn: Der Segelflug. Erklärung des Segelfluges der Vögel; die Möglichkeit des Fliegens ohne Motor. Berichte und Abhandlungen der Wissenschaftlichen Gesellschaft für Luftfahrt (WGL), 5, 1921, 1–26.

**348** Albert Betz: Ein Beitrag zur Erklärung des Segelfluges. ZFM, 3, 1912, 269–272.

**349** Ludwig Prandtl: Bemerkungen über den Segelflug. ZFM, 12, 1921, 209–211. Reprinted in Ludwig Prandtl Gesammelte Abhandlungen, 1, 520–525.

**350** Friedrich Ahlborn: Zur Methode des Segelfluges. ZFM, 12, 1921, 337–338.

**351** Theodore von Kármán: Mechanische Modelle zum Segelflug. ZFM, 12, 1921, 220–223.

**352** Wolfgang Klemperer: Theorie des Segelfluges. Berlin: Springer, 1926.

**353** W. Hoff: Der Segelflug und die Rhön-Segelflug-Wettbewerbe. ZAMM, 2, 1922, 207–218.

**354** Ludwig Prandtl: Lehren des Rhönflugs 1922. ZFM, 13, 1922, 274–275. Reprinted in Ludwig Prandtl Gesammelte Abhandlungen, 1, 526–529.

**355** ZFM, 16, 1922, 232.

**356** Hans Zacher: Studenten forschen, bauen und fliegen. 60 Jahre Akademische Fliegergruppe Darmstadt e.V., 1981.

**357** Ludwig Prandtl: Bemerkungen über den Segelflug. ZFM, 12, 1921, 209–211. Reprinted in Ludwig Prandtl Gesammelte Abhandlungen, 1, 520–525. Correspondence on Prandtl's involvement in the Rhön glider contests is found in GOAR 3667.

**358** Otto Mader: Betrachtungen über den Flugzeugbau. ZVDI, 68, 1924, 1041–1046.

**359** Wolfgang Wagner: Hugo Junkers. Pionier der Luftfahrt – seine Flugzeuge. Bonn: Bernard und Graefe, 1996.

**360** Sef-Gesellschaftsübernahme und Handelsregister-Eintragung, 24 April 1924. DMA, Junkers-Archiv, Luftfahrt und Verwertung, 304/5/21.

**361** Vereinbarung, 5 April 1923. DMA, Junkers-Archiv, Luftfahrt und Verwertung, 304/3/20.

**362** Besprechungsprotokoll, 23 October 1923. DMA, Junkers-Archiv, Luftfahrt und Verwertung, 304/3/47.

**363** Besprechungsprotokoll, 22 December 1924. DMA, Junkers-Archiv, Luftfahrt und Verwertung, 304/5/40.

**364** Zusammenfassende Niederschrift, 8 January 1925. DMA, Junkers-Archiv, Luftfahrt und Verwertung, 304/5/40.

**365** Correspondence from Kármán to Junkers, 22 March 1922. DMA, Junkers-Archiv, Luftfahrt und Verwertung, 304/7/4.

**366** Wolfgang Klemperer: Ein Beitrag zum Spaltflügelproblem. ZFM, 12, 1921, 305–308.

**367** G. Lachmann: Neuere Versuchsergebnisse mit Spaltflügeln. ZFM, 15, 1924, 109–116, 173–176, 181–185.

**368** Theodore von Kármán: Bemerkungen über die Lage der Flugtechnik in Japan mit besonderer Berücksichtigung einer eventuellen Betätigung der Junkers-Werke, 22 June 1928. DMA, Junkers-Archiv, Luftfahrt und Verwertung, 705/4.

**369** Correspondence from Junkers (Hauptbüro) to Kármán, 12 October 1928. DMA, Junkers-Archiv, Luftfahrt und Verwertung, 705/4.

**370** Correspondence from Junkers to Prandtl, 11 January 1923. DMA, Junkers-Archiv, Luftfahrt und Verwertung, 306/3.

**371** Correspondence from Mader to Junkers, 4 August 1919. DMA, Junkers-Archiv, Luftfahrt und Verwertung, 306/3.

**372** Memorandum ("Unterstützung der Göttinger Modellversuchsanstalt für Aerodynamik"), 11 October 1919. DMA, Junkers-Archiv, Luftfahrt und Verwertung, 306/3.

**373** Correspondence from Prandtl to Conzelmann, 25 July 1948. DMA, Nachlass Junkers (NL 11), Junkerschronik.

**374** A list of contractors is preserved in GOAR 3226; files to individual contractors are in GOAR 88 (Arado), GOAR 108 (Dornier), GOAR 75–B11 (Junkers), GOAR 1405 and 1740 (Focke-Wulf), and GOAR 1416–1417 (Heinkel).

375 Otto Schrenk: Systematische Untersuchungen an Joukowsky-Profilen. ZFM, 18, 1927, 225–230; Otto Schrenk: Über die Theorie der Joukowsky-Profile. ZFM, 18, 1927, 276–284.

376 Max Munk, Elton W. Miller: Model Tests with a Systematic Series of 27 Wing Sections at Full Reynolds Number. NACA, TR 221, 1925.

377 F.A. Louden: Collection of Wind-Tunnel Data on Commonly Used Wing Sections. NACA-Report No. 331, 1930, 589–633.

378 Eastman N. Jacobs, Kenneth E. Wards, Robert M. Pinkerton: The Characteristics of 78 Related Airfoil Sections From Tests in the Variable-Density Wind Tunnel. NACA Report 460, 1933, 299–354.

379 Correspondence from Reid to Lewis, 26 November 1932. NaPhil, RG 255, Entry 2, AV400–1 (1921–1944), Box 1025, General Correspondence, 1928–1934.

380 David Bloor: Why Did Britain Fight the First World War with the Wrong Theory of the Aerofoil? 8 May 2000. I am grateful to David Bloor for a draft of this unpublished lecture.

381 L.W. Bryant, D.H. Williams: An Investigation of the Flow of Air Around an Aerofoil of Infinite Span. Philosophical Transactions of the Royal Society London A 225, 1926, 199–245.

382 A. Fage, H.L. Nixon: The Prediction on the Prandtl Theory of the Lift and Drag for Infinite Span from Measurements on Aerofoils on Finite Span. In: Technical Report of the Aeronautical Research Committee, Vol. 1, Reports and Memoranda, 1923–1924, No. 903.

383 Correspondence from Glauert to Prandtl, 21 January 1921. MPGA, Abt. III, Rep. 61, No. 536.

384 Correspondence from Prandtl to Glauert, 23 October 1923. MPGA, Abt. III, Rep. 61, No. 536.

385 Correspondence from Marsh (Secretary of the Royal Aeronautical Society) to Prandtl, 28 April 1922; correspondence from Prandtl to Marsh, 9 May 1922. MPGA, Abt. III, Rep. 61, No. 1401.

386 L. Prandtl: The Generation of Vortices in Fluids of Small Viscosity. Wilbur Wright Memorial Lecture, 16 May 1927. Royal Aeronautical Society, 31, 1927, 720–743. Reprinted in Ludwig Prandtl Gesammelte Abhandlungen, 2, 752–777.

387 Correspondence from Ahlborn to Prandtl, 23 July 1927. MPGA, Abt. III, Rep. 61, No. 15.

388 Correspondence from Ahlborn to Prandtl, 1 December 1924. MPGA, Abt. III, Rep. 61, No. 15.

389 Discussion with F. Ahlborn: Die Ablösungstheorie der Grenzschichten und die Wirbelbildung. Jahrbuch der WGL, 1927, 171–188.

390 Correspondence from Prandtl to Ahlborn, 23 October 1924. MPGA, Abt. III, Rep. 61, No. 15.

391 Correspondence from Prandtl to Ahlborn, 10 November 1924. MPGA, Abt. III, Rep. 61, No. 15.

392 Correspondence from Prandtl to Ahlborn, 3 November 1937. MPGA, Abt. III, Rep. 61, No. 15.

393 W. Birnbaum: Die tragende Wirbelfläche als Hilfsmittel zur Behandlung des ebenen Problems der Tragflügeltheorie. ZAMM, 3, 1923, 290–297.

394 Ira H. Abbott, Albert E. Von Doenhoff: Theory of Wing Sections. New York: Dover, 1959.

395 H. Blenk: Der Eindecker als tragende Wirbelfläche. ZAMM, 5, 1925, 36–47.

396 Bruno Eck: Neuartige Berechnung der aerodynamischen Eigenschaften eines Doppeldeckers. ZFM, 16, 1925, 183–194.

397 H.B. Helmbold: Über die Berechnung des Abwindes hinter einem rechteckigen Tragflügel. ZFM, 16, 1925, 291–294.

398 Herbert Wagner: Über die Entstehung des dynamischen Auftriebes von Tragflügeln. ZAMM, 5, 1925, 17–35.

399 R. Fuchs: Beiträge zur Prandtlschen Tragflügeltheorie. ZAMM, 1, 1921, 106–115.

400 E. Trefftz: Prandtlsche Tragflächen- und Propeller-Theorie. ZAMM, 1, 1921, 206–218.

401 M. Munk: Elements of the wing section theory and of the wing theory. NACA-Report No. 191.

402 Irmgard Lotz: Berechnung der Auftriebsverteilung beliebig geformter Flügel. ZFM, 22, 1931, 189–195.

403 A. Betz: Die Aerodynamiche Versuchsanstalt Göttingen. Ein Beitrag zur

Geschichte. In: Karl Stuchtey, Walter Boje (eds.): Beiträge zur Geschichte der Deutschen Luftfahrtwissenschaft und - technik. Berlin: Deutsche Akademie der Luftfahrtforschung, 1941, 3–166.

**404** Julius C. Rotta (ed.): Dokumente zur Geschichte der Aerodynamischen Versuchsanstalt in Göttingen, 1907–1925. In: DLR-Mitteilungen 90–05, Göttingen, 1990, 149–171.

**405** L. Prandtl: Über die stationären Wellen in einem Gasstrahl. Physikalische Zeitschrift, 5, 1904, 599–601. Reprinted in Ludwig Prandtl Gesammelte Abhandlungen, 2, 891–896.

**406** L. Prandtl: Beiträge zur Theorie der Dampfströmung durch Düsen. ZVDI, 48, 1904, 348–350. Reprinted in Ludwig Prandtl Gesammelte Abhandlungen, 2, 897–903.

**407** L. Prandtl: Strömende Bewegung der Gase und Dämpfe. In: Enzyklopädie der mathematischen Wissenschaften, V, 1905, 287–319. Reprinted in Ludwig Prandtl Gesammelte Abhandlungen, 2, 904–934.

**408** E. Magin: Optische Untersuchung über den Ausfluss von Luft durch eine Laval-Düse. VDI-Forschungsheft No. 62, 1908, 1–29.

**409** Th. Meyer: Über zweidimensionale Bewegungsvorgänge in einem Gas, das mit Überschallgeschwindigkeit strömt. VDI-Forschungsheft No. 62, 1908, 31–67.

**410** A. Steichen: Beiträge zur Theorie der zweidimensionalen Bewegungsvorgänge in einem Gase, das mit Überschallgeschwindigkeit strömt. Dissertation, Universität Göttingen, 1909.

**411** L. Prandtl: Neue Untersuchungen über die strömende Bewegung der Gase und Dämpfe. Physikalische Zeitschrift, 8, 1907, 23–30. Reprinted in Ludwig Prandtl Gesammelte Abhandlungen, 2, 943–956.

**412** Aurel Stodola: Dampf- und Gasturbinen, 4th edition. Berlin: Springer, 1910.

**413** L. Prandtl: Gasbewegung. In: Handwörterbuch der Naturwissenschaften, Vol. 4. Jena: Fischer, 1913, 544–560. Reprinted in Ludwig Prandtl Gesammelte Abhandlungen, 2, 957–985.

**414** Correspondence from Prandtl to Stodola, 30 July 1924. MPGA, Abt. III, Rep. 61, No. 1623.

**415** Correspondence from Stodola to Prandtl, 10 October 1922. MPGA, Abt. III, Rep. 61, No. 1623.

**416** Correspondence from Prandtl to Stodola, 13 October 1922. MPGA, Abt. III, Rep. 61, No. 1623.

**417** Kompressible Flüssigkeiten, MPGA, Abt. III, Rep. 61, No. 2251, 42.

**418** L. Prandtl, A. Busemann: Näherungsverfahren zur zeichnerischen Ermittlung von ebenen Strömungen mit Überschallgeschwindigkeit. Festschrift zum 70. Geburtstag von Prof. Dr. A. Stodola. Zürich: Füssli, 1929, 499–509. Reprinted in Ludwig Prandtl Gesammelte Abhandlungen, 2, 988–997.

**419** A. Busemann: Zeichnerische Ermittlung von ebenen Strömungen mit Überschallgeschwindigkeit. ZAMM, 8, 1928, 423–424.

**420** A. Busemann: Drücke auf kegelförmige Spitzen bei Bewegung mit Überschallgeschwindigkeit. ZAMM, 9, 1929, 496–498.

**421** A. Busemann: Gasdynamik. In: Handbuch der Experimentalphysik, Vol. 4 (Hydro- und Aerodynamik), part 1 (Strömungslehre und allgemeine Versuchstechnik). Leipzig: Akademische Verlagsgesellschaft, 1931, 343–459.

**422** A. Busemann: Flüssigkeits- und Gasbewegung. In: Handwörterbuch der Naturwissenschaften, Vol. 4, 2nd edition, Jena, 1933, 244–279.

**423** Correspondence from Stodola to Prandtl, 21 March 1921. MPGA, Abt. III, Rep. 61, No. 1623.

**424** J. Ackeret: Gasdynamik. In: H. Geiger, K. Scheel (eds.): Handbuch der Physik, Vol. 7 (Mechanik der flüssigen und gasförmigen Körper). Berlin: Springer, 1927, 289–343.

**425** The ETH archive has an Internet exhibit on Ackeret at http://www.ethbib.ethz.ch/exhibit/ackeret/.

**426** Correspondence from Glauert to Prandtl, 5 October 1927. MPGA, Abt. III, Rep. 61, No. 536.

**427** Correspondence from Prandtl to Glauert, 22 October 1927. MPGA, Abt. III, Rep. 61, No. 536.

**428** H. Glauert: The Effect of Compressibility on the Lift of an Airfoil. Proceedings of the Royal Society, London, A, 118, 1928, 113–119.

**429** J. Ackeret: Luftkräfte auf Flügel, die mit größerer als Schallgeschwindigkeit bewegt werden. ZFM, 16, 1925, 72–74.

**430** J. Ackeret: Über Luftkräfte bei sehr grossen Geschwindigkeiten insbesondere bei ebenen Strömungen. Helvetica Physica Acta, 1, 1928, 301–322.

**431** J. Ackeret: Der Luftwiderstand bei sehr grossen Geschwindigkeiten. Schweizerische Bauzeitung, 94, 12 October 1929, 179–183.

**432** N. Rott: J. Ackeret und die Geschichte der Machschen Zahl. Schweizer Ingenieur und Architekt, 21, 1983, 591–594.

**433** A. Busemann: Profilmessungen bei Geschwindigkeiten nahe der Schallgeschwindigkeit (im Hinblick auf Luftschrauben). Jahrbuch der Wissenschaftlichen Gesellschaft für Luftfahrt (WGL), 1928, 95–98.

**434** A. Busemann, O. Walcher: Profileigenschaften bei Überschallgeschwindigkeit. Forschung auf dem Gebiet des Ingenieurwesens, 4, 1933, 87–92.

**435** A. Busemann: Widerstand bei Geschwindigkeiten nahe der Schallgeschwindigkeit. In: Verhandlungen des 3. Internationalen Kongresses für technische Mechanik, Stockholm, 1930, Teil 1, 282–286 mit Bildtafeln.

**436** L. Prandtl: Aufgaben der Strömungsforschung. VDI-Nachrichten, 5, 1925, No. 32. Reprinted in Ludwig Prandtl Gesammelte Abhandlungen, 3, 1538–1544.

**437** J. Ackeret (ed.): Leonhardi Euleri Opera Omnia, Serie 2, Vol. 15. Lausanne: Füssli, 1957.

**438** Carsten Östergaard: Schiffspropulsion. In: Lars U. Scholl (ed.): Technikgeschichte des industriellen Schiffbaus in Deutschland. Vol. 2. Hamburg: Ernst Kabel, 1996, 65–129.

**439** W. Wagenbach: Beiträge zur Berechnung und Konstruktion der Wasserturbinen. Zeitschrift für das gesamte Turbinenwesen, 4, 1907, 276–277.

**440** Correspondence with the "Inspektion des Unterseebootwesens über Versuchsschrauben," 1917–1918. MPGA, Abt. III, Rep. 61, No. 2101.

**441** H. Föttinger: Fortschritte der Strömungslehre im Maschinenbau und Schiffbau. JSTG, 25, 1924, 295–344.

**442** D. Thoma: Die Kavitation bei Wasserturbinen. Wasserkraft-Jahrbuch, 1, 1924, 409–420.

**443** Die Wasserkraft, 20, 1925, 119–120, 369.

**444** VDI-Beirat (ed.): Hydraulische Probleme. Berlin: VDI-Verlag, 1926.

**445** H. Föttinger: Untersuchungen über Kavitation und Korrosion bei Turbinen, Turbopumpen und Propellern. In: VDI-Beirat (ed.): Hydraulische Probleme. Berlin: VDI-Verlag, 1926, 14–64.

**446** J. Ackeret: Zum Vortrag Föttinger. In: VDI-Beirat (ed.): Hydraulische Probleme. Berlin: VDI-Verlag, 1926, 101–105.

**447** J. Ackeret: Kavitation (Hohlraumbildung). In: Handbuch der Experimentalphysik, Vol. 4 (Hydro- und Aerdynamik), part 1 (Strömungslehre und allgemeine Versuchstechnik). Leipzig: Akademische Verlagsgesellschaft, 1931, 463–485.

**448** J.W. Strutt (Lord Rayleigh): On the Pressure Developed in a Liquid During the Collapse of a Spherical Cavity. Philosophical Magazine, 34, 1917, 94–98.

**449** J. Ackeret: Experimentelle und theoretische Untersuchungen über Hohlraumbildung (Kavitation) im Wasser. In: Technische Mechanik und Thermodynamik. Monatliche Beihefte zur VDI Zeitschrift, vol. 1, No. 1, January 1930, 1–22; No. 2, February 1930, 63–72.

**450** H. Mueller: Kinematographische Aufnahme der Kavitation an einem Tragflügel. In: G. Kempf, E. Foerster (eds.): Hydromechanische Probleme des Schiffsantriebs. Hamburg: Hamburgische Schiffbau-Versuchsanstalt, 1932, 311–314.

**451** J. Ackeret: Kavitation und Kavitationskorrosion. In: G. Kempf, E. Foerster (eds.): Hydromechanische Probleme des Schiffsantriebs. Hamburg: Hamburgische Schiffbau-Versuchsanstalt, 1932, 227–240.

**452** A. Betz: Einfluss der Kavitation auf die Leistung von Schiffsschrauben. Verhandlungen des 3. Internationalen Kongresses für technische Mechanik, Stockholm, 1930, Teil 1, 411–416.

**453** O. Walchner: Profilmessungen bei Kavitation. In: G. Kempf, E. Foerster (eds.): Hydromechanische Probleme des Schiffsantriebs. Hamburg: Hamburgische Schiffbau-Versuchsanstalt, 1932, 256–267.

**454** R.T. Knapp, J.W. Daily, F.G. Hammit: Cavitation. New York: McGraw-Hill, 1970.

455 Ch. E. Brennen: Cavitation and Bubble Dynamics. Oxford: Oxford University Press, 1995.

456 Correspondence from Prandtl to Bjerknes, 5 July 1922. MPGA, Abt. III, Rep. 61, No. 139.

457 E. Kuhlbrodt: Über die Polarfronttheorie nach Bjerknes und die neueren Anschauungen von den atmosphärischen Vorgängen. In: Die Naturwissenschaften, 10, 1922, 495–503.

458 Meteorologisches. MPGA, Abt. III, Rep. 61, No. 2275.

459 L. Prandtl: Bemerkungen zu der Bjerknesschen Wellentheorie der Zyklone. Verhandlungen der DPG, 3, 1922, Vol. 2, 60.

460 Correspondence from Exner to Prandtl, 14 December 1922, MPGA, Abt. III, Rep. 61, No. 1061.

461 Correspondence from Prandtl to the Meteorological Office, 16 February 1923, MPGA, Abt. III, Rep. 61, No. 1059.

462 H. Koschmieder: Die Arbeiten des Messtrupps während des Rhönsegelflug-Wettbewerbs 1923. ZFM, 15, 1924, 3–8; Die Arbeiten des Messtrupps während des 2. Segelflug-Wettbewerbs in den Rossitten 1924. ZFM, 15, 1924, 236–242, 257–258; Die Arbeiten des Messtrupps während des Rhönsegelflug-Wettbewerbs 1924. ZFM, 16, 1925, 235–244.

463 Correspondence between Koschmieder and Prandtl during 1924. MPGA, Abt. III, Rep. 61, No. 874.

464 V.W. Ekman: Dynamische Gesetze der Meeresströmungen. In: T. von Kármán, T. Levi-Civita (eds.): Vorträge aus dem Gebiet der Hydro- und Aerodynamik (Innsbruck 1922). Berlin: Springer, 1924, 97–115.

465 V. Bjerknes: Bibliographie mit historischen Erläuterungen. In: V. Bjerknes, J. Bjerknes, H. Solberg, T. Bergeron: Physikalische Hydrodynamik, mit Anwendung auf die dynamische Meteorologie. Berlin: Springer, 1933, 777–790.

466 Correspondence from Prandtl to Ekman, 27 September 1924, MPGA, Abt. III, Rep. 61, No. 396.

467 Correspondence from Ekman to Prandtl, 14 November 1924, MPGA, Abt. III, Rep. 61, No. 396.

468 L. Prandtl, W. Tollmien: Die Windverteilung über dem Erdboden, errechnet aus den Gesetzen der Rohrströmung. Zeitschrift für Geophysik, 1, 1925, 47–55. Reprinted in Ludwig Prandtl Gesammelte Abhandlungen, 3, 1071–1080.

469 B.D. Etling: Strömungen in der Atmosphäre und im Ozean. In: H. Oertel, Jr. (ed.): Prandtl – Führer durch die Strömungslehre. Grundlagen und Phänomene, 11th edition. Braunschweig/Wiesbaden, 2002, 547–592.

470 L. Prandtl: Erste Erfahrungen mit dem rotierenden Laboratorium. Die Naturwissenschaften, 14, 1926, 425–427. Reprinted in Ludwig Prandtl Gesammelte Abhandlungen, 3, 1304–1308.

471 Correspondence from Prandtl to Schmidt, 11 June 1926. MPGA, Abt. III, Rep. 61, No. 1476.

472 Heimo Fette: Strömungsversuche im rotierenden Laboratorium. Zeitschrift für technische Physik, 14, 1933, 257–266.

473 W. Schmidt: Der Massenaustausch in freier Luft und verwandte Erscheinungen. Hamburg: Verlag von Henri Grand, 1925.

474 Correspondence from Schmidt to Prandtl, 17 June and 2 July 1926. MPGA, Abt. III, Rep. 61, No. 1476.

475 Correspondence from Prandtl to Schmidt, 6 July 1926. MPGA, Abt. III, Rep. 61, No. 1476.

476 L. Prandtl: Einfluss stabilisierender Kräfte auf die Turbulenz. In: A. Gilles et al. (eds.): Vorträge auf dem Gebiete der Aerodynamik und verwandter Gebiete, Aachen 1929. Berlin: Springer, 1929, 1–7. Reprinted in Ludwig Prandtl Gesammelte Abhandlungen, 2, 778–785.

477 L. Prandtl: Modellversuche und theoretische Studien über die Turbulenz einer geschichteten Luftströmung. Deutsche Forschung, 1930, 14–15. Reprinted in Ludwig Prandtl Gesammelte Abhandlungen, 2, 777–778.

478 S. Goldstein: On the Stability of Superposed Streams of Fluids of Different Densities. Proceedings of the Royal Society London, A 132, 1931, 524–548. See also correspondence between Goldstein and Prandtl in MPGA, Abt. III, Rep. 61, No. 561.

**479** J. Lighthill: Sydney Goldstein. Biographical Memoirs of the Fellows of the Royal Society, 36, 1990, 174–197.

**480** Correspondence from Prandtl to Taylor, 29 April 1930. MPGA, Abt. III, Rep. 61, No. 1653.

**481** L. Prandtl: Meteorologische Anwendungen der Strömungslehre. Beiträge zur Physik der freien Atmosphäre (Bjerknes-Festschrift), 19, 1932, 188–202. Reprinted in Ludwig Prandtl Gesammelte Abhandlungen, 3, 1081–1097.

**482** W. Tollmien: Luftwiderstand und Druckverlauf bei der Fahrt von Zügen in einem Tunnel. ZVDI, 71, 1927, 199–203.

**483** O. Flachsbart: Beitrag zur Frage der Berücksichtigung des Windes im Bauwesen. Jahrbuch der deutschen Gesellschaft für Bauingenieurwesen, 4, 1928, 160–169.

**484** Tätigkeitsbericht der Kaiser Wilhelm-Gesellschaft. Die Naturwissenschaften, 20, 1932.

**485** Tätigkeitsbericht der Kaiser Wilhelm-Gesellschaft. Die Naturwissenschaften, 21, 1933.

**486** H. Hergesell: Die Strömungsforschung in der Atmosphäre. Deutsche Forschung. Aus der Arbeit der Notgemeinschaft der Deutschen Wissenschaft, 14, 1930, 7–13.

**487** R. Siegmund-Schultze: Mathematiker auf der Flucht vor Hitler. Quellen und Studien zur Emigration einer Wissenschaft. Braunschweig, Wiesbaden: Vieweg, 1998.

**488** A. Beyerchen: Scientists under Hitler. Politics and the Physics Community in the Third Reich. New Haven: Yale University Press, 1977.

**489** N. Schappacher: Das Mathematische Institut der Universität Göttingen. In: H. Becker, H.-J. Dahms, C. Wegeler (eds.): Die Universität Göttingen unter dem Nationalsozialismus, 2nd edition. München: Saur, 1998, 523–551.

**490** U. Rosenow: Die Göttinger Physik unter dem Nationalsozialismus. In: H. Becker, H.-J. Dahms, C. Wegeler (eds.): Die Universität Göttingen unter dem Nationalsozialismus, 2nd edition. München: Saur, 1998, 523–551.

**491** Constance Reid: Courant. New York: Springer, 1996.

**492** C. Tollmien: Das Kaiser-Wilhelm-Institut für Strömungsforschung verbunden mit der Aerodynamischen Versuchsanstalt. In: H. Becker, H.-J. Dahms, C. Wegeler (eds.): Die Universität Göttingen unter dem Nationalsozialismus, 2nd edition. München: Saur, 1998, 684–708.

**493** Gerhard Rammer: Die Nazifizierung und Entnazifizierung der Physik an der Universität Göttingen. Dissertation, Universität Göttingen, 2004.

**494** Ulrich Kalkmann: Die Technische Hochschule Aachen im Dritten Reich (1933–1945). Aachen: Verlag Mainz, 2003.

**495** Kreisleiter Göttingen an das Amt für Technik, Gauamtsleitung, 28 May 1937. Reprinted as document No. 45 in Helmuth Trischler (ed.): Dokumente zur Geschichte der Luft- und Raumfahrtforschung in Deutschland 1900–1970. München, 1993.

**496** Correspondence from Prandtl to Telschow, 11 February 1933. Reprinted as document No. 30 in Helmuth Trischler (ed.): Dokumente zur Geschichte der Luft- und Raumfahrtforschung in Deutschland 1900–1970. München, 1993.

**497** Correspondence from Baeumker to Prandtl, 21 February 1933. Reprinted as document No. 31 in Helmuth Trischler (ed.): Dokumente zur Geschichte der Luft- und Raumfahrtforschung in Deutschland 1900–1970. München, 1993.

**498** Adolf Baeumker: Ein Lebensbericht. Bad Godesberg: Selbstverlag, 1966.

**499** Katharina Hein: Adolf Baeumker (1891–1976). Einblicke in die Organisation der Luft- und Raumfahrtforschung von 1920 bis 1970. Göttingen: Deutsche Forschungsanstalt für Luft- und Raumfahrt, 1995.

**500** K.-H. Völker: Die deutsche Luftwaffe 1933–1939. Aufbau, Führung und Rüstung in der Luftwaffe sowie die Entwicklung der deutschen Luftkriegstheorie. Stuttgart: Deutsche Verlags-Anstalt, 1967.

**501** Luftwissen, 3, 1936.

**502** A. Baeumker: Zur Geschichte der deutschen Luftfahrtforschung. Ein Beitrag (1944). München und Freiburg: Selbstverlag, 1945.

**503** Tätigkeitsberichte der Kaiser Wilhelm-Gesellschaft. Die Naturwissenschaften, 21–25, 1933–1937.

**504** Correspondence between Prandtl, Hunsaker and Gardener, 9 March to 10 July 1933, MPGA, Abt. III, Rep. 61, No. 724.

**505** Luftwissen, 1, 1934.

**506** Luftwissen, 2, 1935.

**507** Reale Academia d'Italia (ed.): Convegno di Scienze Fisiche, Mathematiche e Naturali, 30 Settembre–6 Ottobre 1935, "Le Alte Velocita in Aviazione" (Volta-Congress). Rome: Reale Accademia d'Italia, 1936.

**508** W. Heinzerling: Wozu braucht man Windkanäle. In: L. Bölkow (ed.): Ein Jahrhundert Flugzeuge. Geschichte und Technik des Fliegens. Düsseldorf: VDI-Verlag, 1990, 304–333.

**509** H. Winter: Der Überdruck-Windkanal der Aerodynamischen Versuchsanstalt Göttingen. Luftwissen, 3, 1936, 237–241.

**510** S. Hoerner: Bauarten, Eigenschaften und Leistungen von Windkanälen. ZVDI, 80, 1936, 949–957.

**511** Moritz Epple: Rechnen, Messen, Führen. Kriegsforschung am Kaiser-Wilhelm-Institut für Strömungsforschung (1937–1945). In Carola Sachse (ed.): Ergebnisse. Vorabdrucke aus dem Forschungsprogramm "Geschichte der Kaiser-Wilhelm-Gesellschaft im Nationalsozialismus." Max-Planck-Gesellschaft, Präsidentenkomission, 2002.

**512** Correspondence from Prandtl to Baeumker, 14 June 1940, MPGA, Abt. III, Rep. 61, No. 73.

**513** Luftwissen, 6, 1939.

**514** Correspondence from Prandtl to Hunsaker, 25 April 1934. MPGA, Abt. III, Rep. 61, No. 724.

**515** Correspondence from Prandtl to Göring, 6 February 1935. MPGA, Abt. III, Rep. 61, No. 541.

**516** Correspondence from Prandtl to Göring, 12 December 1936. Reprinted as document No. 43 in Helmuth Trischler (ed.): Dokumente zur Geschichte der Luft- und Raumfahrtforschung in Deutschland 1900–1970. München, 1993.

**517** Luftwissen, 4, 1937.

**518** Luftwissen, 5, 1938.

**519** Correspondence from Prandtl to Baeumker, 3 June 1937. MPGA, Abt. III, Rep. 61, No. 1967.

**520** Correspondence from Prandtl to Durand, 10 July 1937; Durand to Prandtl, 6 August 1937; correspondence from Prandtl to Baeumker, 24 August 1937; correspondence from Prandtl to Durand, 26 August 1937. MPGA, Abt. III, Rep. 61, No. 1967.

**521** Correspondence from Warner to Baeumker, 4 April 1938. Copy in MPGA, Abt. III, Rep. 61, No. 2049.

**522** Correspondence from Durand to Prandtl, 2 December 1938. MPGA, Abt. III, Rep. 61, No. 1969.

**523** Correspondence from Boje to Prandtl, 29 December 1938. MPGA, Abt. III, Rep. 61, No. 1969.

**524** Correspondence from Prandtl to Boje, 3 January 1939; correspondence from Prandtl to Lewis, 11 January 1939; MPGA, Abt. III, Rep. 61, No. 1969.

**525** Correspondence from Prandtl to Knight, 15 May 1937. MPGA, Abt. III, Rep. 61, No. 839.

**526** Mark Walker: Physics and Propaganda: Werner Heisenberg's Foreign Lectures under National socialism. HSPS, 22(2), 1992, 339–389.

**527** Frank-Rutger Hausmann: "Auch im Krieg schweigen die Musen nicht:" Die Deutschen Wissenschaftlichen Institute (DWI) im Zweiten Weltkrieg (1940–1945). Jahrbuch des Historischen Kollegs, 2000, Munich, 2001, 123–163.

**528** Correspondence from Groh to Prandtl, 2 July 1937, MPGA, Abt. III, Rep. 61, No. 2148.

**529** Correspondence from Milch to Rust, 18 February 1938, MPGA, Abt. III, Rep. 61, No. 2149.

**530** Correspondence from the Ministry of Science to Prandtl, 14 March 1938. MPGA, Abt. III, Rep. 61, No. 2149.

**531** Correspondence from the Deutsche Kongress-Zentrale to Prandtl, 22 June 1938. MPGA, Abt. III, Rep. 61, No. 2149.

**532** Correspondence from Mises to Kármán, 4 April 1938. TKC 47.3–8.

**533** Correspondence from Kármán to Mises, 9 June 1938. TKC 47.3–8.

**534** Correspondence from Hunsaker to Burgers, 29 July 1938. TKC 47.3–8.

**535** Minutes, Meeting of the International Congress Committee, 13 September 1938. TKC, 47, 3–8.

**536** Correspondence from Hunsaker to Prandtl, 22 October 1938. MPGA, Abt. III, Rep. 61, No. 724.

**537** Correspondence from Taylor to Prandtl, 27 September 1938. MPGA, Abt. III, Rep. 61, No. 1654.

**538** Correspondence from Prandtl to Taylor, 29 October 1938. MPGA, Abt. III, Rep. 61, No. 1654.

**539** Correspondence from Prandtl to Mrs. Taylor, 5 August 1939. MPGA, Abt. III, Rep. 61, No. 1654.

**540** Correspondence from Prandtl to Hunsaker, 3 November 1939. MPGA, Abt. III, Rep. 61, No. 724.

**541** Correspondence from the Kongresszentrale to Prandtl, 10 October 1939; Prandtl to Kongresszentrale, 14 October 1939; Kongresszentrale to Prandtl, 18 October 1939. MPGA, Abt. III, Rep. 61, No. 297.

**542** Correspondence from Taylor to Prandtl, 15 November 1935. MPGA, Abt. III, Rep. 61, No. 1654.

**543** Theodore von Kármán: Address [at Ackeret's 60th birthday]. Zeitschrift für angewandte Mathematik und Physik (ZAMP), 9b, 1958, 55–56.

**544** Theodore von Kármán, Erich Trefftz: Potentialströmung um gegebene Tragflächenquerschnitte. ZFM, 9, 1918, 111–116.

**545** A. Gilles, L. Hopf, Th. von Kármán (eds.): Vorträge aus dem Gebiete der Aerodynamik und verwandter Gebiete (Aachen 1929). Berlin: Springer 1930.

**546** A. Gilles: Beschreibung des Aerodynamischen Instituts der Technischen Hochschule Aachen und seiner Versuchseinrichtungen. Abhandlungen aus dem Aerodynamischen Institut der Technischen Hochschule Aachen, 10, 1931, 1–13.

**547** Correspondence from Wieselsberger to Kármán, 19 September 1925. TKC, 32.28.

**548** C. Wieselsberger: Die aerodynamische Waage des Aachener Windkanals. Abhandlungen aus dem Aerodynamischen Institut der Technischen Hochschule Aachen, 14, 1934, 24–26.

**549** Correspondence from Baeumker to Kármán, 3 December 1934. TKC, 1.40.

**550** Correspondence from Kármán to Baeumker, 25 November 1935. TKC, 1.40.

**551** Paul A. Hanle: Bringing Aerodynamics to America. Cambridge: MIT Press, 1982.

**552** Correspondence from Prandtl to Sommerfeld, 28 October 1938. MPGA, Abt. III, Rep. 61, No. 1538.

**553** Freddy Litten: Mechanik und Antisemitismus: Wilhelm Müller (1880–1968). München: Institut für Geschichte der Naturwissenschaften, 2000.

**554** Michael Eckert, Karl Märker: Arnold Sommerfeld. Wissenschaftlicher Briefwechsel, Band 2: 1919–1951. Berlin, Diepholz, München: Verlag für Geschichte der Naturwissenschaften und der Technik und Deutsches Museum, 2004.

**555** Carl Wieselsberger: Discussion of Ackeret's report on "Windkanäle für hohe Geschwindigkeiten," in Reale Academia d'Italia (ed.): Convegno di Scienze Fisiche, Mathematiche e Naturali, 30 Settembre–6 Ottobre 1935, "Le Alte Velocita in Aviazione" (Volta-Kongress). Rom: Reale Accademia d'Italia, 1936.

**556** Carl Wieselsberger: Die Überschallanlage des Aerodynamischen Instituts der Technischen Hochschule Aachen. Luftwissen, 4, 1937, 301–303.

**557** Michael J. Neufeld: The Rocket and the Reich: Peenemünde and the Coming of the Ballistic Missile Era. New York, 1995.

**558** Tätigkeitsbericht des Aerodynamischen Instituts der Technischen Hochschule Aachen. Jahrbuch 1939 der deutschen Luftfahrtforschung, 45–46.

**559** Richard P. Hallion: Legacy of Flight. The Guggenheim Contribution to American Aviation. Seattle and London: University of Washington Press, 1977.

**560** History of Aeronautics at the California Institute of Technology, undated typescript. TKC, 73.10. (Presumably the first draft of the "History of the Daniel Guggenheim Graduate School of Aeronautics at the California Institute of Technology." TKC, 73.11.

**561** A. Toussaint: Experimental Methods – Wind Tunnels, Part 1. In: William Frederick Durand (ed.): Aerodynamic Theory, Volume III. Berlin, 1935, 252–319.

**562** Clark Millikan: Über aerodynamische Forschungsergebnisse und ihre Auswertung für den Flugzeugbau. Luftwissen, 3, 1936, 277–287.

**563** Th. von Kármán: A tribute to Walter Tollmien. In: Manfred Schäfer (ed.): Miszellaneen der Angewandten Mechanik. Festschrift für Walter Tollmien zum 60. Geburtstag. Berlin: Akademie-Verlag, 1962, 103–106.

**564** Memorandum on the Status of the Guggenheim Aeronautical Laboratory, California Institute of Technology, April 1936. TKC, 73.13.

**565** William R. Sears: Stories from a 20th-Century Life. Stanford: Parabolic Press, 1994.

**566** Minutes concerning "Die Zusammenarbeit der E.T.H. auf dem Gebiete der Aviatik mit den in Frage kommenden eidg. Instanzen," 31 March 1930. ETHA, Akten des Schweizerischen Schulrats, S3, 1933, Syst. No. 322.

**567** Minutes, "Wissenschaftliche Kommission für Flugwesen," 17 September 1932. ETHA, Akten des Schweizerischen Schulrats, S3, 1933, Syst. No. 322.

**568** H. Sprenger, contribution to the "Stodola-Kolloquium" in 1980, based on personal recollections of H. Bosshardt ca. 1968. Internet exhibit "Jacob Ackeret (1898–1981)" from ETHA – see http://www.ethbib.ethz.ch/exhibit/ackeret/.

**569** Budget report, 15 February 1934. ETHA, Ackeret-Nachlass, HS 552:265.

**570** Correspondence from Ackeret to the Direktion für eidgenössische Bauten, 19 May 1933. ETHA, Ackeret-Nachlass, HS 553:179.

**571** Correspondence from Ackeret to the Direktion für eidgenössische Bauten, 3 October 1933. ETHA, Ackeret-Nachlass, HS 553:181.

**572** J. Ackeret: Das Institut für Aerodynamik im neuen Maschinenlaboratorium der E.T.H. Schweizerische Bauzeitung, 111, 1938, 73–79, 89–93.

**573** C. Seippel: The Development of the Brown Boveri Axial Compressor. Brown Boveri Review, 27, 1940, 108–113.

**574** Correspondence between Ackeret und Crocco, June–December 1934. ETHA, Ackeret-Nachlass, HS 553:143–154.

**575** Correspondence from Ackeret to Prandtl, 20 February 1935. ETHA, Ackeret-Nachlass, HS 553:693.

**576** A complete list is reprinted in the Ackeret Festschrift in Zeitschrift für angewandte Mathematik und Physik (ZAMP), 9b, 1958, 17–23.

**577** Correspondence from Prandtl to Ackeret, 25 May 1939. ETHA, Ackeret-Nachlass, HS 553:700.

**578** Correspondence from Kármán to Ackeret, 21 December 1935. TKC, 1.2.

**579** Correspondence from Millikan to Kármán, 3 March 1936. TKC, 1.2.

**580** Correspondence from Dätwyler to Kármán, 15 May 1934. TKC, 6.37. See also the list of foreign research fellows in TKC, 73.10.

**581** F. Schultz-Grunow: J. Ackeret: Persönliche Erinnerungen. Schweizer Ingenieur und Architekt, 21, 1983, 587–594.

**582** Correspondence from Prandtl to Ackeret, 29 April 1939. ETHA, Ackeret-Nachlass, HS 553:698.

**583** Schulratsakten 1941, ETHA, No. 3294/221.0. Quoted in http://www.ethbib.ethz.ch/exhibit/ackeret/.

**584** Michael H. Gorn: Harnessing the Genie: Science and Technology Forecasting for the Air Force, 1944–1986. Washington: Office of Air Force History, 1988.

**585** R. von Wattenwyl: Professor Ackeret und die Landesverteidigung. ZAMP, 9b, 1958, 31–33.

**586** A. Gerber: Aus der Zusammenarbeit des Institutes für Aerodynamik mit der privaten Rüstungsindustrie. ZAMP 9b, 1958, 37–46.

**587** A. Betz: Aufgaben und Aussichten der Theorie in der Strömungsforschung. Jahrbuch 1936 der Lilienthal-Gesellschaft für Luftfahrtforschung. München and Berlin: Oldenbourg, 1937, 21–28.

**588** A. Betz, F. Keune: Verallgemeinerte Kármán-Trefftz-Profile. Jahrbuch 1937 der deutschen Luftfahrtforschung, 1938, I-38–I-47.

**589** F. Riegels: Neuere Untersuchungen über Profileigenschaften. Luftwissen, 6, 1939, 299–304.

**590** K. Tank: Welchen Nutzen hat der Flugzeugkonstrukteur aus den neueren Profiluntersuchungen gezogen? Jahrbuch 1940 der deutschen Luftfahrtforschung, 1940, I-1 - I-9.

**591** K. Tank in Luftfahrt-Forschung, 18, 1941, 39.

592 J. Stüper: Die heutigen Aufgaben der aerodynamischen Forschung. Luftwissen, 4, 1937, 199–200.

593 A.W. Quick: Die aerodynamische Entwicklung des Ju 86–Tragflügels. Jahrbuch 1936 der deutschen Luftfahrtforschung, 1937, 157–170.

594 H. Blenk: Bericht der Fachgruppe für Aerodynamik. Jahrbuch 1936 der deutschen Luftfahrtforschung, 1937, 597–601.

595 H. Multhopp: Die Berechnung der Auftriebsverteilung von Tragflügeln. Jahrbuch 1938 der deutschen Luftfahrtforschung, 1939, I-101–I-117.

596 A. Betz: Applied Airfoil Theory. Section J in W.F. Durand (ed.): Aerodynamic Theory. Vol 4. Berlin: Springer, 1935, 1–129.

597 Th. von Kármán, J.M. Burgers: General Aerodynamic Theory – Perfect Fluids. Section E in W.F. Durand (ed.): Aerodynamic Theory. Vol 2. Berlin: Springer, 1935, 165–200.

598 H. Multhopp: Die Berechnung der Auftriebsverteilung von Tragflügeln. Luftfahrtforschung, 15, 1938, 153–169.

599 M. Schwabe: Rechenschema zur Berechnung der Auftriebsverteilung über der Flügelspannweite nach Multhopp. Luftfahrtforschung, 15, 1938, 170–180.

600 H. Multhopp: Zur Aerodynamik des Flugzeugrumpfes. Luftfahrt-Forschung, 18, 1941, 52–66.

601 O. Schrenk: Ein einfaches Näherungsverfahren zur Ermittlung von Auftriebsverteilungen längs der Tragflügelspannweite. Luftwissen, 7, 1940, 118–120.

602 H.B. Helmbold: Einige einfache Formeln aus der Tragflügeltheorie. Luftfahrtforschung, 18, 1941, 363–366.

603 L. Prandtl: Beitrag zur Theorie der tragenden Fläche. ZAMM, 16, 1936, 360–361; Über neuere Arbeiten zur Theorie der tragenden Fläche. Proceedings of the Fifth International Congress for Applied Mechanics, Cambridge, MA, 1938. 1939, 478–482. Reprinted in Ludwig Prandtl Gesammelte Abhandlungen, 1, 562–574.

604 Th. v. Kármán: Neue Darstellung der Tragflügeltheorie. ZAMM, 15, 1935, 56–61.

605 H.G. Küssner: Zusammenfassender Bericht über den instationären Auftrieb von Flügeln. Jahrbuch 1937 der deutschen Luftfahrtforschung, 1938, I-273 - I-287.

606 H. Försching: Ludwig Prandtl's grundlegende Beiträge zur instationären Aerodynamik schwingender Auftriebsflächen. In: Gerd E.A. Meier (ed.): Ludwig Prandtl, ein Führer in der Strömungslehre. Biographische Artikel zum Werk Ludwig Prandtls. Braunschweig, Wiesbaden: Vieweg, 2000, 147–171.

607 I.E. Garrick, W.H. Reed III: Historical Development of Aircraft Flutter. Journal of Aircraft 18, 1981, 897–912.

608 James Hansen: The Bird Is on the Wing. Aerodynamics and the Progress of the American Airplane. College Station: Texas A&M University Press, 2004.

609 H. Eick: Der Mindestwiderstand von Schnellflugzeugen. Luftfahrtforschung, 15, 1938, 445–462.

610 E. Heinkel: Vortrag at the Hauptversammlung 1938 der Lilienthalgesellschaft. Luftwissen, 5, 1938, 407–408.

611 L. Prandtl, H. Schlichting: Das Widerstandsgesetz rauher Platten. Werft, Reederei, Hafen, 15, 1934, 1–4. Reprinted in Ludwig Prandtl Gesammelte Abhandlungen, 2, 649–662.

612 Th. v. Kármán, C.B. Millikan: The Use of the Wind Tunnel in Connection with Aircraft-Design Problems. Transactions of the American Society of Mechanical Engineers, 56, 1934, 151–166. Reprinted in CWTK, 3, 49–80.

613 E. Gruschwitz: Die turbulente Reibungsschicht in ebener Strömung bei Druckabfall und Druckanstieg. Ingenieurarchiv, 2, 1931, 321–346.

614 W. Mangler: Einige Nomogramme zur Berechnung der laminaren Reibungsschicht an einem Tragflügelprofil. Jahrbuch 1940 der deutschen Luftfahrtforschung, 1941, I-16–I-25.

615 Theodore von Kármán: Turbulence. In: Aeronautical Reprints No. 89, The Royal Aeronautical Society, London, 1937. Reprinted in CWTK, 3, 245–279.

616 G.I. Taylor: Statistical Theory of Turbulence, Parts I–V. Proceedings of the Royal Society, A151, 1935, 412–478, A156, 1936, 307–317. Reprinted in SPGIT, 288–356.

617 Thedore von Kármán: The Fundamentals of the Statistical Theory of Turbulence.

Journal of Aeronautical Sciences 4, 1937, 131–138. Reprinted in CWTK, 3, 228–244.

**618** Theodore von Kármán: On the Statistical Theory of Turbulence. Proceedings of the National Academy of Sciences, 23, 1937, 98–105. Reprinted in CWTK, 3, 222–227.

**619** H.L. Dryden, G.B. Schubauer, W.C. Mock, H.K. Skramstad, Jr.: Measurements of Intensity and Scale of Wind-tunnel Turbulence and Their Relation to the Critical Reynolds Number of Spheres. NACA Report 581, 1938, 109–140.

**620** K. Wieghardt: Zusammenfassender Bericht über Arbeiten zur statistischen Turbulenztheorie. Luftfahrtforschung, 18, 1941, 1–7.

**621** G.I. Taylor: Eddy Motion in the Atmosphere. Philosophical Transactions of the Royal Society, A215, 1915, 1–26. Reprinted in SPGIT, 1–23.

**622** G.I. Taylor: Diffusion by Continuous Movements. Proceedings of the London Mathematical Society, ser. 2, 20, 1921, 196–212. Reprinted in SPGIT, 172–184.

**623** G.I. Taylor: The Transport of Vorticity and Heat Through Fluids in Turbulent Motion. Proceedings of the Royal Society, A135, 1932, 685–705, with an Appendix by A. Fage, V.M. Falkner: Note on Experiments on the Temperature and Velocity in the Wake of a Heated Cylindrical Obstacle. Reprinted in SPGIT, 253–270.

**624** Correspondence from Prandtl to Taylor, 25 July 1932. MPGA, Abt. III, Rep. 61, No. 1653.

**625** Correspondence from Prandtl to Taylor, 5 June 1934. MPGA, Abt. III, Rep. 61, No. 1653.

**626** Tätigkeitsbericht des KWI für Strömungsforschung verbunden mit der AVA, April 35 bis September 35. Die Naturwissenschaften, 24, 1936.

**627** G.I. Taylor: Turbulence in a Contracting Stream. ZAMM, 15, 1935, 91–96. Reprinted in SPGIT, 280–287.

**628** Correspondence from Taylor to Prandtl, 2 March 1935. MPGA, Abt. III, Rep. 61, No. 1654.

**629** H. Reichard, L. Prandtl: Einfluss von Wärmeschichtung auf die Eigenschaften einer turbulenten Strömung. Deutsche Forschung, 21, 1934, 110–121.

**630** Correspondence from Taylor to Prandtl, 21 April 1935. MPGA, Abt. III, Rep. 61, No. 1654.

**631** Correspondence from Prandtl to Kármán, 14 February 1934. MPGA, Abt. III, Rep. 61, No. 793.

**632** S. Goldstein: Modern Developments in Fluid Dynamics. 2 vols. Oxford: Clarendon Press, 1938.

**633** Theodore von Kármán: I fondamenti della teorica statistica della turbolenza. L'Aerotecnica 17(7), 1937.

**634** Correspondence from Kármán to Prandtl, 3 November 1937. TKC, 47.3–8; see also the correspondence during the preparation of the Congress in MPGA, Abt. III, Rep. 61, No. 2148–2157.

**635** L. Prandtl: Beitrag zum Turbulenzsymposium. In J.P. den Hartog, H. Peters (eds.): Proceedings of the 5th International Congress of Applied Mechanics, Cambridge, Massachusetts, September 12–16, 1938. New York: Wiley, 1939, 340–346. Reprinted in Ludwig Prandtl Gesammelte Abhandlungen, 2, 856–868.

**636** J.P. den Hartog, H. Peters (eds.): Proceedings of the 5th International Congress of Applied Mechanics, Cambridge, Massachusetts, September 12–16, 1938. New York: Wiley, 1939, 294–435.

**637** L. Prandtl: Führer durch die Strömungslehre, 1st edition and 3rd edition of Abriss der Strömungslehre. Braunschweig: Vieweg, 1942.

**638** L. Prandtl: Theorie des Flugzeugtragflügels im zusammendrückbaren Medium. Luftfahrtforschung, 13, 1936, 313–319. Reprinted in Ludwig Prandtl Gesammelte Abhandlungen, 2, 1027–1045.

**639** H. Schlichting: Tragflügeltheorie bei Überschallgeschwindigkeit. Luftfahrtforschung, 13, 1936, 320–335.

**640** Walter G. Vincenti: Engineering Theory in the Making: Aerodynamic Calculation "Breaks the Sound Barrier." Technology and Culture, 38, 1997, 819–851.

**641** Correspondence from Crocco to Kármán, 23 January 1935. TKC, 52.19.

**642** Correspondence from Crocco to Prandtl, 4 December 1934; correspondence from Prandtl to Crocco, 10. December 1934; correspondence from Crocco to Prandtl, 24 December 1934. MPGA, Abt. III, Rep. 61, No. 2143.

**643** Correspondence from Baeumker to Prandtl, 7 February 1935. MPGA, Abt. III, Rep. 61, No. 2143.

**644** A. Busemann: Aerodynamischer Auftrieb bei Überschallgeschwindigkeit. Luftfahrtforschung, 12, 1935, 210–220.

**645** W. Heinzerling: Flügelpfeilung und Flächenregel, zwei grundlegende deutsche Patente der Flugzeugaerodynamik. Festkolloquium zur 6. Verleihung des August-Euler-Luftfahrtpreises, Technische Universität Darmstadt, 29 May 2002.

**646** H. Blenk: Die Luftfahrtforschungsanstalt Hermann Göring. In: Karl Stuchtey, Walter Boje (Bearbeiter): Beiträge zur Geschichte der Deutschen Luftfahrtwissenschaft und -technik. Band 1. Berlin: Deutsche Akademie der Luftfahrtforschung, 1941, 463–561.

**647** A. Busemann: Ergebnisse und Aufgaben der Aerodynamik. Luftwissen, 5, 1938, 353–356.

**648** Theodore von Kármán, N.B. Moore: Resistance of Slender Bodies Moving with Supersonic Velocities, with Special Reference to Projectiles. Transactions of the American Society of Mechanical Engineers, 1932, 303–310. Reprinted in CWTK, 2, 376–393.

**649** A. Busemann: Die achsensymmetrische kegelige Überschallströmung. Luftfahrtforschung, 19, 1942, 137–144.

**650** W. Hantzsche, H. Wendt: Mit Überschallgeschwindigkeit angeblasene Kegelspitzen. Jahrbuch 1942 der deutschen Luftfahrtforschung, 1943, I-80—I-90.

**651** Edward W. Constant II: The Origins of the Turbojet Revolution. Baltimore, London: Johns Hopkins University Press, 1980.

**652** E. Preiswerk: Zweidimensionale Strömung schiessenden Wassers. Schweizerische Bauzeitung, 109, 1937, 237–238.

**653** E. Preiswerk: Anwendung gasdynamischer Methoden auf Wasserströmungen mit freier Oberfläche. Zürich: Leemann, 1938.

**654** Theodore v. Kármán: Eine praktische Anwendung der Analogie zwischen Überschallströmung in Gasen und überkritischer Strömung in offenen Gerinnen. ZAMM, 18, 1938, 49–56.

**655** Correspondence from Prandtl to Taylor, 30 November 1935. MPGA, Abt. III, Rep. 61, No. 1654.

**656** Theodore v. Kármán: The Engineer Grapples with Nonlinear Problems. Bulletin of the American Mathematical Society, 46, 1940, 615–683. Reprinted in CWTK, 4, 34–93.

**657** Walter G. Vincenti, David Bloor: Boundaries, Contingencies and Rigor: Thoughts on Mathematics Prompted by a Case Study in Transonic Aerodynamics. Social Studies of Science, 33(4), 2003, 469–506.

**658** Judith R. Goodstein: Atoms, Molecules and Linus Pauling. Social Research, 51, 1984, 691–708.

**659** Otto Mayr: The science-technology relationship as a historiographic problem. Technology and Culture, 17, 1976, 663–673.

**660** John M. Staudenmaier: Technology's Storytellers. Reweaving the Human Fabric. Cambridge: MIT Press, 1985.

**661** Donald E. Stokes: PasteurÂŠs Quadrant. Basic Science and Technological Innovation. Washington: Brookings Institution Press, 1997.

**662** Bruno Latour: Science in Action: How to Follow Scientists and Engineers Through Society. Milton Keynes: Open University Press, 1987.

**663** Michael Gibbons, Camille Limoges, Helga Nowotny, Simon Schwartzman, Peter Scott and Martin Trow: The New Production of Knowledge. The Dynamics of Science and Research in Contemporary Societies. London, 1994.

**664** Henry Etzkowitz (ed.): Triple Helix. Research Policy, Special Issue 29, 2000, 109–330.

**665** David Wade Chambers, David Turnbull: Science Worlds: An Integrated Approach to Social Studies of Science Teaching. Social Studies of Science, 19, 1989, 155–179.

**666** Ronald N. Giere: Science and Technology Studies: Prospects for an Enlightened Postmodern Synthesis. Science, Technology, and Human Values, 18, 1993, 102–112.

**667** Wendy Faulkner: Conceptualizing Knowledge Used in Innovation: A Second Look at Science-Technology Distinction and Industrial Innovation. Science, Technology, and Human Values, 19, 1994, 425–458.

**668** Terry Shinn: The Triple Helix and New Production of Knowledge: Prepackaged Thinking on Science and Technology. Social Studies of Science, 32, 2002, 599–614.

**669** Gernot Böhme: Autonomisierung und Finalisierung. Gernot Böhme, Wolfgang van den Daele, Rainer Hohlfeld, Wolfgang Krohn, Wolf Schäfer, Tilman Spengler: Die gesellschaftliche Orientierung des wissenschaftlichen Fortschritts. Starnberger Studien 1. Frankfurt a. M.: Suhrkamp, 1978, 69–130.

**670** Gerald Eberlein, Norbert Dietrich: Die Finalisierung der Wissenschaften. Analyse und Kritik einer forschungspolitischen Theorie. Freiburg, München: Karl Alber, 1983.

**671** Th. v. Kármán: Progress in the Statistical Theory of Turbulence. Journal of Marine Research, 7, 1948, 252–264. Reprinted in CWTK, 4, 362–371.

**672** Hans-Jörg Rheinberger: Toward a History of Epistemic Things. Synthesizing Proteins in the Test Tube. Stanford: Stanford University Press, 1997.

**673** Th. v. Kármán: The Role of Fluid Mechanics in Modern Warfare. In: Proceedings of the Second Hydraulics Conference, Bulletin 27, University of Iowa Studies in Engineering, 1943, 15–30. Reprinted in CWTK, 4, 193–205.

**674** Eugene S. Ferguson: The Mind's Eye: Nonverbal Thought in Technology. Science, Vol. 197, No. 4306, 1977, 827–836.

**675** Eugene S. Ferguson: Engineering and the Mind's Eye. Cambridge, Mass: MIT Press, 1992.

**676** Jobst Broelmann: Intuition und Wissenschaft in der Kreiseltechnik, 1750–1930. München: Deutsches Museum, 2002.

**677** Walter G. Vincenti: What Engineers Know and How They Know It. Baltimore: Johns Hopkins University Press, 1990.

**678** John D. Anderson, Jr.: The Evolution of Aerodynamics in the Twentieth Century: Engineering or Science? In: Peter Galison, Alex Roland (eds.): Atmospheric Flight in the Twentieth Century. Dordrecht: Kluwer Academic Publishing, 2000, pp. 241–256.

**679** Kurt Magnus: Forscher an Prandtls Weg. In: Gerd E.A. Meier (ed.): Ludwig Prandtl, ein Führer in der Strömungslehre. Braunschweig/Wiesbaden: Vieweg, 2000, pp. 277–291.

**680** Editorial in Physics of Fluids, 1, 1958, 1.

**681** Siegried Grossmann in an interview in Physik in unserer Zeit, 32, 2001, 233.

**682** Clayton T. Crowe, Donald F. Elger, John A. Robertson: Engineering Fluid Mechanics, 7th edition. New York: Wiley, 2001.

**683** T.E. Faber: Fluid Dynamics for Physicists. Cambridge: Cambridge University Press, 1995.

# Author Index

*The Dawn of Fluid Dynamics: A Discipline between Science and Technology.* Michael Eckert
Copyright © 2006 WILEY-VCH Verlag GmbH & Co. KGaA, Weinheim
ISBN: 3-527-40513-5

# Name Index

## a

Ackeret, Jakob 159, 160, 162, 165–167, 195, 196, 208–213, 229–232
Ackermann, Walter 219
Ahlborn, Friedrich 38–42, 44–46, 110, 135–137, 149, 150
Ames, Joseph S. 85–87, 90, 102, 103, 124
Anderson, Jr., John D. 238, 239
Aristotle 1, 3, 4
Arnold, Henry Harley "Hap" 212

## b

Baeumker, Adolf 179–182, 187, 189, 190, 198, 230
Bairstow, Leonard 98, 149
Bateman, Harry 201
Bell, E.T. 201
Bernoulli, Daniel 14, 240
Bernoulli, Johann 13, 14, 19, 25
Bessel, Friedrich Wilhelm 17
Betz, Albert 54, 55, 60, 73, 89, 107, 108, 118, 124, 135, 142, 145, 152, 163, 167, 182, 184, 204, 211, 213–215, 218–221
Birnbaum, Walter 151, 219
Bjerknes, Vilhelm 29, 98, 168–171, 174
Blasius, Heinrich 43, 73, 108–110, 116, 120, 122
Blenk, Hermann 151, 152, 230
Blondel, Francois 11
Blumenthal, Otto 199
Bohr, Niels 123
Born, Max 44, 177, 241
Bosch, Carl 180, 189
Boussinesq, Joseph 25
Burgers, Johannes Martinus 97–99, 101, 109
Buridan, Jean 1
Busemann, Adolf 159, 161, 210, 213, 229–231
Bush, Vannevar 235

## c

Cauchy, Augustin Louis 15
Chamberlain, Neville 192

Chézy, Antoine 24
Coker, W.G. 98
Constant, Edward W. 239
Costanzi, Giulio 94
Courant, Richard 177, 178
Cranz, Carl 61
Crocco, Arturo 229, 230
Crocco, Luigi 231

## d

d'Alembert, Jean Baptiste le Rond 13, 15, 18, 32
Dätwyler, Gottfried 209, 210
Desaguliers, Jean Théophile 8
Descartes, René 5, 6
Dirichlet, Gustav Lejeune 18, 32, 34
Doepp, Philipp von 78
Dönch, Fritz 110
Douglas, G. 229
Dryden, Hugh L. 124–128, 223, 226
Du Buat, Pierre Louis Georges 25
Dubuat, Louis Gabriel 17
Durand, William Frederick 85, 189, 190, 218

## e

Eiffel, Gustave 49–52, 84, 86, 87, 203
Ekman, Vagn Walfrid 170–172
Engelbrecht, Walter 184
Epstein, Paul 201, 202
Euler, Leonhard 13–15, 18, 19, 25, 32, 37, 162, 240
Exner, Felix M. 173

## f

Ferguson, Eugene S. 238
Flettner, Anton 175
Fokker, Anthony 76, 81
Föppl, August 32
Föppl, Otto 49
Forchheimer, P. 98
Föttinger, Hermann 164, 165
Foulois, Benjamin C. 85

*The Dawn of Fluid Dynamics: A Discipline between Science and Technology.* Michael Eckert
Copyright © 2006 WILEY-VCH Verlag GmbH & Co. KGaA, Weinheim
ISBN: 3-527-40513-5

# Subject Index

The terms "fluid mechanics" and "fluid dynamics" are used synonymously. They are not included in the index because they appear so frequently.

*The Dawn of Fluid Dynamics: A Discipline between Science and Technology.* Michael Eckert
Copyright © 2006 WILEY-VCH Verlag GmbH & Co. KGaA, Weinheim
ISBN: 3-527-40513-5